University Coeducation in the Victorian Era

University Coeducation in the Victorian Era

Inclusion in the United States and the United Kingdom

Christine D. Myers

palgrave
macmillan

UNIVERSITY COEDUCATION IN THE VICTORIAN ERA

First published in 2010 by
PALGRAVE MACMILLAN®
in the United States—a division of St. Martin's Press LLC,
175 Fifth Avenue, New York, NY 10010.

Where this book is distributed in the UK, Europe and the rest of the world, this is by Palgrave Macmillan, a division of Macmillan Publishers Limited, registered in England, company number 785998, of Houndmills, Basingstoke, Hampshire RG21 6XS.

Palgrave Macmillan is the global academic imprint of the above companies and has companies and representatives throughout the world.

ISBN: 978–0–230–62237–1

Library of Congress Cataloging-in-Publication Data is available from the Library of Congress.

A catalogue record of the book is available from the British Library.

Design by Newgen Imaging Systems (P) Ltd., Chennai, India.

First edition: July 2010

10 9 8 7 6 5 4 3 2 1

Printed in the United States of America.

This book is gratefully dedicated
to the students whose names appear in its pages and
to their classmates and colleagues, family and friends,
who supported their efforts, and also to my own

CONTENTS

IMAGES

ACKNOWLEDGMENTS

Many thanks need to be extended to the many people who aided me in writing this book. First and foremost I want to thank my doctoral supervisor at the University of Strathclyde, Professor Hamish Fraser, for his guidance and unswerving faith in my research and me. Numerous archivists and librarians have also helped me in person and from a distance. Alphabetically they are Julie Archer, Elgan Davies, and David Fitzpatrick at the Old College Library, Aberystwyth; Marie Boran and Loretto O'Donohoe at the James Hardiman Library, NUI Galway; Frank Cook, Cathy Jacob, David Null, and Bernard Schermetzler at the University of Wisconsin-Madison Archives; Gary Cox at the University Archives, University of Missouri-Columbia; Paul Dzyak and Jackie Esposito at the Eberly Special Collections Library at Pennsylvania State University; Ivan Ewart and Ursula Mitchel at Queen's University Belfast; George Gardner, Kiara King, Lesley Richmond, and Emma Yan at the Glasgow University Archives; Paul Howarth, Curator of the Gilbert and Sullivan Archive; Douglas McCabe and Bill Kimok at the Mahn Center for Archives and Special Collections, Ohio University; Beth Meko at the University of Tennessee Libraries; Laura Schmidt and Heidi Truty at the Marion E. Wade Center, Wheaton College; Michael Stansfield at Durham University Library, Archives and Special Collections; Tess Watts at the Mary Evans Picture Library; Elizabeth Cassidy West at the University of South Carolina Archives; and finally, to the plethora of people who have worked to make materials from the nineteenth century available online, both at the universities mentioned and elsewhere.

CHRISTINE D. MYERS
October 2009

ABBREVIATIONS OF ARCHIVES AND SPECIAL COLLECTIONS USED

GSA	The Gilbert and Sullivan Archive, Boise State University, Boise, Idaho
GUA	Glasgow University Archives, Glasgow, Scotland
GUSC	Special Collections Department, Glasgow University Library, Glasgow, Scotland
OCLA	Llyfrgell yr Hen Goleg/Old College Library, University of Wales, Aberystwyth, Ceredigion, Wales
MEPL	Mary Evans Picture Library, London, England
MEWC	Marion E. Wade Center, Wheaton College, Wheaton, Illinois
NUIG	Special Collections, James Hardiman Library, National University Ireland, Galway, Ireland
OUMC	Ohio University, Mahn Center for Archives and Special Collections, Athens, Ohio
PSU	University Archives, Eberly Special Collections Library, Pennsylvania State University
QUB	Media Services Photographic Unit, Queen's University Belfast, Northern Ireland
UMC	University Archives, University of Missouri-Columbia
UWA	University of Wisconsin Archives, Madison, Wisconsin
WHS	Wisconsin Historical Society, Madison, Wisconsin

CHAPTER ONE

The Process of Inclusion

Image 1 Earliest surviving photograph of Queen's College, Belfast staff and students, c. 1886.
Credit: Media Services Photographic Unit, Queen's University Belfast.

The historical study of higher education in the nineteenth century is prolific. In both the United States and the United Kingdom there were numerous changes in legislation and policy that revolutionized the nature of university education, primarily by opening it to a wider section of the public. Previously, institutions on both sides of the Atlantic were intended for the elite, the sons of wealthy citizens who were born to a social class that included higher education in its expectations. Beginning in the eighteenth century, there were new philosophies of democracy and equality emerging that encouraged access to all levels of schooling for all members of society. Women in both countries who desired higher education often argued that they could make worthy contributions to society outside the home, if they were only given the chance. By the nineteenth century, women in several countries began to take on new societal roles. One of the most significant of these was the introduction of women to higher education. Although many advances were adopted slowly, by the early twentieth century women were allowed into most fields of study and were able to work toward the same degrees as men.

To more fully understand the impact of the people who worked to gain access for women to universities, the use of a comparative study of more than one nation is beneficial. By studying institutions in the United States and the United Kingdom, comparisons can be made between the cultural conceptions of women in higher education and their actual experiences. Because the purpose of this book is to look at the process of integrating women into male universities during the nineteenth century and the responses of society to this decision, comprehensive, state-supported institutions will be the focus of primary research. These colleges and universities were accountable to the citizens and their representatives and provided a wide-ranging curriculum to all their students.[1] The similarities and differences in these institutions, and in the communities in which they are located, illustrates many of the earliest steps toward gender equality in the Victorian Era. As one historian explains it, understanding "the historical role of women in higher education . . . rests upon understanding a series of related changes in both education and the status of women."[2]

This book aims to address three major themes within the field of university coeducation. First, it will consider the evolving perceptions of women and women's place in higher education, as seen by both sexes. Societal views of women in the nineteenth century focused on the issue of "separate spheres" of influence for men and women, thus prohibiting or restricting women's entrance into public life for most of the century.[3] Though these views had altered greatly by the end of this study, in many ways they did not change as a result of women's admission to universities, with women still guided into certain professions associated with traditionally female roles. The second area of concentration relates to the expectations of women's lives after completing a university education. The conscious and unconscious efforts of university officials and students to ensure that women married and entered into society in a suitably traditional role are key to this study. Despite significant changes in the opportunities open to women, they were still bombarded with images of femininity and matrimony that were intended to help them chart the proper course for their lives. The third and final broad issue considered in this book is the shifting power relationships within the institutions themselves. Through the second half of the nineteenth century, administrative control gradually moved away from boards of trustees and the like into the hands of the faculty and then, to an extent, to the students themselves as their voice on campus increased. This shift in the practical control of the institutions is crucial in understanding the evolution of coeducational policies and experiences. Interaction with the "outside world" is also a factor within this area, as changing values in society were often reflected in the relationships among administrators, faculty, and students.

These three broad themes will be developed by examining both the academic and extracurricular lives of students. Whether inside or outside the classroom, the responses of the administration, faculty, and male students to the presence of women were markedly different. At the opening session of Queen Margaret College, Glasgow, in 1891, Professor Ramsay of the University of Glasgow offered the following thoughts on coeducation: "The tendency was to assimilate female education to male education, and the educational feature of the age had been to throw down the intellectual barriers between man and woman, and open to women the intellectual aims and ambitions heretofore confined to men."[4] This throwing down of barriers intellectually was not often matched by similar approaches in other sections of campus life. In the classroom and housing particularly, new restrictions and regulations were placed on both men and women to control their interactions with each other. Gender segregation continued into purely social endeavors as well, with student organizations and publications often taking on decidedly single-sex appearances. This is not to say that the men and women at university did not interact with each other, but they found it was best to do so in appropriate circumstances that would be sure to have the blessing of the university officials and society as a whole.

A final issue to consider is the elite nature of higher education, especially as it existed in the nineteenth century. In the United States in 1870 "the student population included just over 1 percent of the traditional college age group." In addition, since higher education has never been an inexpensive endeavor, it is generally understood that those who went to university were fairly well off financially.[5] Mary Caroline Crawford made note of the costs of the University of Wisconsin in her book *The College Girl of America and the Institutions Which Make Her What She Is* in 1905. Although tuition was free to Wisconsin residents, the additional costs (primarily room and board) made the access to educational opportunities difficult for some students. With this in mind, how valid is a study of higher education in ascertaining the prevailing gender roles? Certainly women living in different areas had different experiences. Nevertheless, the fact remains that the education received by the men and women at university was revered by society in general. Graduates of these institutions became leaders in business and government, and some brought great prestige to the community. So while few people had the opportunity to attend college, it was still looked to as the ideal of society. As one Victorian historian put it: "It is the summit and crown of the state system of public instruction."[6]

While there were differences between the experiences of women in universities in the United States and in the United Kingdom, the issues of equality or equity in education remained much the same throughout the period. One Victorian definition of a coeducational college or university, provided by the U.S. Office of Education, was "those in which the management and the degrees are the same for the men and the women, though the recitations may be conducted separately."[7] I would argue that full coeducation entails a full mixing of male and female students in all aspects of study. Understandably housing and other facilities might not be mixed, but the classrooms of a coeducational institution would be open to all enrolled students. A better definition then would be the one provided by A. Wallis Myers in her discussion of university coeducation in Aberystwyth, Wales: "Here we get men and women attending the same lectures, learning the same lessons, entering into the same social life, and practically playing the same amusements."[8] What courses women were allowed to, or chose to, take would affect the remainder of their lives considerably, as would their extracurricular activities. Their interactions with men, both faculty and students, would also determine much of their outlook on society in general.

Institutions and Methods Used in this Study

Because this book is about the inclusion of women into male, governmentally funded universities in the nineteenth century, there were a limited number of institutions that fit the criteria. There was also a relatively small window of time during the Victorian Era when male colleges and universities made the decision to admit women and before new institutions were established that were coeducational from the outset. Although there were other forms of coeducation available to students during the Victorian Era, this limited scope is beneficial in a number of respects. First, if a private institution decided to admit women, they only had themselves and their alumni to answer to. It was far easier to find concordance with this smaller group of interested people, and it was easier for the public at large to feel that the actions of one private college or university did not affect society as a whole. Second, if an institution of any sort began as a coeducational one, they did not have the struggles of integration that were found in all-male institutions that now needed to accommodate students of a different sex. Something as simple as having both male and female restrooms, which people take for granted today, was a major obstacle to Victorian colleges and universities. Third, the influence of government legislation on colleges and universities, as in other areas, is an indication of what a country wants itself to become. Only a small percentage of the population in both the United States and the United Kingdom was able to pursue higher education, but those who did were looked to as the leaders of their generation.

The decision to focus on women stems primarily from the fact that they were the first excluded group to gain widespread access to what were generally all-white, all-male universities and therefore chip away at the "monopoly of learning" that existed in the nineteenth century.[9] The methodology used in this research could easily be applied to the inclusion of other demographic groups in subsequent years—ethnic or religious minorities, students with physical disabilities, and so on. I hope that other historians take up some of these challenges and do so with an international perspective, reminding us that no country develops in a vacuum. It is not my intention to write a comprehensive history of coeducation at each of the twenty-four institutions studied in this book. There simply would not be space to do that, and there are many works on some of the individual institutions already in existence. The purpose of this book is to show the similarities across institutions and national boundaries that indicate an upsurge of support of, or acquiescence to, women's presence in coeducational colleges and universities with a comprehensive curriculum.

This book is fundamentally based on primary sources produced by the administrators, faculty, and students of the universities (either during their undergraduate years or through alumni reminiscences), or contemporary writers that commented on the education that was taking place at the institutions. More modern, secondary materials are certainly referenced, but every attempt has been made to utilize firsthand accounts of life at coeducational universities in the Victorian Era. To make the introduction of the colleges and universities in this book as clear as possible, a brief summary of their initial inclusion of women will now be provided, grouped alphabetically by nation in the United Kingdom and by region in the United States.

English Overview: The Universities of Durham, London, and Manchester

Higher education in Britain developed in the Middle Ages, with the foundation of colleges in Cambridge and Oxford in England, and Aberdeen, Glasgow, and St. Andrews in Scotland. The youngest of these so-called "ancient" universities, Edinburgh, opened in 1583, followed by Trinity College, Dublin, in 1592.[10] Despite occasional references in the histories of these institutions to women attending lectures, or indeed, disguising themselves as men to gain admittance (Dr. James Barry at Edinburgh), women were not able to receive university degrees until the nineteenth century.[11] Since universities were traditional institutions, not seen to crave change, the groups that were established to promote women's higher education in several British cities needed to structure their approach carefully to gain admittance to courses and degrees. Emily Davies, founder of Girton College, Cambridge, noted considerable opposition that centered on the view that the admission of women to universities would "entail consequences injurious to the University as a place of education for men."[12] Concerns over the increase in students and the need to share funds, facilities, and faculty proved difficult to counteract. This problem was more intense in England where many colleges were residential, but for universities in the rest of the United Kingdom this was not an issue they had to deal with straight away.

England also had a number of newer, more urban or "civic" universities that opened in the nineteenth and early twentieth centuries.[13] Because this book focuses on male institutions that made the shift to coeducation during the Victorian period, only three of these institutions are of interest—the University of Durham, the University of London, and the University of Manchester.[14] Despite conferring degrees on women, mixed classes were not always available in all subjects.[15] The administrators at these institutions, like their counterparts across the Atlantic, would continue to grapple with the progressive ideals that made them open to idea of the higher education of women while at the same time they remained dubious about the practical realities of fully integrating women into their classrooms and other facilities. Initially many officials felt it would be inconvenient and expensive to become coeducational, though in the end "professors [who] first held separate lectures for them... wearied after a time of repeating themselves and threw open the regular classes to the women."[16]

In 1881 the possibility of admitting women to graduation in arts courses at the University of Durham was considered in Convocation (made up of the graduates of the university).[17] There was a petition in support of the measure within this body, but there were concerns that such a change in policy was not within the power of the Convocation to determine, and there was also worry that the community would not receive it well. Katharine Lake, wife of Dean William Lake, warden of the University of Durham at the time, noted that "Any project for encouraging the education of women is always unpopular with undergraduates, and there were also in the Senate those who strongly opposed the proposal."[18] Dean Lake was able to garner enough support to get the resolution passed admitting women to the institution, though it was not acted on. The affirmative vote, having been reported to the public, was incorrectly perceived by some as the date of women's admission to the university. Even Susan B. Anthony, Joslyn Gage, and Elizabeth Cady Stanton, in their *History of Woman Suffrage*, listed 1881 as the year that "Durham University votes that women may become members" without any clarification about the actual status of women at the institution.[19] Lake's successor, Dean George William Kitchin, also supported women's entrance to the institution and was the one to oversee women's eventual entrance to courses and degrees.[20] Finally in 1895 Durham received a supplemental charter that explicitly stated women be admitted, and they promptly were.[21]

The University of London decided in 1868 to open their examinations to women, and the following year, the first group of women sat for them.[22] Degrees were extended to women in 1878, making them the first university-level degrees available to women in the United Kingdom. Just because women were allowed to sit for examinations and earn degrees from the University of London does not mean all of the university's constituent colleges were coeducational however. It was up to each college to determine which students they would allow into their classrooms and whether they would provide coeducational or separate instruction for men and women. The two oldest colleges—University College, London (established in 1826) and King's College, London (established in 1829)—chose to become coeducational, while others preferred to follow a single-sex approach to higher education. Bedford College, Royal Holloway College, and Westfield College were all established for women only and did not become coeducational until the 1960s.[23]

In 1885 King's College, London opened a "Department for Ladies" to maintain a separation between their male and female students, and it also enabled the college to earmark funds for the education of each group.[24] The decision to proceed with separate courses of instruction has usually been attributed to King's affiliation with the Church of England and the fact that they taught theology, a course which women were not able to pursue because they could not have careers in the Anglican clergy.[25] Another argument for providing a separate education for women at King's was that it was thought to be preferable to the women themselves, and with other all-women's colleges in the Greater London area, it might make King's more attractive to prospective female students and their families. University College, London also experimented with offering separate courses for men and women, but most professors "decided in favour of joint classes" that would prevent them from doubling their teaching load.[26] Some of their junior classes were kept separate, but the majority of senior courses were "common to both sexes."[27]

The University of Manchester began its life as Owens College in 1851. It was named for John Owens, a liberal capitalist who supported the idea of higher education for all men and provided an endowment for the college.[28] The officials of the new college were breaking new ground because they were creating an institution that was unlike any other in the United Kingdom at the time. They were similar to London in that they wanted to provide degrees for middle-class students in an urban metropolis. Looking to Durham and the Scottish universities for inspiration as well, they wanted to make every effort to provide courses that would appeal to those "engaged in the pursuit of trade or commerce."[29] A university-level education would not be a university-level education in the nineteenth century without the study of the classics though, so they did not deviate greatly from the standards expected by society at the time.

As early as 1853 Owens officials appealed to Parliament for financial assistance, as the public subscriptions and other donations they received were not sufficient to meet the burdens of the college. Initially

Owens College could not confer degrees either, and served instead as an examination center for the University of London.[30] Formal ties to government would not be established until 1870 when Parliament passed the Owens Extension College (Manchester) Act, which was quickly followed by the Confirming Bill that was passed in 1871 and solidified state support of the institution.[31] This Act also enabled the governors to admit women if they wished to. Attempts were made in 1874 to include women in courses, but established practice maintained that the institution was for male students only and regardless of the best efforts of the supporters of women's higher education in the city, officials would not change their minds about this policy.[32] Much of this stubbornness can be attributed to the desire of administrators to gain the respect of other colleges and universities in England, which were still seen as male domains. The continuing financial limitations in Manchester, and the need for all space to be used by the students they already had, were more often cited as the reasons women were not admitted.[33]

A final change of organization would occur in 1880 when Owens College received a Royal Charter as Victoria University. This new institution would be federated in a similar fashion to the University of London, with the new Yorkshire College, Leeds and University College, Liverpool joining it in 1884 and 1887 respectively.[34] When women were finally admitted on a regular basis to Owens College, it had already become Victoria University. In 1883 the nearby Manchester and Salford College for Women was taken over by the university, and it became a separate department for women within the larger institution. Owens faculty already taught the majority of the courses at the women's college. Very few courses would be taught in coeducational classrooms in the Owens buildings, and most of the women's education would remain separate for the remainder of the century, resulting in a protracted inclusion process, or what historian Mabel Tylecote termed the "Period of Penetration" by women.[35]

Irish Overview: Queen's Colleges in Belfast, Cork, and Galway

The next country to consider within the United Kingdom is Ireland. Having become a part of the United Kingdom with the Act of Union in 1800, the country would be ruled by the Parliament in London until the Irish War of Independence began in 1919.[36] In 1845 the Queen's Colleges (Ireland) Act formed three institutions in the cities of Belfast, Cork, and Galway.[37] Then in 1850 the Queen's Colleges were united under the name of Queen's University for degree-granting purposes. Students could take their courses at any one of the three Queen's Colleges, although some course work could be done at other institutions of higher education in the nation, as long as they were approved of by the terms of the charter.[38] The administrative structure of Queen's University was replaced in 1881 by a new name—The Royal University of Ireland—as stipulated by Parliament in the University Education (Ireland) Act of 1879.[39] This change was not supposed to alter the teaching at any of the Queen's Colleges, since the Royal University was fundamentally an examination body for the purpose of granting degrees. An interesting aspect to the degrees granted through the Royal University was that a candidate did not have to take classes at *any* college or university to get an undergraduate degree; they merely had to pass the examination for it. In this way everyone could be eligible to gain a university degree, whether they were able to afford to take courses or permitted to enter classrooms or not.[40]

The Queen's Colleges were designed as a way to extend the privileges of higher education to the far reaches of Ireland, that could not afford to attend university in Dublin or preferred not to.[41] It was also supposed to be a "mixed education" with Catholics and Protestants attending classes together. The introduction of three new centers of higher learning on the northern, western, and southern points of the island, which were added to the eastern institutions in Dublin, made perfect sense from a geographical standpoint, but it also had political implications as a result of the religious affiliations in each of these areas.[42] Since the main universities in the country, Trinity College and University College, Dublin, were both Protestant, the Queen's Colleges were intended as a place where Catholics could pursue higher education.[43] As a result, outcry from the Catholic clergy and Irish people about these new colleges focused on the secular education they were designed to provide, and as the historian Judith Harford has pointed out, the "issue of access for women students remained peripheral to the overall debate."[44]

Despite the fact that sectarian issues were of paramount concern during the 1800s, one of the key groups that would benefit from the change in academic approach was women, who were previously excluded from all the universities in the nation due in large part to the latter's original emphasis on theology. In 1876 "six lady students, who presented themselves for admission to Queen's College, Galway, were rejected on account of their sex."[45] Because the University Education (Ireland) Act of 1879 opened university examinations to women, enabling them to earn degrees, there became little reason to continue preventing them from attending classes. The Queen's Colleges in Belfast, Cork, and Galway began to admit women in some or all fields in 1882, 1885, and 1888 respectively. All three colleges had a relatively small number of women attend as students during the Victorian Era, though presence of women in the largely Protestant city of Belfast (as seen in the photo at the start of this chapter) was greater than it was at its sister institutions because the college was larger, "having double the number of students on its rolls that stand on either those of Cork or Galway" where Catholics were reluctant to pursue a secular education.[46] Because the Queen's Colleges were non-denominational, there was a strong feeling among many in the country that a religiously affiliated institution, be it Catholic or Protestant, was preferable to the "Godless" ones established by the British Parliament.[47] From the standpoint of a woman, or her family, a single-sex institution was also often preferable to one where women would be mixing with male students.

Midwestern Overview: Indiana University, and the Universities of Michigan, Missouri, and Wisconsin

In the midwestern United States, the majority of colleges and universities moved toward a policy of coeducation midway through the nineteenth century, with state-supported institutions farther to the west virtually all being coeducational from their inception. The most cited reason for this decision was financial necessity. Since these universities were located in developing states, students were at a premium. Admitting women ensured a more substantial student body and survival of new schools. Additionally, state-supported institutions were often the least restrictive in admittance policies because they felt any taxable individual should be able to attend, without regard to race, sex, or religious preference. The popular nature of these institutions with the people of their state was undeniable:

> The young men and women...belonging to every class of society, the well to do in life, the poor, the rich, all thirsting for knowledge, and desiring to better their condition, are...availing themselves of every advantage offered by the State for their social, moral and intellectual improvement.[48]

This desire for education, as explained by the Board of Curators at the University of Missouri, was felt throughout the Midwest where citizens of new states sought to regain the advantages they had left behind by moving west. The local demands of each community would result in different timelines in the admission process of women, but the number of voices in favor of this action would typically outnumber those opposed to it.

Indiana University began as a seminary and soon became Indiana College in 1828. Ten years later their name changed again to Indiana University.[49] After a steady expansion women were finally accepted as students in 1867 when Sarah Parke Morrison appealed to the Board of Trustees to admit her. She began her college career at Mt. Holyoke and Vassar, but being from Indiana (her father, John I. Morrison, was a former member of the faculty and former president of the Board of Trustees) she wished to attend a university in her home state.[50] Interestingly enough, it was Miss Morrison's petition for admission that brought "the question to a focus."[51] Had this turn of events not occurred, it might have been quite some time before women gained a place as students at Indiana because there was no active campaign to change university policy at the time, despite the existence of such pressure in neighboring states. One of the trustees, Isaac Jenkinson, submitted a resolution that would open the university to women with no special distinctions made between them and men.[52] The vote of the board to admit Miss Morrison was split four to three in favor, and she became the only female student, though others soon joined her; in the 1869–1870 academic year there were a dozen young women in attendance.[53] Because women were admitted with relatively little trouble, historians and supporters of the university heralded their innovation and the fact that "no young men or women need to leave their own State in order to secure the best liberal and professional educations in any vocation they may select."[54] Furthermore the inclusion of women in the extracurricular life of Indiana University was seen favorably, especially in comparison with other institutions: "Women are, in a real sense, a part of Indiana life. . . . Socially, the co-ed is a necessity. Here we cannot comprehend such a state of affairs as is said to exist at some universities where the women are looked upon by men students as interlopers."[55]

Despite legislation in 1837 that stipulated opening the University of Michigan to women, they did not gain access until 1870 after considerable debate in all quarters. One petition for entrance was made by several women in 1858. Victorian sources note that there was a general acceptance in Michigan that the women of the state were entitled to higher educational opportunities, but there was also a strong concern by some that the "two sexes could not associate together frankly and freely, as would be necessary if the university should open its doors to women. It was a question of moral and social advisability."[56] In the antebellum period the cautious approach was preferred and the petition was denied by the Board of Regents.[57] A decade later the argument about the government's obligations to the citizens of the state would be raised again. As the historian Burke A. Hinsdale described it in his 1906 *History of the University of Michigan:* "a democratic state like Michigan, which maintained a University at the public cost, could not, logically, deny admission to any class of citizens prepared to receive this instruction."[58] The state legislature therefore suggested in 1867 that the Regents admit women, and they subsequently did.[59]

The civic sentiment that made it possible for women to attend classes and earn degrees at Michigan immediately caused problems financially with the need for additional facilities and faculty. Debates would ensue about how many funds should be allocated to teaching female students, in some instances in separate classes from the men. The president at the time, Erastus O. Haven, stated that the "poverty of the University has begun to be more painfully apparent since the adoption of the resolution by which the institution has been thrown open to women."[60] Unlike some institutions included in this study, Michigan had a large student body that multiplied the problems encountered as their enrollment figures grew. The first female student, Miss Madelon Stockwell, was praised for "convincing the members of her class that she was intellectually the equal of the best of them and that she could easily hold her own in the strenuous work of a university course. Moreover, she demonstrated at once that she could adapt herself gracefully to an environment that was not only novel but in many respects unpleasant and unattractive."[61] Within five years of Stockwell's enrollment there were one hundred women studying at the University of Michigan.[62] As a result of difficulties with space and funding, as well as concerns about men and women being taught in the same classrooms, the integration of women in Ann Arbor and at some other male universities, would become a multistep process that would take several decades to fully accomplish.

The University of Missouri in Columbia was established in 1839, making it the first state-supported university west of the Mississippi River.[63] Coeducation was first introduced in their Normal School, with three women beginning their studies there during the 1868–1869 academic year. Subsequently they were admitted "cautiously . . . to some of the recitations and lectures in the university building" both before and after the state legislature formally admitted them in 1872.[64] Most of the caution, not surprisingly, related to the intermingling of male and female students. Due to the apprehensions of some administrators and concerns about the community reaction, women were not allowed to use the library at the same time as the men, nor were they allowed to attend chapel initially.[65] The reason there was so much resistance in Missouri to women's admission is unclear, and it would have been highly unusual to believe that preventing them from attending chapel was more morally sound than interacting with men. Some attributed decisions like this to a "prejudice against the University, and favor towards Young Ladies' Seminaries" that might better look after their moral well-being. If efforts were being made to keep parents from sending their daughters to the state university, however, they were unsuccessful in the long run.[66]

In a similar course of events, the University of Wisconsin, founded in 1848, first allowed women into its Normal School for teacher training in 1860. Since the state was expanding, additional teachers were needed, so the state legislature took an active interest in increasing higher educational opportunity to meet the demand.[67] University of Wisconsin alumna, Helen R. Olin reported in *The Women of a State University: An Illustration of the Working of Coeducation in the Middle West* that the "general spirit of Wisconsin people toward the education of women was unusually progressive in early days."[68] Two private institutions in the state had accepted women since their foundation—Lawrence University in 1847 and Ripon College in 1853. The resulting competition from these institutions, along with the various normal colleges in the state that were established after the Wisconsin Normal Schools Act was passed in 1866, propelled the University of Wisconsin to provide a comprehensive and more equitable form of coeducation for their students from that year onward.[69]

The state and community support for women's higher education made coeducation at the university easier to accept, but not simple to implement. The Board of Regents and the university's president governed the University of Wisconsin collectively. The position of president held the greatest day-to-day power at the institution in the nineteenth century, as the Regents only met on an annual basis. The progress of coeducation at Wisconsin, then, depended largely on the amount of support the current president had for the concept. While the Board of Regents as a whole were outwardly indifferent to the idea of coeducation and neither hastened nor slowed women's inclusion, individual presidents altered the course of events dramatically. Most notably Paul Chadbourne, the president at the time of women's admission, worked stridently to prevent their progress at the university. He was soon followed in the position by one of coeducation's most ardent supporters, John Bascom, who endeavored to solidify the position of women on campus.[70] The reactions of male students were often guided by their administrative leaders and faculty. In 1938 *The Wisconsin Alumnus* interviewed the university's oldest living alumna, Mrs. Sophie Schmedeman Krueger. Her most vivid recollection was of the reception given to the female students by their male classmates who treated them "with wholesale contempt." She noted that the women students later found out that "they feared... that our participation in classes might lower the standard of the University," and thus lessen the value of their own degrees.[71] The women were also encouraged to "try not to give cause for the trivial complaints which came in occasionally" from male students or from the faculty.

Ohio Valley Overview: Ohio University, Pennsylvania State College, the Western University of Pennsylvania, and West Virginia University

Ohio University in Athens, Ohio, was founded on February 18, 1804, just one year after the state was admitted to the Union, though it did not open its doors until 1808.[72] It has the distinction of being "the first college in the United States founded upon a land endowment from the national Government, and also of being the oldest college in the Northwest Territory."[73] In the first days of the institution it was an academy, becoming a college in 1822. Ohio University remained a small institution throughout the nineteenth century, and it often struggled financially. The years after the Civil War were a challenging time for the United States economically, with both agriculture and industry facing difficulties after the panic of 1873.[74] Ohio University was one of many colleges and universities that struggled to stay open during this time, with the first real financial upswing coming as the result of a decision of the Ohio State Supreme Court in May 1880. The ruling was "a rental case," which favored the institution and would increase its income by $3,000 per year. *The Ohio*

Educational Monthly remarked on this case in particular because they felt it was "refreshing to record a windfall for the University."[75]

Women were formally admitted in 1871, partially because there was a need for students to keep the institution afloat and partially because there was local pressure that they not be excluded from the institution.[76] In 1872 there were only fifty-five students, with a further fifty-four in the preparatory school that was attached to the university.[77] The total numbers of graduates between 1815 (when the first two students were awarded degrees) and 1901 was 529, of whom 471 were men and 58 were women.[78] The first woman graduate of Ohio University, Margaret Boyd, kept a diary during her time as a student beginning with an entry on Wednesday, January 1, 1873.[79] The following Tuesday, January 7, she recorded: "Commenced college today—Fear I will have hard work but I will try and do it well. Not many students out today. I fear for the future of the old O.U."[80] The admission of women was one of the factors that led to the success of the institution because of the increased number of potential students who might choose to attend it. The fact that the state was known for its progressive views on coeducation (Oberlin College being the first institution in the country to open its doors to both men and women in 1837) meant that the inclusion of women at Ohio University came as a natural process of its evolution as a college.[81]

The Pennsylvania State College was incorporated in 1855, though it did not go by that name until 1874.[82] There was considerable debate during the nineteenth century about the status of the institution, and whether or not it was a "state" college, as its name implies. The first incarnation of the college was as "The Farmers' High School," an institution that was chartered by the governor of the Commonwealth of Pennsylvania in 1855.[83] The connection to the government grew stronger when the state legislature decided to make the newly renamed "The Agricultural College of Pennsylvania" its sole recipient of land-grant funds from the federal government under provision from the Morrill Act in 1862.[84] Along with receiving monies for building and maintenance at various points during the century, further federal funds would come after Congress passed the Hatch Act in 1887. When the commonwealth's financial support became inadequate in the early 1890s, the college's president, George Atherton, argued that they were a "state" institution that the government was obligated to fund. His entreaties were unsuccessful at the time, and it was not until the twentieth century that Penn State's position as a state-supported university was clear to all involved.[85]

In 1871 the Pennsylvania State College decided to admit women into a separate Ladies' Department, which was more of an administrative designation than anything else. The first three graduates who completed a full four-year course of study got their degrees in 1884, with most of their predecessors choosing to withdraw from the institution after only one to two years. Because the enrollment and retention figures had remained so low in the first years of coeducation, in 1884 a new "Ladies' Course in Literature and Science" was started as a way to guide women through the first two years of university study so that they would be more likely to stay for a full four years. Women were not required to take this course, and many did opt to enroll in the standard curriculum of the college instead. Because the Ladies' Course was not as popular as anticipated, it was discontinued in 1891.[86] This decision was seen to validate coeducation on campus, and in the future it was argued that "[a]ny policy which looks toward the practical exclusion of women here would seem to be contrary to the spirit of modern progress which is opening the doors of other institutions to young women."[87]

One of the other institutions whose doors were opened to young women in the Commonwealth of Pennsylvania was the Western University of Pennsylvania. The institution began its life as the Pittsburgh Academy in 1787. After several years it reached university status in 1819 and would later become the University of Pittsburgh in 1908.[88] At the Western University of Pennsylvania women were not excluded from admission by the institution's charter, yet in its first hundred years of existence there were "repeated and earnest appeals being made from time to time to allow the attendance of ladies" to this state-funded university.[89] One such request came from Miss Mary Sophia Lynn in 1893, by way of Professor Francis Phillips, asking to take a special course in chemistry.[90] The chancellor, W. J. Holland, agreed to admit her to the class, even though no official statement had been made about coeducation at the university. He later referred to this action on his part as his "connivance" but added that no one had ever "found fault" with his decision to allow women to study at the university.[91]

Miss Lynn's time at the university was short, and her male classmates regretted seeing her leave, having made no "personal acquaintances among the students."[92] They were pleased, however, that she was replaced by two new female students, the Misses Stein, who "revived" the "drooping spirits" of the student body with their presence. The Misses Stein had written directly to Chancellor Holland asking permission to be admitted. In 1895 the chancellor decided to use his discretion to allow them to enroll as students in the classical and scientific courses. The following year the *Catalogue* reported that the shift to coeducation had been "most pleasant and gratifying" and that the new students had "proved themselves very capable."[93]

The final state in what is commonly known as the Ohio Valley that dealt with the inclusion of women during the Victorian Era is West Virginia.[94] When West Virginia became a state in 1863, having separated from Virginia during the Civil War because it did not wish to secede from the Union, it had no university and not much of a public school system.[95] The state legislature moved quickly to fill this important void and founded an institution of higher education in Morgantown in February 1867. Its first name was the West Virginia Agricultural College, though it soon expanded its curriculum to include more departments and schools, and its name was changed the following year to West Virginia University by the legislature.[96] From the time of the inception of the university, coeducation had been considered by its officials, who even discussed with "the principal of the Morgantown Female Collegiate Institute" the possibility of admitting women.[97] The very existence of female academies or seminaries in the area provided some excuse for the university not to provide an education for women, since there were alternative places they could study, even if they could not offer the status that came with a university degree.

William P. Willey, the chair of law and history, wrote a particularly scathing article in favor of admitting women to West Virginia University in which he accused the people of the state of being "immensely unjust."[98] The article, entitled "West Virginia's Wrong to Womankind," focused primarily on the fact that women of the state needed to go elsewhere to get a university education and that this was not only a disservice to the students it was a great detriment to the advancement of the state. He expressed frustration that the university was only half as useful as it could be by only educating half the young people it actually could. Moreover, a move toward coeducation would show that the citizens of West Virginia were "in harmony with the most enlightened judgment of the educational world" which had already accepted the idea of women attending universities. As he progressed through his points, Willey got increasingly impassioned about the plight of young women in the state, using terms like "ungallant" and "positively cruel." Whether intentionally funny or not, one of his final arguments is definitely amusing: "We seem to think we have done our whole duty when we have made for them the same provision as we make for the tabby cat—a warm place on the rug before the fire."[99] After a prolonged debate in both the state Senate and on the Board of Regents, women would finally be admitted to West Virginia University in 1889.[100]

Scottish Overview: The Universities of Aberdeen, Edinburgh, Glasgow, and St. Andrews

In 1889 the British Parliament passed the Universities (Scotland) Act leading to the admission of women to all Scottish universities. Section 14, sub-section 6, of the Act, states that the legislation will "enable each University to admit women to graduation in one or more faculties, and to provide for their instruction."[101] So the universities could admit women if they wanted to, and if they only wanted to admit women to certain fields of study and exclude them from others, this was all right. Due to circumstances and preferences at each of the four Scottish institutions affected, different approaches to coeducation were tried at each during the Victorian Era.

The University of Aberdeen was originally two separate institutions—King's College (established in 1495) and Marischal College (established in 1593). They merged in 1860 to become one university.[102] Having two campuses created difficulties for students when it came to scheduling and attending classes. The Arts Faculty was located at Marischal College after the union, while science classes were taught at King's.[103] Therefore, along with working to include women after the Universities (Scotland) Act was passed, Aberdeen had the additional goal of helping the male students at both of their colleges feel more unified, and they were still working on this when women joined the community. By the time the institution celebrated its 400th anniversary in September 1906, the incorporation of the male students was complete, while there was still some resistance to inclusion of women to their ranks, primarily by the public viewing the festivities. At the opening procession, observers commented that the women received the "generally appreciative criticism of the spectators" that were perhaps a little miffed that the women were not dressed as uniformly as the men. The women's attire apparently "did not always harmonise well with the scarlet college gown and black trencher."[104] Tradition was clearly changing, and the visual representation of the university that now included women was an adjustment that some members of the community had not yet made.

Victorian women were first admitted to study at the University of Edinburgh in 1869 after an initial application made by Sophia Jex-Blake.[105] Special "Regulations for the Education of Women in Medicine in the University" were established by the University Court to try to accommodate the wishes of the women students and the faculty who would be teaching them. These regulations included the university providing "separate classes confined entirely to women" and the students in those classes paying additional fees "in the event of the number of students proposing to attend any such class being too small to provide a reasonable remuneration" for the professor who would be teaching the additional women's class.[106] This setup was consistent with Victorian morals and was intended to ensure that the female students would not feel uncomfortable or embarrassed learning certain medical information in a mixed classroom with men. The cumbersome scheduling led to inconvenience for the faculty members involved, and some soon began to refuse to teach classes twice—once for men and once for women. Additionally, after Sophia Jex-Blake and her fellow classmates had completed their course work, they were denied permission to graduate by university authorities who determined that "the University Court had exceeded its legal powers in admitting them at all."[107] The resultant legal battle will be discussed in Chapter Three, but at this point it is important to note that women would not again be allowed to take courses at Edinburgh until after the 1889 Parliamentary legislation opened degrees to them.

With such a notably unsuccessful attempt by women to earn degrees at the University of Edinburgh, supporters of women's admission to the other Scottish universities were well aware of the challenge in front of them. Principal John Caird, of the University of Glasgow, in considering these struggles, encouraged the members of the Glasgow Association for the Higher Education of Women "to continue 'knocking at the door of the University till we got the honours we desired'."[108] In 1883, owing to financial difficulties, the association chose to incorporate as a college. They chose as their name Queen Margaret College, honoring as they did St. Margaret of Scotland, the wife of King Malcolm Canmore, who was considered to be the earliest patroness of literature and art in Scotland.[109] The college opened its doors in Glasgow in 1884 offering correspondence as well as ordinary courses for women in the west of Scotland. Throughout the transition the women involved in the association, then college, maintained close ties with the faculty and administration at the University of Glasgow, with most courses at the women's college being taught by those officials. This step was unique in the history of women's access to male universities in Scotland and would lead to a slightly different path toward inclusion than was found at the other three institutions.

Soon after Queen Margaret College was established, it found its foremost patron, Mrs. Isabella Elder, the widow of Glasgow shipping magnate John Elder.[110] Mrs. Elder supported the idea of women's higher education and to this end bought North Park House and grounds, on the banks of the River Kelvin near the Botanic Gardens, and presented them to the ladies of Queen Margaret, rent free.[111] In 1890, a Medical School for Women was also founded in Glasgow, which was constructed with the further financial help of Mrs. Elder.[112] The day-to-day running of the college was left largely to Janet Galloway, the honorary secretary. Working free of charge from the beginning of the association until her death in 1909, Miss Galloway was a strong figure without whom Queen Margaret College might have failed. Indeed, the organization and direction she provided, as well as being an inspirational role model to students, made her one of the primary assets of the women's institution. In 1895, as a result of the passage of the Universities (Scotland) Act, the University of Glasgow, which had been established in 1451, absorbed its neighbor, making Queen Margaret College the institution's newly created Women's Department.[113]

The University of St. Andrews is composed of three colleges—St. Salvator's, St. Leonard's, and St. Mary's. The first two were combined in 1747 into the United College of St. Salvator and St. Leonard, while St. Mary's remained separate, to some extent because it was the college of theology.[114] The first woman included on the matriculation rolls at the University of St. Andrews was Miss Elizabeth Garrett (Anderson) in 1862.[115] After petitioning the University of London for entrance to medical studies unsuccessfully in 1861, she made enquiries in Edinburgh, Glasgow, and St. Andrews in hopes of having greater success in Scotland. Her attempts in Scotland would also amount to little. As noted by the University of St. Andrews' librarian James Maitland Anderson, "she was never really a student of the University, as she was prohibited from entering the United College class-rooms, and her matriculation and class tickets were declared to be null and of no effect."[116] Another interesting aspect of the history of the University of St. Andrews was the creation of University College Dundee, which was originally a constituent college of the older institution. Dundee was a coeducational institution from its inception, a stipulation insisted upon by Miss Mary Ann Baxter who, along with her cousin Dr. John Boyd Baxter, founded the institution.[117] From one standpoint then, it can be said that the University of St. Andrews became coeducational when courses began at Dundee in 1883.

Prior to the Universities (Scotland) Act, the Association to Promote the Higher Education of Women in St. Andrews negotiated with the university to offer women a degree of L.L.A.—Lady Literate in Arts or Lady Licentiate—that involved some instruction, but was primarily based on examinations.[118] The women were supposed to be held to the same standard as men who received a Master of Arts degree, and the degree was advertised explicitly that way.[119] In theory this would have satisfied the requests of many Victorian women, but in reality the feminine nomenclature caused the L.L.A. to be viewed with suspicion and a lack of respect by many. As educationist Maria Grey pointed out, the women "were entitled Licentiates, instead of Bachelors of Arts, from some foolish fear of the ridicule attaching to the latter term applied to women."[120] This system was maintained as a form of correspondence courses into the twentieth century, well after St. Andrews was opened to women by Parliament. St. Andrews would remain on the forefront of women's access to university degrees when they allowed women to pursue the degree of Doctor of Laws, making them popular with women in other nations where this was not possible. In 1897 the first woman to earn this degree was Eugenie Sellers from Munich. She was followed by Millicent Garrett Fawcett from London who would go on to be a well-known suffrage campaigner in the twentieth century.[121]

Southern Overview: The Universities of Alabama, Mississippi, South Carolina, and Tennessee

In the southern United States the moves toward coeducation "lagged behind" other regions, as a result of traditional views of women as a gentler, protected class of people.[122] Because university coeducation was seen as a "northern" idea, those who maintained loyalty to "Confederate" sympathies were offended that such a concept was being introduced to their community.[123] The historian Amy Thompson McCandless argues that the "preference for single-sex colleges was closely connected to Southern racial policies" because women's status was often closely tied to that of blacks before the Civil War in a strict hierarchical system. Following the war, it was important for institutions to maintain their white male hegemony, and this meant keeping both African Americans and white women out of positions of authority and away from university education.[124] In the end, state-supported universities in the South opened their doors to women much sooner than their private counterparts, just as in other sections of the country. All of these institutions did so after the Civil War and their readmission to the Union, and not until spurred on by the Morrill Land Grant Act (which was extended to former Confederate states during Reconstruction).[125]

The University of Alabama in Tuscaloosa opened on April 18, 1831, with a grant of land from Congress.[126] When the institution was founded, a provision was included for the eventual formation of a "female branch" of the institution, though one never materialized.[127] The first substantial agitation for coeducation began in the 1870s, but because rebuilding was necessary after the Civil War, the idea of expanding the student body to include women was "too difficult a task" and took more than twenty years to happen.[128] The ultimate pressure put on the University of Alabama to admit women came from Julia S. Tutwiler, daughter of one of the institution's first faculty members, who petitioned the administration in 1892.[129] She called on officials to provide education for all "youth of the state" because, as a public institution, it was their duty to do so. The Board of Trustees invited her to make her case to them in person, and she was apparently "persuasive as well as eloquent." They moved immediately to open their doors to qualified women.[130] The faculty at Alabama voted in favor of the trustees' decision after an investigation was completed by a committee composed of "Professors Wyman, Edgar, Parker and McCorvey."[131] It is worth noting that three of these men—McCorvey, Parker, and Wyman—had daughters who would eventually benefit from the coeducation provided at Alabama, making them not as impartial as other faculty members might have been.[132] The presence of women at Alabama seemed to be embraced by the male students, regardless of how few female classmates they actually had. The male students wrote flattering rhymes about the ladies, who were referred to as "Tuscaloosa Girls": "She is pretty to walk with, And witty to talk with, And pleasant, too, to think on."[133] The fact that the University of Alabama was also a military school during the nineteenth century meant that women would always be in a male domain regardless of a new policy of coeducation.[134]

The University of Mississippi, located near the town of Oxford, was chartered by the state legislature in 1844 and opened in 1848.[135] The first attempt to bring coeducation to the campus was in 1870, when the Board of Trustees of the Columbus Female Institute offered their buildings and grounds to the University of Mississippi with the hope that they would be converted into a women's department in the state university. At the time the university's own Board of Trustees was dealing with the question of racial integration and chose not to add the question of women's admission to the already complex issues it was debating.[136] Two years later the state formed the "Reneau Female University of Mississippi at Oxford." Had it survived beyond its first year, this institution could have remained a female department within the University of Mississippi for decades to come, as a similar approach to coeducation was taken at Duke University in North Carolina and lasted until 1972.[137] As it was, the state legislature repealed the law they had passed establishing the women's university just ten months later.[138] Women were finally admitted to the University of Mississippi, known popularly as Ole Miss, in 1882, but only to the collegiate courses.[139] One of the early supporters of women's admission to the University of Mississippi was its former chancellor Frederick A. P. Barnard.[140] When women were finally admitted, he expressed pride in the accomplishment in a paper he wrote entitled *Should American Colleges Be Open to Women as Well as to Men?* The University of Mississippi, he pointed out, "honorably leads the way" in coeducation by becoming the first "strictly Southern college" to open its doors to female students.[141]

The General Assembly of South Carolina chartered the South Carolina College in 1801, though it did not open until 1805. A century later it would become the University of South Carolina in 1906.[142] In the state there were several small women's colleges in the nineteenth century, but government support for the higher education of women did not come until the 1890s. In 1891 the legislature created "The Winthrop Normal and Industrial College of South Carolina" which provided training for women in "industries suitable for them," including work as teachers.[143] Women were then admitted to the South Carolina College in 1895, after the General Assembly passed an act the previous year, making it a coeducational institution.[144] There was resistance to this move, from the college officials, students, and state politicians alike. Governor Ben Tillman argued that women would lose their femininity if surrounded by men and feared they might even be corrupted by male influences.[145] In an effort to appease the governor and his supporters, initially women were only allowed to enter the junior class, but a subsequent act of the legislature opened to them "any class they might be prepared for, on the same footing as the men."[146]

The final southern university included in this study is the University of Tennessee. Initially, in 1794, the name of this institution was Blount College. It would subsequently undergo three name changes, first to East Tennessee College in 1807, second to East Tennessee University in 1840, and third to the University of Tennessee in 1875, after the state legislature passed an act making it the official institution of higher education in the state. Early records of Blount College show that five women attended classes, including Barbara Blount, the daughter of William Blount, the territorial governor who was also the namesake of the college, and Polly McClung who was the daughter of one of the trustees, Charles W. McClung. All the women did well in their studies and earned "marks of distinction" that were recorded with their names.[147] A century later supporters of further expansion of higher education for women heralded the achievements of these women for their academic success and "discreet conduct."[148] More modern historians have questioned whether these young ladies were taught at the college level, and even if they were, they did not receive college degrees.[149] When women were readmitted to the University of Tennessee in the 1893–1894 academic year, they were often reminded of their female predecessors even choosing to name their new literary society after Barbara Blount.[150]

Welsh Overview: University College of Wales, Aberystwyth

In Wales, the timeline of university education is far different from that in other parts of the United Kingdom. The decision to begin a University College for Wales did not come to fruition until the 1850s, with the first constituent college opening in Aberystwyth in 1872. The University Colleges in Wales were originally based on the Queen's Colleges in Ireland. This initial idea, suggested by Sir Hugh Owen in 1854, was not put into motion until there was a sufficient ground swell of support from the people of Wales. After a great fundraising effort, the college was opened in 1872 in a building that had been designed by J. P. Seddon to be a hotel but was instead purchased for use as an institution of higher education.[151]

The College at Aberystwyth continued to struggle financially during its first decade of operation.[152] Then in 1885 the institution looked doomed to failure when the Old College Building was destroyed by a fire that took the lives of three men.[153] Finally in 1886 Parliament began to give the Welsh college an annual grant that would give it the stability it needed to survive and eventually thrive.[154] The institution had several name changes, and there is often inconsistency among contemporary writers regarding the name by which it was referred to at any specific point in time. The titles include Aberystwyth College; University College Aberystwyth or Aberystwyth University College; the National University of Wales, Aberystwyth; the University College of South Wales, Aberystwyth; and the University College of Wales at Aberystwyth. It is now Aberystwyth University.[155] Each of these name changes indicates a change in the organizational structure of university education in Wales.

In 1883 the Aberystwyth campus was joined by a second Welsh college in Cardiff (known originally as the University College of South Wales and Monmouthshire), with a third following in Bangor in 1884 (known originally as the University College of North Wales).[156] As both these institutions admitted women from their founding, they do not meet the criteria for discussion in this book, but the new options for women in Wales who wished to pursue higher education are still important to keep in mind. Technically Aberystwyth was also coeducational from the outset, but had no women students in the early years it was open: "There was nothing to forbid the admission of women, it was simply a thing no one had contemplated."[157] Women began studying at Aberystwyth in 1883 and initially met little resistance, due, as in Scotland, to the nonresidential nature of the college. The three colleges were later incorporated together as the University of Wales in 1893 after which they worked together to administer matriculation examinations and offered the same degrees.[158]

Historiographical Approaches

Though studies of the shifts in higher educational policy in the nineteenth century are extensive, those of women's admission to universities are less frequent. The innovators—Oberlin College in the United States and Girton College in England—have received due coverage from historians and commentators alike, often focusing on the first women to accomplish a certain goal. As the historian Gillian Sutherland puts it, "[T]he historiography of women's secondary and higher education...has been one of the last bastions of heroic fairy-tale, a story of great women battling against all obstacles."[159] The moves made toward women's admission to medical studies are an example of this type of historiography. As at time, the actions of this handful of women are marked more by their distinctiveness than for being reflections of larger societal forces. Taking a gendered approach to the study of higher education enables a wider perspective on conditions in society and their impact on the subject. The historian Helen Lefkowitz Horowitz argues in her 1995 article "Does Gender Bend the History of Higher Education?" that it is important to eliminate "the boundary separating schools from other key institutions in society, such as the family or the associations of civic life."[160] Moreover, to fully understand the effects of changes in higher education on all the participants, it is necessary to consider the experiences of both men and women, as they increasingly shared the university experience.

The general historiography of women's admission to higher education falls into two basic areas. The feminist movement has influenced many historians leading some to look to the entrance of women to college and university as the first step toward their social and political emancipation, suffrage, and equality with men.[161] Studies of the early female students of higher education have looked at the interests and outcomes of those women during and after university, often focusing on the work of individual women like Emily Davies or Elizabeth Garrett Anderson. These historians see this bottom-up approach as the most personal method, giving the most human portrait of the individuals involved. In terms of higher educational history, this type of explanation reinforces feminist theory by studying the agency of the students themselves. The other approach of this period of higher educational history is more institutionally based. It involves studies of individual institutions and includes the admission of women as a necessary chapter in the progress of the school, not as a feminist watershed. These studies, though less politically charged on the whole, focus on the history of educational policy, rather than focusing on the people most affected (the students). They are primarily approached by studying institutions in a top-down direction. Simply put, the administrators, policies, and legislation are studied as the primary elements needed to make changes.

A primary example of the institutional study is historian Rita McWilliams-Tullberg's *Women at Cambridge*. Reprinted in 1998 on the occasion of the fiftieth anniversary of the admission of women to degrees and membership at Cambridge, this work is often cited as an "indispensable" history of women's higher education in England. In her introduction to the revised edition, historian Gillian Sutherland points out that this is not a "self-congratulatory house history," but is instead a synthesized study of the context and changes of women's admission to the University of Cambridge.[162] McWilliams-Tullberg follows an institutional approach, investigating the separate histories of Girton and Newnham Colleges as they were added into the larger university body. To further facilitate her research, various legislative turning points and a series of failed attempts to gain admittance to membership of the university are used as a framework for the book. These ties to the institutional structure are not eased a great deal to allow for further investigation of the student's lives on campus, primarily because of the nature of Cambridge's history. The length of the struggle to gain full admittance to the institution necessitates McWilliams-Tullberg's choice of format, though it does remain part of the top-down school of educational history.

A similarly groundbreaking study of women in British universities is historian Carol Dyhouse's *No Distinction of Sex? Women in British Universities 1870–1939.* Like McWilliams-Tullberg's work, Dyhouse needed to lay the foundation for studies of women in British higher education by tracing the more general societal evolution of thought on the subject. The challenges to the established structure of education and community belief can be seen in the organizations Dyhouse chooses to investigate and the obstacles they needed to overcome. Unlike McWilliams-Tullberg, Dyhouse includes analysis of the lives of women students (housing and extracurricular activities), providing a bottom-up look at events.[163] Although studies of individual institutions, like Cambridge, are certainly valuable, a better picture of societal affairs may be garnered by the investigation of more universities as Dyhouse does. Additionally, making a comparison between different countries illustrates the transnational similarities in the development of thought about women's place in Victorian society.[164]

My study attempts to bridge the gaps between these main areas of educational historiography, as well as expand the focus to include comparisons across national boundaries. Building on the groundwork laid by my predecessors, it is possible to examine more fully the events and experiences in both the United States and the United Kingdom. Though the conclusions drawn by other historians of women's higher education are by no means incorrect, many studies have been too narrowly focused to develop much of the true nature of women's admission to universities in the Victorian Era. It is undeniably important to recognize the significant contributions made by individual students and groups who fought for women's admission to colleges and universities in the overall course of women's history. This progress was both linked with that of their male counterparts and separate from the history of men's higher education in many ways. What is unwise is to assume that because some noteworthy individuals used the admission of women to universities as a means to revolutionize gender roles, that all, or even most, women students felt the same way. The primary theoretical question posed by this book—whether women's education worked with or against prevailing societal beliefs—has appeared in a handful of previous histories of the subject. Historians continue to debate the intentions of nineteenth-century educators and students, as well as the results of their actions. No consensus has yet been reached, particularly because the variation of opinion is both so slight and so significant.

The historian Barbara Miller Solomon indicates the currently accepted answer to this question in her book *In the Company of Educated Women:* "Academic study became another way to reinforce the differences between men's and women's lives."[165] The line drawn between Solomon's statement and the thesis of this book comes in her use of the word *became.* Solomon argues that women's education, at every level, was intended as a way to change the roles played by men and women in the nineteenth century, but that the educators who were pushed into this course of action eventually found a way to subvert the female students and their quest for equality. While undoubtedly this can be seen in some specific cases, the generalization is unwarranted. More commonly, Victorians in both the United States and the United Kingdom saw the extension of education to women, from the outset, as a method of sustaining separate male and female spheres of influence. Solomon does acknowledge this when she notes that "no woman could forget that she was in a man's world," but her assertions regarding women's "intellectual awakenings" at college detract from a realistic understanding of the community at large.[166] The differences in opportunities for men and women, no matter how slight, continually reinforced their respective roles in society and gently pointed them toward appropriate fields of study and extracurricular activities. To accept Solomon's summary is to assume that nineteenth-century educators and students wanted men and women to become equals in all areas, and that simply was not true.

An alternative approach to the views expressed by Solomon is that of historian Lynn D. Gordon in *Gender and Higher Education in the Progressive Era.* Gordon argues that administrators occasionally discussed how the female students were "shaped" by higher education, "without considering in turn the effect of students on institutions or the larger society."[167] By inverting the question in this fashion, Gordon is better able to study the students themselves, and their experiences at university. Unfortunately, she does not give the educators sufficient credit for understanding the influence of their actions on the communities they worked in. Gordon's work, though very thorough in the issues discussed, inflates the role students had in their own situation, while unduly deflating that of the administrators. In this way she imposes a modern twentieth-century framework on the nineteenth century and gives credit to the students for far more agency than they ever could have had. While this type of study has its benefits, looking solely at the student motivation in attending college ignores the initial question of women's admission to higher education, for without understanding the general decision of Victorian society to allow this to happen, no female student could have made steps toward changing their communities. Much of this is the result of Gordon focusing on institutions that accepted coeducation from their beginnings, and the fact that her timeframe extends well into the twentieth century.[168]

Another criticism of the existing historiography of women's higher education comes from a more recent study of university coeducation, historian Andrea G. Radke-Moss' *Bright Epoch: Women and Coeducation in the American West.* Radke-Moss refers to most previous histories of the subject as "practicing a 'fly-over' approach to women's education, focusing on eastern academia or large urban centers in California and the Midwest" and leaving out numerous colleges and universities in the country that were also significant to the overall progress of women in university. Her focus on land-grant institutions in the western portion of the country, specifically in Iowa, Nebraska, Oregon, and Utah, deals with many of the same topics as this study, but she does not have need to deal with the process of inclusion since all of the colleges and universities she studied were coeducational from their outset.[169]

My proposal in this book is that there was a polarization of the curricular and extracurricular worlds on the basis of gender that was a concerted effort by both institutional officials and students to reinforce the gender expectations of the community at large. Regardless of the idiosyncratic conditions in a particular town, city, or country, the underlying presence of separate male and female spheres in Victorian society helped to direct the activities of both men and women in universities. Despite the new opportunities given to women, the structure of the institutions continued to guide them into traditionally female roles,—particularly those as wives and mothers. The inadvertent reshaping of these established gender roles was precisely that, inadvertent, as the goal of the majority of women who chose to attend a previously male university was not to revolutionize the world; they simply wanted to gain an education and earn a degree. The resulting need to assess and compare both the administrative and legislative positions in regard to women's education, along with the implications for the students themselves, in a transatlantic context, sets this study apart from many of its predecessors in the field.

CHAPTER TWO

Victorian Views of Coeducation

Image 2 Illustration of "Princess Ida" from *Songs of a Savoyard*, 1894.

Credit: W. S. Gilbert and the Gilbert and Sullivan Archive, Boise State University, Boise, Idaho.

In 1847 Alfred Lord Tennyson wrote a poem entitled "The Princess" which featured as its main character Princess Ida. The heroine decided to eschew the company of men and begin a women's college, entirely staffed by women. The purpose of the poem was "to exhibit the mental relation of woman to man" and to provide a satirical look at the movement for women's admission to colleges and universities in the United Kingdom.[1] After he became Poet Laureate in 1850, Tennyson's words held more weight than the average writer, and his use of the phrase "sweet girl graduates" in the poem created a sort of new classification of women and resonated in a time when the higher education of women was debated regularly.[2] The illustration on the preceding page is from the 1894 publication *Songs of a Savoyard* that included the lyrics of a number of Gilbert and Sullivan productions including "Princess Ida" which was based on Tennyson's poem.[3] In it we see Ida refusing the entreaties of her male suitors, as she chose independence over marriage. The possibility that giving women too much education would make them not want to marry or indeed make them unmarriageable was an enduring cause of anxiety during the nineteenth century. The fascination with the story presented by Tennyson and its implications for gender relations in society over the whole of the Victorian Era are unmistakable, and they were present not only in the United Kingdom but also in the United States.[4]

The far-reaching influence of Tennyson's poem and Gilbert and Sullivan's interpretation of it can be seen at the University of Aberdeen, where a student poem, "Trim Little Maids at 'King's'," echoed some of the sentiments in the original poem, as they referred directly to "the song that the Princess spun," while using another of Gilbert and Sullivan's works as inspiration as well.[5] A distinction was drawn between the women's college Tennyson wrote about and women at Aberdeen who were "wise and wary, Tired of a ladies' seminary" and preferred instead to attend a coeducational university. In Aberdeen the ideals of women in the Victorian Era were still important, but many did not feel that receiving higher education would threaten them or lessen a woman student's femininity. In the poem "Trim Little Maids at 'King's' " the author brushes off the concerns of opponents of university coeducation and says that the "feminine instinct" will not necessarily "fail" if a woman attends university.[6]

In this chapter questions regarding the ability of women to undertake university study, raised by opponents and proponents alike, will be considered. Within this issue were many aspects dealing with the inherent strengths and weaknesses in the female constitution. Medical evidence was used to prove women were not able to handle higher education, while other evidence showed the opposite.[7] Women were often thought to lose that which made them distinct from, and compatible with, men as a result of education. Finally, contemporaries often questioned the value of educating women at higher levels. If their only job in life was to be wives and mothers, what use did they have for a university degree? Clearly when several of the professions (particularly teaching and medicine) were opened to women, this question held less weight, but the underlying issue of women's position in society and the desire to maintain established gender roles remained.[8]

"Separate spheres" Ideology

In the nineteenth century, male views of women and the views women had of themselves, greatly affected the progress of women's access to higher education. Both in Britain and America, a primary topic of conversation was the actual physical capabilities of women. As Charles Rosenberg and Carroll Smith-Rosenberg note, the "Victorian woman's ideal social characteristics—nurturance, intuitive morality, domesticity, passivity, and affection" were believed to be under threat by higher education.[9] Victorian sensibilities were relatively universal on either side of the Atlantic, because they were based on similar codes of morals and religion. The fact that the United States had only been separated from Great Britain for one hundred years shows that there was little evolution in women's rights or educational realities in either country during that period.[10] The key feature of women's existence in both countries was the notion of separate spheres of influence for men and women. The increasing influence of the industrial revolution also worked to define more strictly the "separate sphere" in which women functioned, both inside and (to an extent) outside the home. Middle-class divisions of labor were generally applied in conjunction with an ideology of domesticity. A view that women were domestic and subordinate members of society passed from mother to daughter over time.[11] Together these placed limits on women in terms of occupation and, what is more important to this study, of education.

There was a considerable "surplus" of women in the nineteenth century, which made marriage prospects more limited than in the past. The need for the maintenance of social status dictated new solutions for women and their role in society.[12] In this way, education was seen as a means of keeping women from going down the social ladder, while for men it was a means of moving up it. Economic arguments for giving women a university education factored into the question as well. There was an increasing need for employment for middle-class daughters. After the 1850s, the rate of marriage decreased considerably, with the age of marriage postponed until the mid- to late twenties, due in large part to the increase in professional men who chose to marry at a later age.[13] The historian Michael Sanderson notes that, in England, "over a quarter of a million women had little expectation of marriage and the lifetime protection of husband and home."[14] As a result of relatively poor marriage prospects, many young women were forced to look for other avenues for their life to take. Middle-class families were often unable to support unmarried daughters for life, so many young women were required to seek training to become governesses or teachers, the two main "acceptable" occupations for young women prior to marriage. With particular reference to university education, concerns centered on whether women's entrance into the workforce would threaten male occupations.[15] In South Carolina male university students had long been "confined to a more limited and obscure sphere—farmers, country preachers, teachers and doctors, who contributed... to the material and moral uplift of the commonwealth."[16] The more women pushed to become teachers and doctors, the more unsettled society was thought to become.

In a speech given to the student literary societies at West Virginia University in 1888, the industrialist and future West Virginia senator

Stephen B. Elkins discussed the need for appropriate education for women, but stopped short of stating that university coeducation should be in their future: "The age is industrial, commercial, and productive, and men and women should be prepared to live in it by being educated in a way that would fit them for such pursuits."[17] The worry that an inappropriate education, which many believed coeducation was, would make women unfit for adult life permeated many discussions of admitting women to male universities. Students in Pennsylvania also wondered about the relationship between women's education and their role in society. In an editorial in *The University Courant* in Pittsburgh, the long-term consequences of this evolution in women's roles was brought into question: "Woman's sphere has of late widened out considerably, but it is a question whether it will continue to broaden or whether a reaction will set in, and her rule will be again confined to the narrow limits of the home."[18] And at the Pennsylvania State College a student identified only as H. R. L. concluded, "there is in woman a grade, a delicacy, a fineness of sensibility, a tenderness and quickness of insight not natural to the stronger sex. These points are . . . shown in the spheres of life which they are to fill. . . . A woman needs what will make her a queen of the household and of society, while man needs what will fit him for the harder, sterner duties of life, to which ladies should never be driven except in cases of exigency."[19] Whether there was an urgent need for women to be admitted to male universities is debatable, but there was certainly a demand that they be.

In Indiana the question of "separate spheres" was also discussed in the 1880s in regard to the efficacy of coeducation in all levels of schooling, but especially as it pertained to university training:

> That the great mass of women do, and always will, find their chief work inside the home circle, is true, and that they need some special training for the various employments they engage in. It is further true that the studies pursued in acquiring a general education should have some reference to such callings. But the great fact remains that the main object in study is not simply the facts learned, but the culture and development gained by the study. The development of mental power and of character are the great ends to be reached. Now whatever is good to develop a boy's mind is good to develop a girl's mind. What will give a boy culture will give a girl culture. A study that will make a man reason will make a woman reason.[20]

Society would be stronger if more of the people, male and female, were educated. The fact that it was likely that a woman was still most likely to remain "inside the home circle," or sphere, was no excuse for not educating her because she could take care of all her responsibilities more effectively if she approached them in a thoughtful and reasoned manner. Adding in a moral component would also appeal to Victorians who sought to raise the level of "character" in society. Surely having well-educated wives and mothers would be a stronger argument in favor of women's higher education than trying to convince people that continuing to hinder female education would help the community progress.

"Republican Motherhood"

In the newly formed United States of the eighteenth century, a concept regarding the status of women, termed "Republican Motherhood," emerged to raise women's position in society. This theory was established for women who were not involved in public life so they "might play a deferential political role through the raising of a patriotic child."[21] In part, this theory was applied as a way to keep women in their societal role (or "separate sphere"), while elevating that role considerably.[22] Supporters argued that women should be provided with education in order to become teachers, both in the home and outside it. Campaigners for women's admission to higher education in Britain utilized the theory of Republican Motherhood as well. Women were expected to produce "strong enterprising men," and an education would be extremely valuable in this aspect of women's traditional societal role.[23] At the Scottish Institution for the Education of Young Ladies in Edinburgh, organizers cited the "universally acknowledged" fact that the "education of women does not influence the tastes and opinions of present society alone,—it affects those of succeeding generations."[24] This argument was used in this instance as an incentive for parents to send their daughters to the institution, but it also reinforced the belief that girls needed a good education if they were to be good mothers.

Factoring into the idea of women's traditional societal role of motherhood was the introduction of women to the teaching force. In both the United States and the United Kingdom teachers were predominantly male before the nineteenth century. The drain on manpower caused by military conflicts (most notably the U.S. Civil War) is often cited by historians as the reason women were initially called on to teach. Practical concerns, such as the fact that women could be paid smaller salaries, were more likely the cause of the shift. Theories of Republican Motherhood also played a part in women's relatively easy acceptance into the teaching profession. As women's primary role was to raise the next generation, it made sense for them to do so in the classroom as well as in the home. Most of the young ladies who taught in schools did so only until they married; this made the profession one of discretion and respectability. Employment as a governess was viewed similarly, as good practice for girls who would eventually become mothers. As a result of these two job opportunities, much of women's early higher education was focused on the skills and subjects needed by teachers and governesses.[25]

Care needed to be taken to guarantee that a woman's education would be appropriate to these professions, and would also help to make her desirable to men. The higher education of women was seen by some to make them more attractive, rather than less so. In one article published in *The Westminster Review* in 1881, the author argued that a woman's education could make her more marriageable:

> "Jane Eyre" is a typical and profoundly true embodiment of one of the best established laws of human nature—viz., that a woman of highly developed intellect, who is ever so plain and unattractive in person, can command the passionate and lifelong devotion of men who are far enough from being either saints or heroes.[26]

Whether one agrees with the author's assessment that Mr. Rochester fell in love with Jane Eyre because of her education or believes it was because of her love for him is questionable, but for the purposes of an argument in favor of women's education, it would have been compelling to people who were familiar with the novel. For the concept of Republican Motherhood to reach fruition, it was not enough to educate girls and women. It was also necessary for them to get married so that they could pass their knowledge on to subsequent generations.

The historian W. Gareth Evans, in his study of *Education and Female Emancipation* in Wales, attributes a passive acceptance of the role of education as a mechanism of social control in the Victorian Era: "the education of girls and women came to be viewed as a means of strengthening the female's influence in society at large, while still maintaining the status quo of male supremacy in the workplace."[27] The first half of this comment indicates more thought on the part of those who pushed for women's acceptance to schools and universities. They knew that by educating women, they were also educating future generations. In this way, women's traditional role was being reinforced and simultaneously strengthened by greater experience and further academic training. The second half of Evans' statement, regarding male supremacy in the workplace, is also worthy of comment. In this area, he is entirely correct; the male domain of the workplace was to be preserved at all costs.[28] As noted above, the topics women were directed toward studying at university were only those which would refine positive female virtues like the arts, languages, literature, and so on, all of which would help them raise future generations appropriately.

Moral and Religious Concerns

Women's role in Victorian society was most often associated with issues of virtue. The possibility of a lady soiling her nature by undertaking tasks to which she was not suited led to a significant portion of the opposition to women's access to higher education. Proper women were not to consort with strange men, and the social atmosphere seen on male university campuses would inevitably offer the chance for male-female interaction. The religious undertones of this set of arguments are undeniable. One writer suggested that Christ himself had set women in "the highest place in the hierarchy of virtues, leading to women's continued piety throughout history."[29] Through this placement on a pedestal, women were constantly seen as the weaker or gentler sex, who needed protection from their male counterparts. The evolution of separate spheres ideology is a basic outcome of this theory. Additionally the idea that women should be virtuous and kind led to new roles for women in the nineteenth century. Religious groups looked to women to reach out to the community in various forms of charity or philanthropy, and women saw a way to mix with the wider public spheres of society.[30]

Social commentators saw education as one of the key ingredients of moral and social improvement, which was needed by all members of society, particularly women. Maria Grey, writing in 1871 *On the Special Requirements for Improving the Education of Girls*, argued for the improvement of all education, both male and female, as it would provide "sound judgment" and "moral discipline" to all who received it. She further concluded that the first object in improving the education of girls should be "to fit them morally and intellectually, not for the matrimonial market, but to do their duty wisely and faithfully in that station of life whatever it may be, to which it may please God to call them."[31] The religious aspect of Grey's argument remained strong throughout the nineteenth century, whether at private or state-funded institutions. Colleges and universities in each country had historic connections with specific religious groups, and most maintained, as a primary purpose, the education of future clergymen. This did not necessarily lead to religious restrictions being placed on students, and many more practical subjects were introduced in the course of the century as the needs of society required people to be trained in new fields at a high level.

Education was a key area of focus for social reformers during the nineteenth century because, as Josephine Butler commented, it was "the sure road to emancipation. It is to education that we must first look for the emancipation of women."[32] The evolution of such arguments came from abolitionist groups to women's rights groups in both the United States and the United Kingdom during the nineteenth century. Women worked through their Christian duty and moral sensibilities to promote human rights and, in the process, honed their political and public-speaking skills. Eventually these women began to apply those talents to their own lives, with human rights issues eventually focusing on women's rights in particular.[33] Women's activism was encouraged in the United States by the Second Great Awakening of religious fervor in the country.[34] This movement brought with it a great desire for women to be educated to spread the word of God to the public. The reformers in the 1830s were not the same women pushing for women's entrance to higher education in the 1870s and 1880s, however. While there were often familial or other links with the earlier movement, this new group of women was truly a second generation who, while influenced by their predecessors, were driven by different personal experiences.

From a modern perspective, women's entrance into public debates over abolition, or health and housing issues should have proven them worthy of a chance at higher education. Formal and traditional religious sensibilities often prevented this from taking place. It was thought by many that women who strove for further education were tempting fate and would surely displace nature as God intended it. "Lectures and sermons to 'females' upon their God-given limitations and the shamefulness of it if mental activity should carry them beyond, were designed to fill up any conscious gap in their lives."[35] Coupled with this idea was the belief that women, as complements to men, should do their best to cultivate the talents they were imbued with:

> The man has always been regarded as the rightful lord of the woman, to whom she is by nature subject, as both mentally and physically the weaker vessel; and when in individual cases these relations happen to be inverted, the accident becomes a favourite theme for humorists—thus showing that in the general estimation such a state of matters is regarded as incongruous.[36]

This leads easily into a discussion of the moral concern for educated women, but the more significant conclusion to be drawn from this quotation regards the idea of the status quo. Since women had always been of a status inferior to that of men, many opponents of coeducation felt that they should always be. The possibility of admitting women to men's institutions of higher education put under threat the entire way of life for society in general, and the innate fear many comparatively well-educated men acted through was often one of the most difficult obstacles for women to overcome.

Danger to the moral well-being of women was used as an argument to keep them out of universities in the nineteenth century. In practice, the female societal role as the protector of the community's virtues had different implications for the campus environment than many might have expected. Many universities found that the introduction of women to their institution meant that the male students were more restrained than previously. Codes of "chivalrous" behavior, which persisted into the nineteenth and twentieth centuries, meant that men had to show an amount of respect to their new female counterparts, whether they wanted to or not. An historian of the University of Wisconsin, J. F. A. Pyre, noted, "The difficulties of student discipline were rather mitigated than increased by coeducation.... Upon the usual peccadilloes of the college student, the presence of women acted as a restraint."[37] Pyre was not the only commentator who was impressed with the "civilizing" effect the introduction of women had on the male student body and their overall behavior. Lynn D. Gordon reports the use of the same argument at the University of Chicago where the university president desired "to minimize the rowdier aspects of male undergraduate life" through the introduction of women to the institution.[38]

The religious influence through education reached further in the United Kingdom than it did in the United States. At the Universities of Cambridge and Oxford students had to take an oath to the Church of England to be able to matriculate. One of the reasons the University of London was formed was to eliminate this religious requirement in higher education.[39] The newer civic institutions, like Owens College in Manchester, which opened later in the nineteenth century, followed London's lead by not requiring a religious test for employment or admission. This meant that the

> students, professors, teachers, and other officers, and persons connected with the institution shall not be forced to make any declaration as to or submit to any test of their religious opinions, and that nothing shall be introduced in the matter or mode of education or instruction in reference to any religious or theological subject which shall be reasonably offensive to the conscience of any student, or of his relations, guardians, or friends under whose immediate care he shall be.[40]

To allay fears that this secular approach to higher education would be amoral, all candidates for admission also had to provide testimonials of their "good character" from former instructors or school principals before they were allowed to matriculate.[41]

In Scotland the church, or Kirk, was largely responsible for children's schooling prior to 1872, when Parliamentary legislation gave more control to the state (in the form of the Scotch Education Department [SED]). Primarily where parish schools existed, they conducted Scottish education, with adventure, burgh, and other endowed schools providing training for some in the middle and upper classes.[42] The Disruption of 1843, which divided the Scottish church into three sections (the Free Church, the Established Church, and the United Presbyterians), also spurred the development of additional schools, with each group wanting an equivalent number to the others. The curriculum in these schools served several purposes, the first of which was the moral education of the younger generations. The other main function of the education system was to prepare Scottish children for their place in society. Girls received "education for dependence," for example. This strict protestant philosophy was modernized to a certain extent at the end of the seventeenth century. What remained constant was a belief that the education system should play a positive role in society and provide guidance for the nation.[43]

When the Queen's Colleges were established, they were intended to make higher education available to all men in Ireland, regardless of their religion. Belfast was a city dominated by the Protestant Scots-Irish community, who did not want to travel to Dublin to study for a university degree. "The Queen's Colleges at Cork and Galway were intended for Catholics from Munster and Connaught, but, as it transpired, were as likely to have attracted Protestants as Catholics."[44] This was due primarily to the opposition by the Catholic Church to the teachings at the Queen's Colleges. As a result, all three colleges suffered from a lack of students during much of the nineteenth century. Many women would graduate under the auspices of the Royal University of Ireland, but few of those women did their course work at one of the Queen's Colleges.[45] Instead most preferred to attend a women's college, one that was religiously affiliated, or one that was both; the most popular were Alexandra College, Dominican College, Loreto College, and St. Mary's University College in Dublin and Magee College and Victoria College in Belfast.[46] Many families considered the higher education of their daughters to be a potential danger to their moral development, and adding a questionable secular curriculum to the mix made it far less likely that they would consider sending their daughters to any of the Queen's Colleges.[47]

Questions of religion and character were raised by many U.S. universities during the nineteenth century just as they were in the United Kingdom, even if the same level of hostility to secular education was not present. At some institutions, like the University of Missouri and the Western University of Pennsylvania, students were expected to attend chapel services regularly.[48] In other states, like Alabama, Michigan, and Tennessee, arguments were made that as governmentally funded institutions, their universities should not be religious in nature and therefore "could give no preference to the creed of any religious sect."[49] To reassure citizens that they were maintaining proper morals when coeducation began, in Alabama they stipulated that only "women of good character, who have attained the age of eighteen, may be admitted to the University."[50] Bessie Parker was described in her senior class yearbook as "A maiden never bold; of spirit so still and quiet."[51] Such praise, of both male and female students, was made regularly by commentators, officials, and the students themselves who expressed particular pride in their charitable work and the lack of irresponsible behavior that was found at other institutions of higher education.[52] In *The Corolla* in 1896 the editors noted that a "goodly sum" of students' expenditures went to "home missions" and that "two-fifths of the students" were members of the Young Men's Christian Association.[53]

The admission of women often went hand in hand with the evolution of state-supported colleges and universities that followed a secular approach to teaching and learning, emphasizing a general concept of "morals" rather than any particular set of denominational teachings. Some faculty members still incorporated religious concepts into their lectures whether they were supposed to or not. Margaret Boyd noted one such lesson given by Professor William Scott, Ohio University's president: "It was about manners, our associates, books, secret thoughts, Marriage etc. I think I never saw him so much in earnest. How he did speak of those who had evil thoughts—lust. . . . I wish O so earnestly that I was a better stronger girl. Strength of character."[54] Boyd's praise of this lecture indicated how intensely she believed in what she was being taught and how hard she felt she should work to live up to the expectations society had for her. One can question if Professor Scott would have discussed manners and marriage to a group of all-male students or if this were done for Miss Boyd's benefit, but it appears his words had the desired effect on her if on no one else.

The shifts in curriculum in the second half of the nineteenth century also included a move toward learning more practical and less esoteric subjects.[55] University curricula in the sixteenth and seventeenth centuries focused on theology, making the admission of women a nonissue (as the study of theology at this time led to a position in the clergy or ministry). With the expansion of university teaching to include more arts and literature, it became more conceivable to include women in that education.[56] It was not a coincidence, for example, that the Supplemental Charter that permitted the admission of women to the University of Durham in 1895 also included provisions for the first degrees in Letters.[57] Opponents of one transformation were more than likely opponents of both, because they saw a secular, coeducational university as containing "all the features of Sin personified."[58] The passionate attacks made on these institutions, directed at either the admission of women or the lack of religious teachings, would continue throughout the nineteenth century in both the United States and the United Kingdom. At times one topic would overshadow the other, and at other times the rhetoric was so interchangeable that it was not always clear if the reason for the attack was sectarian or gender related.

Medical Concerns

During the nineteenth century there were many medically based ideas about what women's role in society should be and what effects higher education would have on that role. If women were to be wives and mothers, then their physical health in producing healthy children was of great importance. In regard to this view, many people saw women as intellectually inferior and far too physically delicate to handle the rigors of a university course.[59] Physicians and educators wrote about the significant physical demands women faced that men did not in terms of reproduction. They stated that if women used energy to develop mental strength, they would not have enough left to properly develop the physical strength needed to have children. Eventually suggestions emerged that university-educated women would become infertile.[60] The associated belief, which received much notoriety on both sides of the Atlantic, was the concept of "race suicide." As noted by the historian Cynthia Eagle Russett in her book *Sexual Science: The Victorian Construction of Womanhood*, women "it seemed, had no right to self-fulfillment that could stand for a moment against the claims of society on their wombs."[61] Russett terms this concern about women's reproductive capabilities being threatened as "obsessive," reinforced by scientific and other demographic information that showed a decline in marriage and birth rates in the second half of the nineteenth century. The introduction of pseudo-scientific and sociological theories to explain this phenomenon increased throughout the period, so the new higher education of women often served as an easy target of writers.

Following the publication of Darwin's *On the Origin of Species* in 1859, white men were presented with the argument from Social Darwinists that they were superior to other races, as well as to women. The preservation of this status relied greatly on the propagation of strong offspring, created by the most suitable parents. The argument that followed held that "the higher education of women is surely extinguishing her race," with college studies making women too weak for childbearing. This theory was two-fold, incorporating medical opinions and societal expectations. Women who attended university tended to delay marriage; this shortened their period of fertility and possibly reduced their chances of having strong, healthy children.[62] Educated women were also expected by many in society to lose their beauty as a result of poor health, brought on by the sedentary nature of university study. This idea was reinforced with the belief that educated women would not be able to have children due to this loss of femininity, which made them unattractive to men, and thus not marriageable. So, although women might be capable of undertaking higher education, they could only do so at the expense of their attractiveness and maternity.[63]

In 1887 an article entitled "Mental Differences Between Men and Women" was issued in Great Britain.[64] It stemmed from Darwin's study of the *Descent of Man* (1871) and included some comments made by the evolutionist. The author, George J. Romanes, was a friend and colleague of Darwin at Cambridge.[65] In this piece he made references to females as "secondary sexual characters," clearly indicating that males were a more evolved group of the human species. Along with this he described females as being a separate psychological species, with distinctly different abilities and needs. The crux of the argument in the terms of education came with the discussion of an indisputable fact:

> Seeing that the average brain-weight of women is about five ounces less than that of men, on merely anatomical grounds we should be prepared to expect a marked inferiority of intellectual power in the former.[66]

Women were basically incapable of the mental activities of men, because they lacked the mental capacity needed to undertake them. The historian Joan Burstyn notes that, along with determining that women's brains weighed less than men's, scientists of the day had concluded "that the bone structure of their heads was less mature—in short, that evolution had passed them by."[67] This use of Darwinism to illustrate the physical deficiencies of women and their inability to learn caused writers to argue that the idea of women taking part in higher education was fanciful at best. Women were also thought to be less mentally robust, lacking the energy needed to sustain "serious or prolonged brain action." This was thought to lead to significantly weaker powers of acquisition and creativity, while women were also more emotional and less impartial. He argued that women crave entertainment and society and any attempts at education up to that point in time had only been (and could only be) superficial. Romanes did grant that there were exceptions to the rules laid out, particularly in the area of musical ability in which women were generally equal to men. The only benefit he could see to education for women was a basic refinement of the senses and "of nervous organisation."[68]

The argument laid out by Romanes set itself up as being indisputable. Clearly women could not succeed in the male world of higher education; their brains could not handle the mental strain it required.[69] Opponents of women's entrance to universities would have taken such an argument as proof of their claims and attempted to either stop further advancement or, more likely, redirect the course of women's higher education as it stood in the 1880s. The question of women's educational aptitude was not easily determined by scientific evidence. Though women's brains are lighter than men's on average, that does not necessarily equate with less mental capacity. In 1889, just two years after Romanes' article was printed, the *Encyclopedia Britannica* contained a section entitled "WOMEN. Position of American."[70] This entry discussed a wide range of issues, from legal to social, and interestingly, it included a counterargument to that made by Romanes. It acknowledged that men and women are biologically different and might benefit from different types of education, but it disputed the idea that women were incapable of tackling the education provided for men. Along with this argument was a different angle on the medical evidence noted earlier:

> Much stress has been laid upon the supposed fact of women's smaller brain weight. Since, however, physiologists have declared... that so far as investigations have gone the relative brain-weight of women is about one-forty-fifth of the body's weight, while that of men is one-forty-sixth.[71]

This tends to point to the idea that women are in fact *more* capable than men when it comes to intellectual ability, but this is not the reasoning applied in the article. Instead, the *Encyclopedia Britannica* merely notes that there is no great difference between the relative brain weights of males and females and, beyond this, there was no proof to support the correlation between brain size and intellectual power. Therefore each side of the debate on women's higher education was able to cite medical "evidence" to support their claims, and neither was sufficiently strong enough to sway large portions of their audience. More than likely, those who supported women's entrance to higher education would continue to, and those who opposed it would not change their stance either.

Writing in *Popular Science Monthly* in 1905, A. Lapthorn Smith offered a different explanation of the race suicide theory, showing that it was not resolved by the end of the Victorian period. He included many of the traditional arguments about women's health, saying that the "duties of motherhood are direct rivals of brain work, for they both require for their performance an exclusive and plentiful supply of phosphates." But, he went on to argue that women in higher education became more independent, more self-confident and, as a result, chose to marry later in life or perhaps not at all. Smith believed that, "in the cultivation of the powers of analysis and criticism," no educated woman could ever be satisfied with her status in life, especially if she lived only as a wife and mother. Educated women also had a knowledge of their abilities which created "an aggressive, self-assertive, independent character, which renders it impossible to love, honor and obey the men of their social circles who are the brothers of their schoolmates."[72] These conclusions had wide implications for the progress of women in colleges and universities. There was a perception that women could become too educated. Smith also mentioned that there were many "failures of marriage, directly due to too great a cultivation of the female intellect, which results in the scorning to perform those duties which are cheerfully performed, and even desired, by the uneducated wife."[73] If Smith's statements proved true and educated women were becoming dissatisfied with their expected role in society, then institutions would need to be carefully structured in an attempt to preserve the existing society.

The view that higher education was physically debilitating and detrimental to women's reproductive processes carried beyond the issue of motherhood. Some contemporaries felt the mental strain was deemed unnecessary for all women, irrespective of the possible benefits of education in raising children. In many areas on either side of the Atlantic, recommendations for improvements in women's education included retaining a separate educational system. Since women were physically smaller, weaker, and less capable, they could not study in the same ways that men could. Though such a statement seems derogatory to readers today, these arguments did at least give women credit for being intelligent. The difference in biology was simply seen as evidence that access to education itself and to certain courses of study in particular should be restricted to preserve a woman's safety and well-being.[74]

Coeducation Versus Separate Education

An initial question raised by nineteenth-century educators (once higher education was within women's grasp) was whether men and women should attend the same institutions. In 1872 Edward H. Clarke wrote a book entitled *Sex in Education; or, a Fair Chance for the Girls* that addressed the issue of women's education. He was an eminent Harvard Medical School professor whose argument was so effective and popular that his book went through seventeen printings. Clarke's argument spurred much debate in both the United States and Britain. He offered a definition of coeducation and was careful to indicate that schools may offer the same education to men and women, but that it was not necessary. He supported his arguments with medical "evidence" that indicated that brainwork unnecessarily drained energy from other bodily functions, such as development of the reproductive organs. So, although education was valuable for girls, it was also potentially dangerous and should be approached with great caution.[75] Female students were well aware of Dr. Clarke's book, and discussed his ideas. One student at the University of Michigan, Olive San Louie Anderson, concluded in her semi-autobiographical novel, *An American Girl and Her Four Years in a Boys' College,* that Clarke "makes a great ado about nothing, and fails to hit the point."[76] The point was that women were perfectly capable, in Anderson's opinion and experience, of handling the same university education as men.

In 1874 Henry Maudsley of the University College, London wrote a book entitled *Sex in Mind in Education* which built on Clarke's ideas in a British context.[77] He agreed with Clarke's basic premise that mental exertion by women, particularly during menstruation, would damage their prospects of reproduction or at the very least of having strong and healthy children.[78] These arguments fed the debate over coeducation with a primary concern being whether an equal education was also an identical one. The differences between the sexes were the focus of much of the controversy when the first women were admitted to universities. There was a belief that any type of education for women might neglect the needs of women. Many who supported women's education felt that they needed protection from the outside world and saw higher education as a way to safely prepare young ladies for their future roles as wives and mothers.[79] Numerous all-female colleges were established on both sides of the Atlantic, supporting the notion that separate societal spheres of influence would be best served by separate educations.[80]

Popular literature regularly commented on the notion of university coeducation versus single-sex education during the Victorian Era. The use of the term *co-ed* to refer to women students only was seen by many as irreverent, though it was a quick and easy way to convey specifically which young women someone was referring to. Many works of fiction popularized the term; some meant it to be insulting, while others just used it as a form of shorthand. One such book was *A Man for a' That*, written by George Van Derveer Morris. In it the life of students at the fictional "Darnforth College" are discussed. According to the author the "Co-eds . . . were from the good solid families of the State. They were the daughters of ministers and pious laymen; they were the young women who could be trusted and who had ambition to study what their brothers were studying."[81]

This nonjudgmental and fair appraisal of the women students and their position within society was not held by all, as the author went on to note that some girls in "the town looked down with something approaching scorn upon the college society, and especially upon the poor 'Co-ed.' No girls from that lofty set had ever brought shame upon the family name by attending a mixed school." The concern was not the pursuit of higher education; it was the decision to attend a coeducational institution that some saw as bringing "shame" on one's family.[82] The narration continued with an acknowledgement by the "Co-eds" that "they knew that they were looked upon as outcasts" who might never fit in with society's expectations of what a woman should be. This does reflect Olive Anderson's views in her novel, but a modern reader of both of these books cannot help but wonder if these works were intended to portray absolute reality or if tension over women's place in coeducational universities was simply an effective literary device during the nineteenth century.

The desire of women and their supporters to debate the benefits and drawbacks of coeducation was approached in more formal ways also. In 1887 Mr. Francis H. Underwood, L.L.D., U.S. consul in Scotland, began a series of talks on the subject of women's higher education.[83] Two of these talks were given to the women of Queen Margaret College in Glasgow. Mr. Underwood discussed the effectiveness of different systems of women's inclusion in higher education in the United States, as well as the press response to them. He said that the subject of the higher education of women in the United States had received great attention within the last twenty years and that several excellent institutions existed. As a resident of Cambridge, Massachusetts, the first institution he discussed was Harvard, and its annex, Radcliffe. He argued that these schools had the same standards of admission as the Universities of Edinburgh or Glasgow, and he used this as evidence that they were of equal stature to the European institutions. He also noted that the new presence of women under the administration of Harvard University had in no way lessened the reputation of the university.[84]

From a more theoretical standpoint, Underwood believed that the education of women and the development of female minds were equivalent to the growth of democracy, because both were necessary tasks at the end of the nineteenth century. This section of his speech received much applause and a great deal of local press coverage. Underwood recognized the financial and ideological pressure Queen Margaret College was under at the time and reassured them that the ladies in Massachusetts "could stand" the criticism they received from the public and press alike. The support of international figures, like the U.S. consul to Scotland, was clearly thought to strengthen the likelihood that Parliament would include provision in the Universities (Scotland) Act for women, a topic they were debating at the time. Hearing the basic reinforcement of the ideals of women's higher education was also a great comfort to those in

Glasgow, as they felt more confident in their struggle knowing the same goals had already been achieved in the United States.[85]

Another American who traveled to the United Kingdom to discuss the prospect of university coeducation was Nathan Sheppard, who spoke at the Universities of Aberdeen, Edinburgh, and St. Andrews about how to persuade audiences. Because one of the obstacles women had to overcome in the nineteenth century was the belief that they should not speak publicly in political or other forums, Sheppard used his platform to encourage those in attendance to throw their support behind the admission of women to their universities. Sheppard even addressed his talks to "any man or woman who seeks a living or renown by means of the most perplexing and elusive of the arts—the art of public speaking."[86] His more progressive, open-minded view that both men and women could become accomplished speakers was not well received by some of the students at Edinburgh, due largely to the fact that they were in the process of fighting the admission of women to medical courses at the time.[87] Sheppard recalled being hissed by the male students when he commented on the status of women in U.S. universities where "if a woman passed the examination, she was given the degree or admitted to the class."[88] When Sheppard challenged the visceral response of the students, the remainder of the audience, who did not feel that the students had responded appropriately regardless of the heated atmosphere on campus at the time, applauded.

A third commentator who compared the decisions made about university coeducation between the two nations was David Staars, who wrote a book called *The English Woman: Studies in her Psychic Evolution*, which looked at women in English-speaking nations. In discussing Aberystwyth College, the coeducational practices were of particular interest and specific reference was made to Wales having followed "the example of American universities" in this respect. The author noted that women students were well looked after and that "young men and young women meet on a footing of perfect equality" both inside and outside the classroom. This would lead "to the great advantage of the natural evolution of moral and psychic relations between the sexes."[89] Psychic in this instance presumably meant mental or intellectual, not clairvoyant. The larger argument that coeducation provided men and women the chance to get to know each other before choosing a husband or wife was a common selling point for coeducational institutions that their single-sex competitors could not offer to the same extent.

In the United States almost all women's colleges were private ones, which made them in most cases more expensive and often farther away than state-supported institutions. The first exception to this rule came in Mississippi in 1884, two years after women were admitted to the collegiate courses at the university, making it coeducational.[90] Proponents of separate educations for men and women were still so in favor of that approach that the institution that would later become the Mississippi University for Women was opened, making it the first state-supported college for women in the country.[91] The existence of private and normal colleges additionally meant that there was a wide variety of options for women in the state, so that they were provided with many types of higher education that they could pursue, depending on what type of future they wanted to have and what method of university teaching they preferred.

Conclusions

With all the varying ideas about women's need for education and ability to undertake it, an observer during the Victorian Era could not have easily concluded that university coeducation would ever be a success. While government legislation invariably contributed to women's entrance to higher education, the most crucial vote of confidence came with the assistance of forward-thinking male administrators. Despite the relative autonomy of universities in determining the course of coeducational study they chose at this time, changes in official policy would not have succeeded without a general consensus of support from the public at large who supported institutions financially. Having leading university faculty and officials lend their support to the admission of women greatly increased the public's confidence in the notion. Families of prospective university students needed to accept the forms of coeducation adopted by individual institutions as well, or they would have opted to send their children somewhere else. In hindsight, despite the many obstacles it faced, the admission of women to male colleges and universities happened in a relatively short period of time. This was only the first step, as more would have to be done than simply admitting women to lectures in order for them to have full access to higher education.

The simplest representation of both the public's reluctance to accept women's higher education and their acquiescence to women's admission to universities can be seen in the writing of George Romanes. His medical concerns about women's mental capacity have already been examined closely, but the conclusions he draws in the same essay show a different resolution than one might expect. After discussing the many hereditary faults of the female species, Romanes states that they cannot be changed by education. He also argues that women's desire for higher education is "preparing for the human race a second fall."[92] The biblical reference notwithstanding, his opposition to coeducation is apparent. He concedes that women's progress in attaining higher education in Britain has been significant and admits his arguments may do little to slow this advancement. He has decided, then, that the best he (and those who agreed with him) could do was help to direct "the flood into what seem likely to prove the most beneficial channels." Educated women *could* become good complements to well-educated men, if they were guided properly. In addition, the society as a whole must choose to either support or oppose women's entrance to universities, so that no harm came to those involved. As Romanes saw it, the decision to admit women to universities needed to be unanimous for there to be a possibility of success.[93]

CHAPTER THREE

Administration and Legislation

Image 3 "The College Government," a student illustration from *La Vie*, the Pennsylvania State College Yearbook, in 1893.

Credit: Penn State University Archives, Pennsylvania State University Libraries.

Decisions about who should be admitted to a university were never in the hands of the students, despite the fact that the students were the ones most affected by the change to their environment. The students at the Pennsylvania State College showed their view of administrative and government control over them in their yearbook, *La Vie,* in 1893 with the illustration shown above.[1] The handle on the press reads discipline, and the officials were grinding the students into shape with it. There are no women visible in the illustration even though they had been part of the student body for more than two decades. This was probably out of a sense of delicacy but also because the male students were far more likely to be in need of discipline than the women. In no case was the admission of women to a university the direct result of agitation on the part of the male students. On many campuses the idea was welcomed with little to no difficulty from the men, but in some cases there was loud disapproval of the decision of their administrators. Students were well aware of the debates about women's ability to undertake higher education, and there were mixed reactions to the prospect of having them do so in mixed classrooms.[2]

Due to the fact that the institutions studied in this book are state funded, or as D. I. Mackay put it in his study of Aberdeen University, "a large part of the expenditure on university education is met from public funds," government legislation could be used to compel or enable colleges and universities to change their admissions policies.[3] This chapter will examine the decisions made by national and state governments that contributed to the admission of women to higher education in the United States and the United Kingdom and the reactions of the students and the wider community to these changes. Finally the role of administrators in opening the doors of some universities to women will be considered, since it was often the support of individual faculty, chancellors, presidents, or principals that brought university coeducation into practice. During the nineteenth century colleges and universities were governed on a daily basis by faculty members, some of whom had administrative responsibilities. The idea of having full-time administrators was not present, and there would not have been funding for such an option at most institutions. The rules and regulations regarding student conduct on campus, or discipline as emphasized by the students in *La Vie*, were in the control of the faculty and were "conducted with large dependence upon each student's sense of honor and moral responsibility."[4]

Early Efforts at University Coeducation

The historian Jane Rendall notes that women's push for greater equality in the United Kingdom was "encouraged by the accession of Queen Victoria in 1837" although, ironically, the monarch did not believe women should receive university degrees.[5] Her Majesty's appreciation for complementary roles for men and women, and for the upholding of tradition, was not held by all of her subjects. Numerous individuals and groups worked on behalf of women's admission to universities, either targeting specific institutions or promoting the admission to degrees generally. The first women's colleges that were established—Queen's College, Glasgow (1842), Queen's College, London (1848), and Bedford College, London (1849)—were not able to grant degrees to their students that were on par with those awarded at male universities. Clearly this earlier period of women's higher education is of great importance, and as the historian Sarah J. Smith argues in her study of the Scottish institution, women's historians ignore this earlier period of women's higher education because graduation was not an issue for students (whether male or female) and because concern over women's access to the professions did not emerge until the late nineteenth century.[6] Indeed this distinction between access to courses and the granting of degrees was of the utmost importance to advocates of women's entrance to universities, as they did not believe themselves to have real or complete higher education without the distinctions received by men.[7]

In the second half of the century groups whose sole purpose was advancing the opportunities available for women's higher education were formed in most British cities, including the Edinburgh Association for the University Education of Women, the Glasgow Association for the Higher Education of Women, the Ladies' Association for the Promotion of Higher Education of Women in Cork, the Ladies' Education Association of London, the Leeds Ladies' Educational Association, and the Manchester Association for Promoting the Education of Women.[8] The popularity of such groups extended to the United States, with the formation of the American Woman's Educational Association as well as several local groups.[9] The members of the ladies' educational associations were primarily the wives, widows, or daughters of wealthy citizens who had disposable income of their own. Professors' spouses were also key figures in the Edinburgh Association, as they were the most sympathetic to the higher education of women. Although the majority of members in these organizations were women, men did take part in their activities. For instance, in Glasgow the university principal, John Caird, became the association's chairman, and Mrs. Campbell of Tullichewan its vice president.[10] The title of president was given to HRH Princess Louise, Marchioness of Lorne, one of Queen Victoria's daughters, though it was an honorary distinction.[11] The organizations intended to offer women teaching similar to that which men received at universities and "to promote generally the higher culture and education of women."[12]

Lectures were given by university faculty who generally supported the idea of women's higher education, though some did not believe that university coeducation was the best means of accomplishing this goal. A few years before the Glasgow Association (or the G.A.H.E.W.) was formed, Professor John Nichol, chair of English at the University of Glasgow, offered a series of lectures on English literature for women in the Corporation Galleries to an audience of both men and women.[13] Professor Nichol held progressive views on women's equality largely due to the influence of his stepmother, Elizabeth Pease, who had been a member of the radical Quaker network and an active antislavery campaigner. Despite this, he continued to doubt the ability of women to work at the same level as men in universities. He warned the ladies of the danger of "an over-stimulus in the direction of competition—to which he had occasion to know the minds of women were still more liable than those of men."[14] This sentiment did not deter the work of the ladies who desired access to the university, and ultimately the inaugural lecture of the G.A.H.E.W. in November 1877, given by logic professor John Veitch in his rooms in the university, was seen as a great popular success.

The lectures arranged by the ladies' educational associations became so noteworthy in the press that other similar talks were arranged, even in cities where no such group existed. Women in Durham availed themselves of any educational opportunities they could, including public lectures given by members of the faculty. One such talk was given on January 31, 1883, at the College of Physical Science demonstrating the use of "lime-light and other apparatus." The presentation was attended by more than one hundred "spectators...amongst whom were several ladies."[15] Clearly if the author of this piece felt it was important enough to mention the women in the audience, this was not a typical occurrence, possibly because of the topics at hand which included the anatomy of beetles and moths among other things.[16] The desire for women to further their education was unmistakable, and the fact that all fields were of interest, not just the English literature taught by Professor Nichol, was an indication that limiting women to gender-specific subjects of study would not be possible. The push for university admission would also build in strength as more and more lectures were given and attended because women found that they wanted more knowledge and they found that they could, in fact, learn it.[17]

As noted earlier, some advocates of women's higher education focused their attention on specific institutions. One of these was the University of Edinburgh, more specifically its medical degrees. Indeed, the admission of women to medical degrees would prove to be one of the most difficult barriers to break through. Jessie Meriton White first asked the University of London if women could take the examinations for a medical degree in 1856. She was followed by Elizabeth Garrett in 1862, but both were told that medical degrees for women were out of the question.[18] As discussed in Chapter One, Sophia Jex-Blake made the first application in Edinburgh and was initially accepted to study there in 1869. The male students, who were described as having a "ruffianly element" in their number, acted out after this decision was made. Descriptions indicated that the inclusion of women went smoothly at first, but after women started to do better than their male counterparts on examinations, and thus be in competition for scholarships, tensions started to increase. In one incident the male students "mobbed the women at the entrance to Surgeons' Hall, where the lectures were given, pelting them with mud and assailing them in the streets with foul language."[19] The concern over admitting women to universities in the medical field in particular was acute because many felt that having men and women in the same classroom while studying anatomy was not only unwise; it was immoral.

University officials decided to reconsider their decision to admit women to medical courses (even though the women were taught separately from the men) and questioned whether they had overstepped their power to have taken this action in the first place, and denied permission for the women to graduate. This new ruling caused consternation amongst the British medical community, many of whom had not previously made statements on whether they believed women should be trained as doctors. Most could agree that charging the women students for courses leading toward a degree and then not letting them complete it was unfair at best and "conspicuous injustice" at worst. The case was taken to court and won by the ladies, but on appeal the verdict went in favor of the university.[20] A thorough accounting of the legal case can be found in Sophia Jex-Blake's *Medical Women: A Thesis and a History.* She concludes, quite magnanimously, that she will leave it to "the public to judge how far such a course would have been more prudent and more commendable than that which they actually followed, and for which they have been so bitterly punished."[21] What the legal question boiled down to was whether the University Court, at any institution, had the power to grant degrees to women because the right to do so had "never been exercised or even claimed by any University in these kingdoms."[22]

It was determined after events in Edinburgh that Parliament would need to clear up the situation. They made their first attempt in April 1874 when "A Bill to Remove Doubts as to the Powers of the Universities of Scotland to Admit Women as Students and to Grant Degrees to Women" was introduced by Mr. Cowper-Temple, Mr. Russell Gurney, Mr. Orr-Ewing, and Dr. Cameron.[23] Though this Bill received its first reading in Parliament, it was withdrawn prior to its second reading on 11 May without any discussion. The reasons for this action are unclear, but regardless of the failure of the 1874 Bill, the idea of opening Scottish universities to women was now finding a legislative voice. A second similar bill, the Universities (Scotland) (Degrees to Women) Bill, was introduced the following year by the same four MPs. The historian Lindy Moore reports that the primary opposition to this bill was not its provision for the higher education of women in general, but was specifically related to the "vested interests of the medical profession."[24] Once the second bill was under consideration in Parliament, those who supported it made their opinions known. Sixty-five petitions were sent in an effort to convince the members to pass the legislation, coming from leading faculty and administrators at Edinburgh, Glasgow, and St. Andrews, together with the town councils of Aberdeen, Edinburgh, and Linlithgow.[25]

Despite this visible show of support from the academic and nonacademic communities in Scotland, the press reported overwhelming opposition to the idea of women studying in the universities of Scotland. In 1875 *Punch* claimed that the "weight of Scotch opinion—above all, of Scotch University opinion—is dead against the Bill. The Scotch Universities do not see their way to mixed classes of both sexes in Anatomy and Pathology, and cannot undertake to provide separate classes for Ladies."[26] The opposition noted in *Punch* could have been a reference to the male students who, though not unanimously opposed to the admission of women to the University of Edinburgh, did have a number of extremely vocal opponents in their ranks. Modern readers may assume that the strong feelings against the admission of women to university study were the result of sexism or misogyny, but it is far more likely that the male students at the time did not wish to share their educational resources with anyone. Mr. Noel, MP for Dumfries, also questioned whether it was an enabling bill that gave the Scottish universities the option to admit women or a bill that was forcing them to do so. This concern over the wording of the legislation and the possible problems with implementing it as a result led to Noel stating that "[m]uch as he regretted appearing to oppose the higher education of women, he felt bound to vote against the Bill on the grounds he had stated."[27] These issues placed a considerable amount of doubt in the minds of many MPs, leading to the defeat of the bill by 194 votes to 151.

Through the debates over the Scottish legislation, it became clear that the structure of universities would first need to change if the admission of women was to go smoothly the next time it was attempted. The new university courts, established in the Universities (Scotland) Act of 1858, were thought to be too weak to affect the needed changes, while it was felt that the Lord Rector's position was far too strong, with their personal view capable of deciding the issue single-handedly.[28] As a result, the following year another royal commission was appointed to investigate Scottish higher education (there had been commissions previously in 1828 and 1870). Concurrently, in 1876, the passage of the Russell Gurney Enabling Act permitted medical examining bodies to admit women to their examinations on the same terms as men.[29] Mr. Gurney had previously been one of four MPs who put forward the unsuccessful 1874 and 1875 Bills, indicating a particular interest on his part in women's right to higher education and degrees.[30] This did not ensure that women would be admitted to male universities, but it did remove one of the arguments being used to prevent it. Soon after the passage of the Enabling Act, the University of London opened its doors to female students in 1878, becoming the first male institution in Britain to do so.[31]

The Morrill Land Grant Colleges Act (1862)

The role of legislation in the progress of women's admission to higher education was significantly different in the United States and the United Kingdom. The primary distinction comes in the lack of national policies in the United States since education was not included in the Constitution, leaving it instead under the control of individual states.[32] It is true that the "founding fathers" believed that U.S. citizens should be educated because "an informed populace was a prerequisite for the healthy functioning of the democratic system."[33] Suggestions were made that a national university be founded, but the idea was determined to be an impractical use of the limited resources of the new country. As the historian Merle Curti notes in his *Social Ideals of American Educators*, America's revolutionary leaders had professed the necessity of illuminating "the minds of the people at large... without regard to wealth, birth or other accidental condition of circumstance," but the employment of the ideal was often absent.[34] Curti goes on to argue that in the early years of the nineteenth century, U.S. educators began to adopt the philosophy that public education prevented social disintegration and was therefore necessary in maintaining a balanced society.[35] The need for the education of citizens, as discussed in Chapter Two, was combined with a "belief that the citizenry, through its elected government, should provide the material support such a system of formal education required."[36] Taxes, in other words, could and should be used to ensure that education took place and that the United States could take its place on the international stage. Throughout the nineteenth century each U.S. institution was allowed to determine their own policies of admission and the federal government had no power to change those policies, though they could encourage changes by offering funds with strings attached to them.

The primary example of this approach is the Morrill Act, in 1862, during the U.S. Civil War.[37] Officially known as the Land Grant Colleges Act, this piece of legislation included provisions for the introduction of scientific and technical subjects, such as agriculture and engineering, to university curriculums that would be more useful to the average student. Congressman Justin Morrill of Vermont suggested that these institutions should "lop off a portion of the studies established centuries ago as the mark of European scholarship and replace the vacancy... by those of a less antique and more practical value."[38] This helped women in two ways. First, changing the curriculum to include more practical subjects in place of the classics would cause institutions to rethink what a university education was. This in turn would affect their view of what a university student needed to be, and it would also eliminate many of the entrance requirements (particularly those in Latin and Greek) that were not always taught to girls in the country. Second, it was unlikely that the government would send funds to any school that was likely to fail due to an insufficient number of students.[39] This pressure, though indirect, helped to support the case of women's admission to higher education in developing states with smaller populations.

Justin Morrill had made a previous attempt to open colleges to students "of average means" in 1857. That bill was passed by Congress, but vetoed by President James Buchanan on the grounds of unconstitutionality.[40] Once the Civil War started, the Morrill Act became one of several pieces of legislation designed to spread northern ideals and limit the spread of the Confederate states, should their rebellion succeed.[41] In practical terms the federal government gave 30,000 acres of federal land for each member of Congress in a state to a college or university, which they could use or sell to help them establish new departments of study. In many cases this resulted in the establishment of an agricultural research farm, since one of the stipulations was that buildings (construction or repairs) not be paid for with the land grant.[42] Institutions that benefited from this Act included the Universities of Michigan and Wisconsin and the Pennsylvania State College. They were required to invest money from the sale of any of the land grant in U.S. stocks or bonds, the interest of which could be used to hire faculty and purchase books or other materials. One of the most important stipulations in the Act was the requirement for military training, not a surprising requirement for a country during wartime.[43]

Evidence of concern about the passage of the Morrill Act and the desire for these newly accessible funds can be found at many colleges and universities in the Union during the 1860s because only one institution per state was to receive a land grant. Competition resulted in some cases, with administrators making pleas that their institution was "the state college" and should therefore become a land-grant college. Indiana University did not receive land-grant funds, with Purdue University in Lafayette being founded instead; it opened its doors in 1874.[44] Other educators saw the legislation as a "full-scale attack on the time-honored study of the 'classics'" and did not wish to lower the content of their curriculum to meet the needs of middle- and working-class Americans.[45] The Morrill Act's stipulations were not always interpreted in the same way by universities either. In West Virginia some members of the state Senate and House of Delegates believed that the military component of the legislation was a requirement for all students and if women were not part of the military training required by the Act they could not be admitted to the university at all. Despite plenty of evidence to the contrary in other states, some members of the state government continued to use this argument as their reason for not adopting a policy of coeducation at the university.[46] At Wisconsin the Act was used as an impetus to admit women. As the historian Jean Droste points out, "it was chiefly practical needs rather than theoretical considerations about the need for women's education" that opened the institution to female students.[47]

After the Civil War ended, the Morrill Act was extended to southern colleges and universities as they were readmitted to the United States. New complications would arise after the Thirteenth, Fourteenth, and Fifteenth Amendments to the Constitution freed the slaves and granted them equal rights with whites. The federal support for land-grant institutions that had racially restrictive admissions policies became controversial, especially after more federal funds were added to the initial Morrill Act through the Hatch Act of 1887.[48] As with the earlier legislation, officials decided which institution in the state would receive the federal funds. In South Carolina, Clemson College (later University) was the designated institution, not South Carolina College.[49] Eventually Congress also passed a Second Morrill Act in 1890 that required states that wanted the new round of federal funds to either admit black students to their land-grant institutions or establish a segregated land-grant college for them.[50] For instance, the "West Virginia Colored Institute" was "for the benefit of agriculture and the mechanic arts," the practical courses that Morrill argued for in his initial legislation.[51] It was not until the latter half of the twentieth century that the federal government began to forcibly open the doors of state-supported colleges and universities, like those in Mississippi and Tennessee, to excluded groups in the country, with private institutions remaining autonomous from much of this legislation.[52]

The Owens College Act (1871)

In 1870 the British Parliament passed the Owens Extension College (Manchester) Act that was quickly supplemented the following year by a "Confirming Bill," titled the Owens College Act.[53] These pieces of legislation first established the Owens Extension College in Manchester, where there was already an Owens College, and second "amalgamated the two colleges and their resources as The Owens College."[54] In addition, the legislation established government ties with the newly reconfigured institution, which would no longer be governed strictly under the terms of John Owen's will. As stated in Chapter One, the Owens College Acts also enabled the governors to admit women if they wished to do so.[55] The standard arguments about the practicability of educating women for professions that they could not yet enter, when a lecture scheme or other means could be organized instead, was made in Manchester, though the primary issue that prevented their admission was the belief that all resources of the college were needed for the men of the city.[56] Looking back on this provision at the end of the century, Christina Sinclair Bremner noted that the "attitude of Owens to women students in the past can hardly be styled cordial. It was expressly founded to instruct and improve 'young persons of the male sex.' Even in 1871, when it obtained the power to open the College classes to women, it carefully protected the young male person by its famous clause: 'conditionally upon adequate provision having been made for the instruction of male applicants.' "[57]

In 1874 Latin and comparative philology professor A. S. Wilkins attempted to bring university coeducation to Manchester by letting women attend his class.[58] Approximately seventy women took him up on his offer, showing the great demand for access to higher education among the ladies of the city. Soon after, the Manchester Association for the Higher Education of Women

> passed a resolution to the council and senate expressive of their warm thanks for the accommodation afforded [by Prof. Wilkins]...and at the same time respectfully entreated the authorities of the college to consider how far similar accommodation might be made available for some other, or for all the classes held under the sanction of the association.[59]

The administration at Manchester maintained the view that there were not enough funds, or enough space, at Owens College to admit women alongside men in the classroom, or indeed to offer separate classes for women only. Even if women had been admitted in the 1870s, they would not have been able to earn degrees from Owens College, since the institution could not confer degrees on any of its students. Instead they took the examinations of the University of London until a new charter was granted, after a "strenuous effort has been made to secure from Parliament a recognition of its deserts by clothing it with the privilege of conferring degrees, and transforming it into a university," creating the Victoria University in 1880.[60] Even though Victoria University could grant degrees of its own, many students still chose to sit for examinations for the degrees of the University of London.[61]

The University Education (Ireland) Act (1879)

The Queen's Colleges in Belfast, Cork, and Galway were established in 1845 by the Queen's Colleges (Ireland) Act. These institutions were intended as a means of bringing higher education to areas outside Dublin, and also to people whom, for financial or religious reasons, could not attend the colleges there. The socioeconomic level of Cork and Galway, in particular, meant that having colleges in these cities would mean that higher education was no longer the exclusive domain of the wealthy. One commentator noted, "Galway is essentially a poor man's college—and the better for that."[62] Arguments were made in the late 1840s that women should also be included in this extension of university teaching because they were essential to retaining the Irishness in the country. A letter to the editor of *The Nation* in 1847 explained, "The hopes of Ireland must mainly rest on the rising generation."[63] Women, as future wives and mothers, were primarily responsible for childcare so they too should be educated about what it was to be Irish. "There is a species of patriotism which may be at once firm and feminine."[64] The patriotic element was a direct result of the desire of many Irish to regain their independence from Great Britain, and having control over the Irish education system was seen as a way to control the future of the country. Others viewed the higher education of women as a waste of resources.[65]

When the Queen's Colleges were united as Queen's University in 1850, women were still not admitted as students. Their desire for entrance would be sidelined by sectarian concerns, and religion in the Irish universities would remain the "dominant political issue for the remainder of the nineteenth century."[66] An attempt was made in the mid-1870s by Mr. Cowper-Temple, MP for Hertford and stepson of Lord Palmerston, to make provision for women's admission to the three institutions because the "Council of Professors" of the three colleges had "thwarted and rendered inoperative" attempts by women to matriculate.[67] Finally in 1879 the University Education (Ireland) Act made it possible for women to earn degrees through examination with the new Royal University of Ireland (RUI), an examining body based on the University of London. This new configuration was meant to be "an olive branch to the Catholic hierarchy" because students could either study the secular curriculum provided at the Queen's Colleges or they could attend a religiously affiliated institution, but all would be awarded the same degrees.[68] Because of the continuing Catholic opposition to secular education, particularly in the Catholic counties in the southern part of the country, the enrollment at Belfast, and especially Cork and Galway, for both men and women, would remain small throughout the century.[69]

As in other countries, having legislation that made university coeducation possible did not make it happen in Ireland. Pressure still needed to be placed on the individual institutions to admit women to classes. For instance, women were still not admitted to Trinity College, Dublin, or University College, Dublin, the main institutions of higher education in the country, except for an occasional lecture.[70] A drawback of the RUI was that no residency was required for students to earn degrees through their exams, except in medical studies, so it was possible for women to be privately trained and receive a university degree.[71] This fact was used by opponents of coeducation to say that there was no reason to admit women to the Queen's Colleges, since they could get degrees without studying in the same classrooms as men. Isabella Tod of the Belfast Ladies' Institute petitioned Queen's University in 1873 and 1882 to admit women.[72] After the later request, and having little grounds for denying it once degrees were available to women through the RUI, women were admitted to arts courses at Belfast in 1882. The following year science courses were opened to them, and in 1889 medical courses as well.[73] Women gained admission to Cork in 1885 and Galway in 1888, all subsequent to the University Education (Ireland) Act.[74]

The Universities (Scotland) Act (1889)

In the early 1880s Parliament made numerous attempts to pass legislation intended to fine-tune the Scottish universities. These draft bills are covered in sufficient detail in the historian Lindy Moore's book *Bajanellas and Semilinas: Aberdeen and the Education of Women*, but are worthy of a quick summary here. First the 1883 Universities (Scotland) Bill, that made no reference to women, was withdrawn due to university criticism of financial provisions included. In 1884 all four Scottish institutions openly opposed a similar bill. In both 1885 and 1886 bills failed, the significant difference between these two drafts and their predecessors was that they would have given commissioners the power to consider the issue of women's admission.[75] In 1887 this new clause was omitted, and then in 1889 it was restored. Moore argues that after 1887, "it was harder for women to argue that the Commissioners' general powers included the power to admit women to the universities once two previous drafts had implied that such powers required a specific clause."[76] Surely if any of the supporters of higher education for women began to lose hope after 1887, they could have looked as easily to the 1885 and 1886 drafts of the bill as positively addressing the question of women. It is also important to note that the failure of these various bills does not owe itself entirely to the question of women's higher education, but in fact had more to do with the general lack of Parliamentary action on any Scottish legislation at the time.

By 1889 the Universities (Scotland) Act passed in Westminster with relative ease, marking the opening of Scottish universities to women, although there was a question about what Parliament intended the institutions to do in order to make this happen. The subsequent ordinances that were released in 1892 were somewhat clearer, but there remained debate over the wording of the government's position. The key issue revolved around whether Parliament was *forcing,* or *compelling,* the universities to admit women or whether they were *empowering* them to do so.[77] Much time can be wasted debating terminology, but in this case the issue of what women were entitled to under the law is based on the specific powers set out by the government. The report issued by the 1889 Commissioners stated that "the University Court in each University may admit women to academic instruction and graduation in any Faculty" and that it "lies with the Court to determine what subject shall be taught in mixed classes or to men and women separately."[78] The ordinances that followed in 1892 specifically used the term *enabled* when speaking of women's entrance to the universities.[79] Parliament saw themselves as giving the schools the power to let women matriculate and graduate on terms equal with men. J. N. Morton, a writer in Victorian Glasgow, commented in his *Analysis of the Universities (Scotland) Act* that some "disappointment was felt that the Act had not made express provision for women studying and graduating at the Universities." Further, women's "right to admission to a University thus seems to depend, first, on the will of the University, and secondly, on an enabling ordinance of the Commissioners."[80] Taking the ordinances into account, it was apparent that one way or another the universities had to begin educating women or they would be in violation of the law. The impetus to find a solution then rested in the hands of the universities, and if they did not fully admit women to their courses, they would have to find an alternative route to educate women.

While it did seem as though the institutions were forced unwillingly into providing education for women, they did have the choice to do so "either by admitting them to the ordinary classes, or by instituting separate classes for their instruction."[81] The responses to this Act at the four Scottish universities varied.[82] Aberdeen was the only Scottish university to immediately admit women to graduation in all faculties, while Edinburgh, Glasgow, and St. Andrews all began admitting women in various fields where, it was argued, they would be the most comfortable.[83] In the "very wide and important" section of the Act that allowed universities to admit women, the commissioners were also able to "institute either an examination for entrance to the University, or one for entrance on a course of study for a degree, or both."[84] At Glasgow newer, more stringent preliminary examinations were introduced in arts and sciences. This increase in entrance standards may at first seem a detriment to women, but the courses offered to them were designed to help them pass these examinations. Additionally, after 1895 all university bursaries were opened to the women students under the provision of Ordinance No. 58.[85] Each institution was required to make any bursary instituted within twenty-five years before the passage of the 1889 Act available to the new women students. Along with this, all new awards needed to be open to both sexes, though like the admission of women itself, the individual institutions were still in control of dispersing bursaries as they saw fit.

Other Considerations Relating to Legislation

One of the earliest, most vocal, and well-known parliamentary advocates of women's equality with men was John Stuart Mill. In his *On the Subjection of Women*, written in 1869, he argued that women should receive the same education as men and that this was the only way for society to be fully civilized. Although he did not specifically mention their admission to universities, it was clear that his call for a "better and more complete intellectual education of women" would include higher education and degrees of some sort if the United Kingdom were to raise women's education to "the level of that of men."[86] Because much Parliamentary legislation is handled on a country-by-country basis, it was common for national interests to come to the fore in government debates and community reactions to them. One article in 1890 reviewed the Universities (Scotland) Act and, at that point, anticipated the forthcoming ordinances:

> The lady students hope that ere October our *Alma Mater* will follow the example of London University, and allow all ladies who have taken the required lectures in them to go for their first professional examination. Why should Scotch women be required to go to London, or Dublin, or the Continent, for a University degree, or be content with obtaining in their own country simply a license to practise?[87]

Though there is little evidence that a large number of Scottish women went abroad to seek degrees, the very fact that they ever did so was a concern to many in the Scottish educational establishment. National and civic pride, in the face of control from Westminster, was evident in Ireland and Wales too. Judith Harford notes that the Royal University of Ireland was "[k]een to justify its contribution to Irish society" as a national institution that was training the next generation of men and women.[88] In Wales the creation of a national university was seen as "a recrudescence of the national sentiment" in the "restoration of Welsh autonomy" with Aberystwyth itself becoming "the pivot around which Welsh national life... has been turning."[89]

In the United States regional and state pride were often just as apparent as national sentiment was in the constituent nations of the United Kingdom. In West Virginia several articles in favor of women's admission to the university centered on the need for women students to be able to stay in the state to earn their college degrees, rather than being forced to go to other states. Once women were admitted in 1889, *The West Virginia School Journal* admonished those who chose to leave their home state to pursue their studies:

> The young men and women who go out of the State to get their higher education make a great mistake. The West Virginia University belongs to the people of West Virginia, and it is eminently worthy of their patronage.[90]

As noted in Chapter One, West Virginia was in a unique position as far as regional loyalties were concerned. The decision of the counties that became West Virginia to split with the rest of the Commonwealth of Virginia over the issue of secession meant that they had considerable ground to make up in establishing themselves as a state in their own right. In addition, the people of the state still harbored sympathies for the southern way of life and southern traditions, especially when it came to gender relations. Coeducation was popular in northern states, not southern ones; this meant the decision to admit women to the state university was a clear choice to become a "northern" institution.[91] This decision was praised by A. R. Whitehill in his *History of Education in West Virginia* in 1902: "Standing at the head of the public-school system, the university has taken the lead in this educational progress."[92] So while sometimes legislation drove universities' coeducational practices, sometimes the institutions themselves helped to bring the state forward in their mindset and practices.

The admission of women to universities for courses and degrees did not guarantee them a chance at a career in their chosen field. Women's careers after graduation will be discussed at length in Chapter Eight, but it is appropriate to discuss some legislative issues at this point. Teaching, in most communities, was open to female graduates without much difficulty, although laws did exist in some areas that restricted conduct or prevented women from keeping their jobs after they married. A textbook used in Mississippi schools in the early twentieth century to study the state's history attributed the admission of women to the university to "a recommendation made by the State Teachers' Association" which felt that women needed to be trained to a higher level in order to be as qualified as possible to become teachers in the education system, because they were in need of more employees in a state with a growing population.[93]

Medical practice, the other key field pursued by early female graduates of male universities, would prove more difficult for them to enter. Legislation was needed to open licenses to female candidates, and hospitals needed to be willing to hire female staff. Even before women were admitted to degrees at universities in the United Kingdom, Mr. Cowper-Temple tried to head off this issue a bit with another piece of legislation he introduced to Parliament that would have allowed women who trained in foreign countries to practice medicine in the United Kingdom.[94] In the United States, the state legislature of Michigan passed an Act "to provide for the employment of women physicians in certain institutions of this State" in 1899.[95] From that point onward at least one female physician needed to be employed by state-supported hospitals or other medical institutions where girls or women were treated. Methods of this sort were needed to smooth the way for professional careers for women and to help society accept women in professional positions. Without these additional steps women's higher education, whether coeducational or not, would be a form of cultural improvement or means of edification and nothing more.[96]

The Role of University Officials

A final issue to consider in regard to legislative decisions about the admission of women and their implementation at universities was the role of university officials. It was common for a university's chancellor, president, or principal, or members of the board of trustees to be cited as the reason women were or were not admitted to their institution, irrespective of legislation from government. Much of the legislation was worded to give some amount of flexibility in how administrators proceeded, with the possibility of restricting the extent of women's inclusion in various courses or aspects of campus life. For instance, President Read was heralded for being "broad minded enough" to support and facilitate the admission of women to the University of Missouri, with a women's residence hall being named for him as a result.[97] In contrast, Paul Chadbourne at the University of Wisconsin was a great obstacle for women's equality on campus. A women's residence hall was also named for him, but it was done so out of irony, rather than thanks. The motivation of administrators who facilitated women's admission and inclusion was sometimes a result of their belief in women's abilities, and sometimes the specific result of knowing a young woman who wanted to become a university student. And often, these young women were the daughters of the administrators in question.

At the University of Wisconsin, Paul Chadbourne, who opposed university coeducation, was followed in the presidency by John H. Twombly. Then, in 1874, Twombly's tenure was abruptly terminated due to "irreconcilable differences of opinion."[98] John Bascom was unanimously elected to the presidency by the end of the same evening. He is described by historians of Wisconsin as a fair, open-minded administrator who always tried to do what was best for the university. His swift response to various concerns about women's ability to handle higher education was also due to the fact that his own daughter, Florence, was a student at the university. It is probable that Bascom's fervent belief in the intellectual abilities of women was fueled by his own erudite and extremely successful daughter.[99] Florence Bascom received degrees of B.L., A.B., and B.S. at the University of Wisconsin and then became one of the institution's earliest graduate students, earning an A.M. degree in 1893. She went on to complete her Ph.D. at Johns Hopkins University in Maryland. This provided a more personal motivation for John Bascom's support of coeducation, one which would prove to solidify women's place at Wisconsin and one which would result in Bascom becoming the university's most honored president.[100]

Family relationships led to the admission of women to Ohio University as well: "Mr. Hugh Boyd, of Athens, was desirous that his sister, Margaret, should have a college education."[101] Although another sibling, William F. Boyd, had recently graduated from the university in 1866, it was apparently Hugh Boyd (Class of 1859) who took it upon himself to discuss the matter with William Adney, a professor of mathematics.[102] The professor agreed to teach Margaret privately since both men felt that a formal application to the institution would probably be denied. After a year of private study Miss Boyd was enrolled as a student in 1868 under the name M. Boyd, with no reference made to her sex, though clearly her classmates knew who she was. The following year the M. was changed to Miss in the records. Margaret Boyd's experience was positive enough that she encouraged her niece, Ella Boyd, to become the second woman to attend Ohio University; she did so and graduated in 1876.[103] The Boyd family, of which Margaret, or "Maggie," was the youngest of nine children, placed a great value on education. Along with Hugh Boyd's request, evidence of the academic success of her other siblings would have strengthened the case for granting Margaret admission because it could be used as "proof" that she would succeed in her studies as well.[104]

A similar skirting of the rules took place before women were admitted to West Virginia University. The debate over coeducation, which was led in many respects by Professor William P. Willey, as noted in Chapter One, was resisted by the Board of Regents. Just before Willey's article "West Virginia's Wrong to Womankind" was published, the subject had been filibustered by a "small minority" of the board, delaying the vote until "train time" when enough members in favor of the measure left and there was no longer a sufficient quorum left to pass the measure. The editors noted these events in the same issue Willey's article was published in, taking it as a good sign that opponents of women's admission had to go to such drastic lengths to stop the change, because the majority of people supported it.[105] Along with publishing articles and making speeches on behalf of women's admission to the university, Willey also took action on campus by allowing the daughters of Professor Franklin Lyon to attend his history courses in 1883.[106] They had already been sitting in on their father's English classes informally and there had been no outrage at this, but attending those of someone they were not related to caused upset on campus and elsewhere in the state. It should be noted that Professor Lyon's contract was terminated two years later, but one of his daughters, Harriet, returned to Morgantown once women were admitted and became the first female graduate of the university.[107] The Board of Regents at West Virginia decided in June of 1889 to admit women to the Collegiate Department and they would be admitted to the entirety of the university in 1897.[108]

Familial ties were also important to the faculty and administration at the University of Alabama as they were even included in the official "Record of Students of the University" published in 1901, with both parents' names given when known.[109] Bessie Jemison Parker, the first female student to enter the University of Alabama in 1893, was the daughter of Professor William Asa Parker and his wife Martha English Foster.[110] Professor Parker was also an alumnus of the institution, from which he graduated in 1858 along with his older brother Osborne.[111] The Parker's family residence was "located on the campus not very far from his class room"; that meant Bessie and her siblings grew up literally as part of the university family.[112] She had five brothers and one sister, Mary Parker, who all attended Alabama during the Victorian Era.[113]

Administrators and faculty members were equally influential in Ireland. When six women tried to matriculate at Queen's College, Galway, in 1876, the council refused to permit them to enter.[114] This decision contradicted the personal views of Sir Thomas William Moffett, who became the president of Queen's College, Galway, the following year.[115] During 1877 he discussed women's higher education saying, "Not only had the ladies excelled in such subjects as music and history, but he found that they were perfectly familiar with the mode and figure of the Aristotleian logic, and adepts in the science of Adam Smith."[116] His public acknowledgement of women's academic abilities would make the inclusion of women at Galway smoother than it was at Belfast where the "male students burned a wad of cayenne-pepper pods" to welcome their new female classmates in Professor Joseph David Everett's physics class.[117] The reaction by the men in Belfast was especially interesting because one of their new female classmates was Alice Everett, the professor's daughter.[118] While at some institutions this would have guaranteed respectful treatment, the opposition of the male students to coeducation was strong enough that they acted out in protest regardless.

Conclusions

At publicly funded colleges and universities, the decision to shift to coeducation was not always in the hands of the people who would have to execute it, and it was definitely not in the hands of those who would experience it. The male students at the Western University of Pennsylvania, one of the last institutions to admit women in the Victorian Era, summed up this situation before coeducation began on their campus:

> Since thoughts rule the world, and our colleges are supposed to turn out trained thinkers, it follows that the opinions of the present crop of college students will largely shape the principles and policy of our government at some time in the future. With this in view, a vote of the students on two important questions would not only be interesting, but also might give us an inkling of the future. The first is the question of co-education. Should girls be admitted to any or all of the schools for boys, and should they follow the same course laid down for boys, or should a special course be arranged for them? Opinions seem at present to be drifting toward co-education.[119]

Although opinions might have been drifting toward coeducation, there would be no formal survey of the male students' opinions themselves. The interconnection between higher education and government, which was necessary in societies where citizens participate in that government, meant that local, state, and federal authorities took seriously the need to legislate the direction that these institutions would take in all areas. In both the United States and the United Kingdom higher education was viewed as a vehicle for social advancement for both men and women.[120] Graduates were expected to set the example for their communities, and "drifting toward" a new status quo of university coeducation would fundamentally alter gender relations for future generations.

By the end of the nineteenth century, women's admission to male universities had become common enough that when the University of Durham received its Supplemental Charter permitting coeducation, it was not greeted with any fanfare. Christina Bremner, in her book *Education of Girls and Women in Great Britain,* commented that the "opening of another university to women excited hardly any attention either at the time or afterwards."[121] Women's entrance to colleges and universities did not guarantee access to all aspects of academic life, however. Questions remained for students, faculty, other officials, and legislators alike about whether "in the long run, society [will be] better for the setting free of so much latent energy and intellectual power in women" or what ramifications there would be for gender roles or higher education.[122] To determine the affects of women's admission to university on the students, the next few chapters will examine different aspects of coeducational university life and the efforts made by officials and students to maintain gender divisions.

CHAPTER FOUR

Academic Student Life

Image 4 Dr. Allen's English Class, c. 1890s.
Credit: Courtesy of University Archives, University of Missouri-Columbia.

The admission of women to male universities in the nineteenth century brought with it many new debates and concerns. While the supporters of women's higher education, in both the United States and the United Kingdom, considered all the reasons a coeducational form of instruction was ideal, they had not fully considered the applicability of such a scheme. As will be shown in this chapter, there were numerous difficulties and disagreements in mixing the male and female students in the classroom. The administrative decisions made on behalf of the student curriculum, in terms of the form that instruction took, were wide ranging. The responses of the male students to the presence of their new female classmates, and the reactions of women to their welcome also varied. At West Virginia University some of the men supported coeducation, while others were rude or impolite to their new female classmates.[1] At the University of Durham the idea of having women attending lectures was seen by some of the men as a very good thing, though not for academic reasons. Writing in *The Durham University Journal* in 1882, an unnamed student remarked that "the presence of the fair sex would also have the great advantages of making the lectures much more attractive to the ordinary undergraduate."[2] Depending on one's perspective and one's confidence in his own abilities, then, it was possible to see the inclusion of women in male universities in a positive or negative light. The fact that coeducation was but one of a number of changes to the traditional university education during the nineteenth century sometimes clouds the response to it, while at other times the reaction was all too clear.[3]

Certificates, Diplomas, or Degrees?

One of the main reasons women gained access to higher education was to have the training needed to enter both the teaching and medical professions. Of course, for women to enter into any male-dominated profession, their education would have to be equivalent to that received by men.[4] The first and most widely followed solution provided was the admission of women to examinations at the University of London. In 1869 matriculation exams were opened to women, and in 1878 examinations for degrees were also opened to women when "the Senate and Convocation . . . agreed to accept from the Crown a supplemental charter, making every degree, honour, and prize awarded by the University accessible to students of both sexes on perfectly equal terms."[5] In the first several years the majority of the women who took these examinations had studied privately because there was no residency or teaching requirement.[6] When women were first admitted to examinations at London, predictions were made that "the percentage of those who will actually avail themselves of this privilege will always be small."[7] This prediction proved incorrect as women from all parts of the United Kingdom availed themselves of the opportunity to gain a university degree by examination, while others would continue to agitate for admission to courses as well. In both Ireland and Wales women's first access to higher education was in the form of London examinations, a practice that would continue even after women's and coeducational colleges were opened in those nations.[8]

The ladies' educational associations in some cities also offered certificates for students who took their classes. In Edinburgh this "certificate formed a valuable link between the Association and the University" in the years leading up to women's full admission to the institution.[9] Students who wished to pursue this certificate needed to complete at least three courses with the Edinburgh Ladies' Educational Association and were also expected to take and pass the local examinations at the Universities of Edinburgh, London, or Oxford; this caused women students around the United Kingdom to take notice.[10] Although the selection of courses was limited in comparison with what was available to the male students of the university, the reading materials were very similar, indicating that the women were expected to master university-level materials to succeed in their studies. For example, women studying logic and mental philosophy (or metaphysics) were told to study Locke's *Essay* and "the Professor's edition of Berkeley's *Philosophical Works*" along with their lecture notes to prepare for their examination in 1878.[11] This is precisely what male students were told to do in preparation for their pass examinations to graduate in arts from the university. The women were also told to have a "knowledge of the History of Philosophy" and study "Hamilton's Lectures... on *Logic* or *Metaphysics;* Mill's *Logic*."[12] These were subjects and readings required of male students who wanted to graduate with honours. The popularity of the "University Certificates in Arts for Women" at Edinburgh can be seen in its placement in the Appendix of the *University Calendar.* In 1878 the system was the seventh to be explained, but only five years later it was the second, leapfrogging ahead of other programs administered by the university like the "Civil Service of India Examinations."[13]

At St. Andrews, the university agreed to grant women the title of Lady Literate in Arts (L.L.A.).[14] This degree was considered more useful than a mere certificate for obtaining jobs, particularly in teaching. The nomenclature was the real issue here, and once Scottish universities began to admit women, even the title of L.L.A. was deemed insufficient, and a full-fledged degree was called for. The regulations for an L.L.A. at St. Andrews were not, in fact, comparable to those of a bachelor of arts. In addition, there were no academic robes issued for students who earned an L.L.A. as there were for all the other degrees of the university.[15] This may seem like a small point, but it does indicate a lack of respect for the degree, that it was somehow less than the others. Criticisms of the L.L.A. followed along the same lines, saying that the degree "would have proved more generally useful, and certainly more attractive, if [the university] had simply offered to candidates of both sexes examinations of the same academic value and under the same conditions."[16] The choice to offer something special for women only meant that they would never be taken as seriously academically as male university graduates. The L.L.A. had expanded to such an extent as a correspondence degree for women by 1900 that examinations were held in seventy-four cities around the world.[17]

In Glasgow the desire of women to be allowed to obtain university diplomas, to replace the less valuable certificates issued by the Glasgow Association for the Higher Education of Women, and later after the association was incorporated in 1883 as Queen Margaret College, illustrated the need for an equivalent education to that of men. The women in

Glasgow further argued that they should receive a similar title "in accordance with all Scottish University tradition... with teaching authorised by the University."[18] Those arguing on behalf of the women continued their appeal on several levels, even targeting the university's desire to be the finest in Scotland. The fact that many Glaswegian women had "gone elsewhere to obtain the desired title" was perhaps the most persuasive for officials of the university, who soon opened most degrees to their new female students.[19] A report of the Lecturers' Committee gave another reason incorporation would be beneficial, stating that "a great incentive will be given to the public to fully endow the College when it is definitely understood that it is to be the first University College for women in Scotland, and not merely a higher secondary ladies school."[20] This distinction between university education and other forms of higher education was crucial, as many at the University of Glasgow, and at Queen Margaret College saw that their institutions were collectively making history as the first in Scotland to offer a full university education to women.

In the United States there was less argument over the granting of degrees to women because admission to a college or university necessarily meant permission to earn a degree. A notable exception to this can be found at the University of Wisconsin where a controversy over issuing degrees to women did take place. While it was agreed women should receive some sort of degree, the name of the degree was at issue. In 1874 President Chadbourne, expressed opposition to the idea of giving women "bachelors" degrees because he considered it to be a male term and, therefore, a male degree. But, after a professor cited a dictionary definition of the term *bachelor* that made no reference to gender, Chadbourne acquiesced. Chadbourne could offer little other opposition after this, so the Board of Regents resolved to confer the same degree on female and male graduates "provided the same courses of study are satisfactorily completed."[21] Looking to other coeducational universities in the United States, most followed the same decision as Wisconsin and provided an equitable curriculum and degree status. This granting of equal degrees actually caused a problem at Ohio University where Margaret Boyd expressed concern when she saw her diploma for the first time because the wording, in Latin, had "masculine endings."[22] Her wish to have a different diploma than that of her male classmates was accommodated by the university's president, William Henry Scott, who agreed the following day to have it "fixed."[23]

Course Selection

As each institution made the decision to admit women to their student body, questions arose about what courses women should be permitted to take, whether they would be mixed or single sex, or if special arrangements needed to be made. In some locations half measures were taken because university officials were unsure what the response to coeducation would be in the community, and there were concerns that male students might choose to attend an institution that did not admit women. The university officials also needed to determine how best to teach their new students. When Parliament opened the Scottish universities to women with the Ordinances of 1892, women at Aberdeen were admitted to the "classes of Botany, Chemistry, Geology and Zoology . . . at once" although they had to wait four more years to gain entrance to "the other classes of the medical curriculum."[24] Female students at Aberdeen also took some of their courses in separate sections from their male counterparts. The topics of anatomy, forensic medicine, and midwifery, as well as "limited parts of [the] course on surgery" were taught to single-sex classes.[25] By 1900 women in Edinburgh were "admitted to the examinations, and to nearly all the classes in the faculties of Arts and Sciences, on the same terms as men."[26] The courses that they did not have access to were botany, chemistry, and zoology, which they took instead at the Medical College for Women on Chambers Street. These courses counted toward their graduation from the university itself and had more to do with a continuing reluctance to educate men and women together in the sciences that had been established during the first admission of women in 1869.

At the University of Glasgow, as specified when Queen Margaret College became the Women's Department, separate courses were offered in most subjects. The women could therefore choose to take courses at either Queen Margaret College or on the main Gilmorehill campus. Annie McMillan, one of the early women students, related a comment by a University of Glasgow lecturer who apologized to his students because the lecture had been "prepared for the weaker intellects of Queen Margaret College."[27] Faculty displeasure over the great inconvenience of traveling the fifteen minutes to the college to present their lectures often exaggerated the continuing struggle against such prejudice. Miss McMillan also commented that the male lecturers were unable to "address an audience of women without being either slightly condescending or slightly facetious—either of which is intolerable to the feminine mind." The prejudices faced by the women students were clear, at least to the women themselves. The male lecturers, on the other hand, stated such opinions as fact, this being an acceptable approach to women's higher education well into the twentieth century.

The issue of separate versus mixed classes was also debated at great length in Glasgow. Not long after Queen Margaret College became a part of the University of Glasgow, the school's favorite patron, Mrs. Elder, began writing letters of complaint to the administration, regarding the education given to female students. She argued that the administration was blatantly ignoring her stipulation of equal education outlined in her "Deed of Gift," despite assurances to the contrary. Mrs. Elder held strongly the belief that the principle of a separate but equal education was "misplaced, misread, or ignored."[28] The historian Carol Dyhouse notes, "Mrs. Elder's advocacy of separatism was based on her belief that this was the best way to earmark resources for women." Mrs. Elder felt so strongly in fact that she brought her solicitors into the conversation to apply added pressure on the university. Throughout 1892, and for the next several years, heated discussion and exchange of correspondence between Mrs. Elder, her solicitors, and the institution continued. Eventually the *Glasgow Herald* even printed reports of "insults" to the lady in January 1896. The basic concern expressed by Mrs. Elder was that although there were courses offered for the women at Queen Margaret College, these did not necessarily qualify for graduation. For example, Mrs. Elder complained that there was no history course specifically designed for women who wanted to graduate in arts. Once the University Court received this complaint, they told the new chair of history, Professor Richard Lodge, that he would need to teach an extra course, at Queen Margaret College. While Lodge had no problem with teaching women, he worried that "the daily duplication of my ordinary lectures to two classes" was "an extraordinarily irksome and intolerable burden."[29] This was the first such complaint by a faculty member, but it was to become a decisive obstacle for the continuation of Queen Margaret College. In March 1897 the University Court placated Mrs. Elder temporarily when they decided "to provide separate teaching at Queen Margaret College in the Classes of Logic and Moral Philosophy to qualify for the Degree of M.A."[30]

In England there was often a division between subjects that women students might study for furthering their cultural knowledge versus those that might prepare them for a suitable career. At King's College, London women could study "divinity, Greek Testament, Church history, moral sciences, history, literature and language, Latin, Greek, modern languages and literature, mathematics, natural sciences, ambulance and nursing, elocution, wood-carving, architecture, art, music" and could take correspondence courses in "harmony, Latin and Greek, French and German."[31] One of the more interesting courses offered for the women at King's College was "Our Navy and its Work," taught by Professor J. K. Laughton. Women studied naval warfare including "piracy" and "raiding" along with the rules of engagement and causes of naval conflict.[32] Britain's status as a naval power makes the subject understandable, though many may have seen the subject as being unfeminine. Courses in experimental physics that included acoustics, electricity, magnetism, and optics were also not feminine, though they were more likely to be preparing a woman for a potential career than studying the navy.[33]

Critics of the teaching offered for women at King's College believed that there was not enough of a systematic approach to their curriculum. At least not in comparison with what was being provided at University College, London. Alice Zimmern, in her book *The Renaissance of Girls' Education in England: A Record of Fifty Years' Progress* which was published in 1898, described the curriculum at King's as "more on the lines of miscellaneous lectures and general culture."[34] This was not an overtly scathing statement, but it is a denunciation of their efforts to provide a university education to their female students. Zimmern had a point, to a certain extent, in claiming that the instruction given to ladies was not as rigorous as that provided for men. The Ladies' Classes at King's started and ended on different dates than those for men; this made them shorter and indicated that contemporary critics had reason for finding that the women were being asked to do less rigorous work.[35] Another difference between the men's and women's classes was that women were only taught during the day, while men were taught in either the day or the evening, depending on when they had time available. It would have been inappropriate for women to attend classes on their own after dark, so this was a distinction of propriety and safety rather than a desire to discriminate.[36] The existence of a separate Department for Ladies also led to some confusion about the admission of women to King's College. In 1896 the *Handbook of Courses Open to Women in British, Continental and Canadian Universities* noted that King's was "for men only" and stated that the typical instruction given for women was "of a very elementary nature, but more advanced classes can in some cases be arranged when desired."[37] The faculty who taught these classes were from King's College in most cases, so it appears that Isabel Maddison, the handbook's editor, also believed the material they were providing to the ladies was watered down to some extent.

Curricular concerns in Ireland had little to do with the sex of the student learning the material and everything to do with their religion. The secular approach to education had ramifications for the curriculum taught at the Queen's Colleges. Rather than studying theology or literature from a moral perspective which might be seen as religious, more emphasis was placed on science and logic. This led to countless politicians, church officials, and average citizens protesting the colleges because they were "Godless" and, worse still, anyone who attended them and studied such inappropriate material as Darwin's theories would be seen by God as a sinner.[38] The desire of parents to protect their daughters, in particular, from a dangerous secular education caused many to send them to religiously affiliated colleges. In some cases families felt assured of their daughter's moral character and knew that no harm would come to them intellectually if they attended one of the Queen's Colleges.

There were only two women who graduated from Queen's College, Galway, during the Victorian Era, both in 1900. Margaret Aimers earned her B.A. with Second Class Honours in modern literature and Margaret Clarke earned her B.A. with First Class Honours, also in modern literature.[39] Clarke was the daughter of a Presbyterian minister, Reverend Dr. Courtney Clarke, who believed in the value of educating women and could see to her religious instruction himself.[40] Flora Hamilton in Belfast, like Margaret Clarke in Galway, had a minister for a father, Reverend Thomas Hamilton. Because both of these fathers were in a profession that provided great service to the community, with limited financial rewards, it was entirely likely that their daughters might be placed in a position where they had to support themselves financially, and a university education would make them better able to do so.[41] As a child Flora was described as being "possessed in an eminent degree of the 'matter of fact-ness' for which her generation of the family was famous."[42] These were qualities that would make her well suited to be a pioneer in the higher education of women in her nation and to withstand the opposition to the Queen's Colleges themselves.

In the United States student choice was also a common answer to questions about their course of study. The amount of choice they were given was greatly restricted. Women at Penn State were offered substitute courses if they did not wish to take subjects that were thought to have little "practical utility" to them in their future lives. Instead of mechanic arts and surveying, they could take a course in music.[43] As the historian Michael Bezilla summarizes these changes, "Music, literature, and mental and moral philosophy were thought to form a more appropriate core of higher learning for women than surveying, chemistry, or botany."[44] Another possible change was the creation of a course in industrial art, focusing on "designs for wallpaper, carpets, house decoration, and fabrics," in place of the mechanical drawing courses taken by men.[45] The instructor hired to teach these courses, Anna E. Redifer, also included other artistic and popular material in her curriculum, like designing stained glass, something that was of high interest during the 1890s. These courses were open to both sexes, as either might enter into a career in interior design following graduation.[46] Advertisements for the Pennsylvania State College highlighted the "special advantages" that they offered to women in history, modern languages and literature, and philosophy in the late 1890s, though this was dropped from the ads by 1901.[47]

The University of Wisconsin had similar alternatives for female students in the early years of coeducation. As women had already been admitted to the university's Normal School in 1863, there were stipulations for their place within the university, somewhat apart from their male counterparts. In 1869 the *Annual Report of the Board of Regents* expressed a desire for state funds to build a "Female College" at Wisconsin, and the state legislature agreed.[48] It was "designed expressly to segregate university women from the mainstream of campus life."[49] The existence of a separate college for women within the University of Wisconsin did not last long.[50] The Board of Regents decided "that the distinctive features of a Female College be maintained by furnishing a separate education to females, when preferred."[51]

In 1871 students in the Female College were allowed to take elementary rhetoric in place of agriculture and meteorology, elementary English literature in place of calculus, German literature in place of analytical chemistry, or other elective study that was approved by the faculty.[52] Certainly all the female students did not choose to substitute these courses, but the very possibility of making substitutions indicates a softening of standards for a portion of the student body. So, while it was thought that once women "got their foot in the door" of higher education things only got better, the University of Wisconsin is an example of regression in equality, not improvement. The Board of Regents later summarized these "modifications in the courses of instruction" for the female students by saying that women "may prefer less exacting mental labor, and a minor degree of culture." This was further illustrated the following year when new substitutions were made that, from a modern perspective at least, seem even easier than those only one year earlier. In 1872 members of the Female College could exchange surveying, navigation, agriculture, analytical geometry, calculus, chemistry and analytical chemistry with Latin or drawing.[53] Despite the offer of "easier" courses, there is little evidence to indicate that a significant number of women took the option.

At West Virginia the belief that the women entering in the early years of coeducation would not be prepared to enter into a full collegiate course in certain subjects caused administrators to direct female students to less challenging topics of study. The classes thought by officials to be too advanced were Greek, Latin, and mathematics. Instead they were instructed in the *Catalogue* to take anatomy, chemistry, field botany, history, junior English, physics, and zoology. These open courses had further stipulations noted as well. For instance, "in order to enter the class in History, the applicant must have sufficient age and general culture to pursue the study profitably," whatever that meant.[54] For the women to gain entrance to any or all of the courses, they had to provide testimonials proving that they had completed the requisite preparatory studies, and also had to pass the same entrance examinations that were given to the male students.[55] It should be noted that the special directions to women entering the university had been removed from the *Catalogue* by 1891, and it was simply stated that they were to follow the same admissions guidelines as the men.[56]

Not all U.S. institutions provided different courses for their male and female students, though all increasingly offered choice to their students in their catalogues. Part of the reason women had a smoother transition to life at Indiana University may be because of the "principles" the administration worked with in the 1890s. The three fundamentals were:

(1) No two minds are alike, and different minds require different discipline; hence, after the completion of certain studies deemed essential to all culture, great freedom in the choice of studies should be granted.
(2) The thorough study of any subject is conducive to mental discipline; hence all departments should be placed on the same footing.
(3) The beginnings of any study are easy compared with the difficulties the student meets after going beyond the mere elements of his subject; hence a better mental training can be obtained from the study of one subject for several years than from the study of a number of subjects for a short period of each.[57]

Aside from the use of the word *his* in reference to students, the emphasis placed on equitability of subjects would ensure that women as well as men were seen as valuable members of the university community. The acknowledgment that all minds were different, not just male versus female, was also forward thinking for the time period and would have allowed for the success or failure of all students in their courses regardless of their sex. The modernizing of the curriculum at Indiana extended to students selecting a "specialty" that is the Victorian equivalent of a major in modern U.S. colleges and universities. Every student was required to choose one that they would study for all four years.[58] A look at the graduates of 1890 shows women with specialties in English, German, Greek, history, and Latin. It may not be surprising to modern readers that there were none in the sciences or mathematics, but their absence from the "Specialty of Pedagogics" or the study of teaching, is, perhaps, unexpected.[59]

At Ohio University in Athens the only reference to coeducation in their *Annual Catalogue* in 1875 was this statement: "Ladies are admitted to all departments of the University on the same terms and under the same conditions as those prescribed for young men."[60] The lack of distinction between male and female students was possibly the result of there being so few women students at Ohio in the early years of coeducation that no further comment was needed. Margaret Boyd's graduating class had six students getting bachelor of arts' degrees and one getting a bachelor of science.[61] Some of the subjects Margaret Boyd studied included astronomy, elocution, Greek, international law, logic, "Mental Science", political economy, and theology.[62] All of her courses were mixed, and all were of at least some interest to her personally, although mental science appears to have been her favorite. The lack of distinction between the education available to male and female students at Ohio carried over into their advertisements to recruit new students. In one placed in *The Review of Reviews* in 1895 the text read: "Young people of either sex who desire to obtain an education may find something of interest in our catalogue."[63] Permitting students to choose their courses of study and offering subjects that would be of interest to prospective students were becoming the best means institutions had of recruiting new students.

At Michigan statements made in the university's *Calendar* were somewhat contradictory in respect to the extent of coeducation on campus. In 1880 it was noted in one sentence that the "course of instruction for women is in all respects equal to that for men." And in the next sentence readers were told:

> Practical Anatomy is pursued by the two sexes in separate rooms, and such of the lectures and demonstrations as it is thought by each member of the Faculty not desirable to be presented to the combined classes, are given separately; but in most of the lectures, in the public clinics, in the chemical laboratory, and in various other class exercises, it is found that both may with propriety be united.[64]

Despite being written well before the U.S. Supreme Court ruling of *Plessy v. Ferguson* in 1896, this appears to have been a "separate but equal" system of education. By 1895 the *Calendar* reported that this separation continued in the study of practical anatomy because concerns about men and women dissecting and examining a naked corpse, of either sex, remained strong in society.[65]

In the United Kingdom as well many people believed "there were a number of classes in which no man would declare that it was right and proper that men and women should be taught together."[66] In Glasgow this debate spread to the editorial columns of the local newspapers. One letter, written anonymously by "A Member of Council" (of Queen Margaret College) made what they felt to be a practical suggestion to end the controversy:

> But it has occurred to me that this difficulty at any rate might be got over by making two separate entrances to the class-room, and dividing it by a partition 8 or 10 feet high, from the back wall down to the professor's desk, one side for women, other for men. It would be impossible for students on different sides to see one another, while the professor would have both equally fully in view. Both men and women would in this way get equal benefit from lectures or demonstrations.[67]

Although separate entrances were used for some time at Glasgow, the idea of incorporating a partition in the room was not put into use. Victorian sensibilities may have supported such a suggestion, with certain topics of instruction such as anatomy (or, at institutions with agricultural programs, animal husbandry) deemed too sensitive for male and female students to learn together. But the idea of a literal partition was not conducive to learning.

The combination of Victorian sensibilities and limited fiscal resources led to various classes being taught to men and women separately at Manchester also. This included some of the junior classes in the medical school and early courses in what would become the Department of Education.[68] Accounts of the latter courses indicate that both the male and female students had limited resources at their disposal:

> The men's class had been housed in the Studio, a room fraught with many memories, but on the advent of women students (who began their collegiate existence in a back room in Brunswick Street, with a cracked blackboard as sole apparatus), this room was afterwards (1894) vacated in their favour, the men migrating to the Old Court Room, now the Bursar's Office. Such were the small and unostentatious beginnings from which sprang the large Department of Education whose majority we are now celebrating.[69]

The male students had been asked to vacate their rooms so that the new women students could be more comfortable and so that they did not have to remain in an inappropriate "back room" that would have been a poor selling point to prospective students. As the historian Sarah V. Barnes notes, "[i]n part their reluctance involved logistical and financial difficulties, in part philosophical ones. First, there was the problem of 'benches and chairs'" which literally meant there was not enough space for all the students, male and female, who wished to take a course to fit in the same classroom together. As in Scotland suggestions were made to physically separate men and women in mixed classrooms "perhaps by drawing an imaginary line down the centre, with male students on one side and female students on the other."[70] Logistical constraints (or excuses) made the moral question of coeducation a secondary concern at Manchester. In the twentieth century both of these areas would finally be addressed and a fuller coeducation would eventually be possible.

Similar worries about the limited amount of funds available triggered discussions about the need to share those resources between male and female students at the University of Durham. In this case the debate over the admission of women to Durham dragged on for several years, due in particular to the distribution of university prizes (fellowships and scholarships) that would need to be opened to women if they were granted full student privileges. This stumbling block was cited in *The Durham University Journal* in an article, which argued that admitting women to competition for these honors "would be to rob the male undergraduates for whom those prizes were instituted."[71] Those who supported university coeducation regularly stipulated that women should be given the same "advantages as men" which would include awards, degrees, and other opportunities, but the institutions themselves often found different ways to interpret this type of statement that did not result in the full coeducation many advocates wished for.[72] Until new sources of funding were made available, and new university prizes instituted, this opposition to women's presence on campus would linger.

As governmentally funded institutions, it was also possible that elected officials would consider the choice of studies made available at universities. As noted in the previous chapter government could incentivize the teaching of certain subjects, as the Morrill Act had, but during the nineteenth century nothing was done to wrest control of the curriculum from university officials. Typically all government officials did was comment on the effectiveness of the education provided and such praise or censure might result in changes at individual institutions. The practicality of the education received by female students at Aberystwyth was raised in a discussion of the Royal Commission on Land in Wales and Monmouthshire in 1896. The agriculture courses dealing with experimentation with various crops and methods for planting and fertilizing were considered by the Members of Parliament on the commission. In particular respect to women, Mr. Richard Jones (an MP from Montgomeryshire) noted, "we are first of all practical in Aberystwyth. We say that we have to give instructions to these young women that are already doing the dairy work."[73] This interaction between the college and the country was described by one commentator as the institution exercising "a kind of superintendence over" the agricultural interests of the nation.[74] Along with training the future leaders of Wales, efforts were made to offer extension courses that local citizens could benefit from, even if they were not able to take a full degree at the university college itself.

Courses Added after Coeducation

Concerns over the enrollment figures for women at Penn State persisted into the 1890s with accusations being made that it was the "policy of the institution to crowd out young women."[75] As a state-supported college many believed that more should be done to appeal to prospective female students with courses of study designed especially for them, beyond just the substitutions noted earlier. At several institutions new courses were added after the admission of women that were particularly thought to appeal to feminine interests. Some new subjects, like domestic science, or domestic economy, could be paid for with funds from the Morrill Act since they met the criteria imposed by Congress.[76] In this way the admission of women became a mechanism to gain monies for institutions, rather than always being seen as a drain on resources. This was only one new subject of study that found favor with many colleges and universities in the late nineteenth century, some of which became whole degree programs or even departments.

The study of the domestic sphere was not an innovative idea, as courses and schools had existed for women to learn about new advances in the area in Britain for some time. For instance, in Scotland the Glasgow Association for the Higher Education of Women, in a combined effort with the Glasgow School of Cookery, offered courses on domestic economy that included the topics of cooking, health, housekeeping, and thrift beginning in 1879.[77] Textbooks for the course included *Huxley's Lessons in Elementary Physiology* and the *Post Office Savings Bank Guide Book*, along with other manuals on housekeeping and thrift that were popular at the time. What was new in the United States was the idea that these topics could be considered worthy of a place in a university curriculum and degree program. The elevation of what would one day become home economics as a formal field of academic study was brought on by women's entrance into universities. Ironically the increase in domestic technology was one of the factors that led to women's increasing desire to enter universities. Emily Davies noted at the time that there was an "increase of wealth, and the supply of domestic wants by machinery" which freed up time for women to pursue other interests like higher education.[78]

The academic subject of domestic science, or domestic economy, was a more formal means of teaching household skills required by the advances of the industrial age. Harriet McElwain described the discipline as follows:

> The Department of Domestic Economy... would prove an invaluable means of fitting young ladies for the performance of the ordinary duties of domestic life, not as a drudgery, but as an attractive exercise in the application of clear and intelligent knowledge of the sciences upon which such economy is founded. The department would not aim to make housekeepers, but would aim to teach botany, chemistry and physiology and domestic management, by means of the exercises of the class room carried on as a practicum in these sciences.[79]

Clearly there was more going on in the addition of such courses than just drawing women to the campus. By elevating a woman's traditional role to the university level, administrators were validating women's presence in their institutions. They could in turn use the fact that they were teaching women to be better wives and mothers as an argument for society to accept women's presence in universities. In one article in *The Westminster Review*, the author stated that higher education would develop a woman's domestic side because it would be "a guarantee of thoroughness in everything."[80] And if the subjects being studied were feminine in nature, there was no reason to fight against women's higher education.

Further to the south the same type of courses were called domestic science at West Virginia University and the University of Tennessee.[81] When the latter institution added the courses just three years after women were admitted, *The American Kitchen Magazine* extolled the move. The publication regularly reported on the courses for women at tertiary institutions around the country because they felt the move toward more formal training in "Household Arts" was a great step forward for society.[82] After women were admitted to the University of Tennessee in 1893, President Charles W. Dabney reported "No special concessions were made to them and no new courses offered to attract them. Every woman applying for admission met the same requirements as men and when admitted she took one of the same courses of study."[83] This was not an entirely truthful statement, as domestic science was not likely to appeal to male students. Another new course at Tennessee was "Floriculture: A Special course in the propagation and culture of flowering plants and the laying out and management of flower gardens. Especially for women students." It was accompanied by the coeducational "Plant propagation and gardening" both being taught by Professor Keffer.[84]

The University of Michigan took a somewhat different approach to the increased interest in having domestic studies as part of their curriculum. They developed a course entitled "Domestic Relations." The *Calendar* in 1880 listed the textbooks to be used as a reference by students which included "Schouler on the Domestic Relations; Schouler on Husband

and Wife; Bishop on Marriage and Divorce; Bishop on Married Women; Cord on Married Women; Macdonell on Master and Servant; Simpson on Infants."[85] Even without examining each book individually, the titles alone give some information about the students who were taking the subject. The socioeconomic level was expected to be high, or there would be no need to study about masters and servants. The inclusion of a book about divorce also stands out, perhaps as an indication that the topic was prevalent enough to warrant debate among the students. Presumably this course was geared more for women than men, as it would have been less likely that a man would take a course to study infants. There was no restriction on enrollment, however, so men could have taken the course if they thought it would be of value to them.

Some universities earmarked subjects as ones they thought both men and women could benefit from. At the University of Mississippi the School of Pedagogy targeted all students, stating that "all intelligent men and women should be interested in the study and solution of educational problems, since all are to have more or less to do with the education of the children of the State."[86] This reads to some extent like it is written for prospective parents, rather than to prospective teachers, but in other years the *Catalogue* descriptions show that the intention was that men and women should pursue careers as teachers. In 1894 Professor James Underwood Barnard stated that the "demand was never stronger than at this time for thoroughly qualified teachers, for men and women in the school-room who combine full and accurate scholarship with professional skill and knowledge."[87] The increasing standards in the teaching profession led to a need for colleges and universities to step up the level of their own teaching of people entering into those fields. The University of Mississippi did this by offering a degree of bachelor of pedagogy for a number of years at the turn of the century.[88] With normal schools in abundance at the time, the faculty and administrators were conscious of the need to establish their teacher-training program as offering something that students might not find elsewhere, in an effort to attract them to enroll at the university instead.

Teacher training was included at other universities as well. Pedagogy was added to the curriculum at South Carolina College in 1894. This course was not specifically targeted at women. Two men from each county in the state earned scholarships to study to become teachers at the college.[89] In 1901 only four of the eleven students in the spring teacher's course were women.[90] St. Andrews took a different approach to the training of teachers, who were referred to as "Queen's Students." These students could attend any Scottish university approved by the "Scotch Education Department" and might live anywhere in the country while completing "their practical training in a school approved by the Department." In 1899 there were fifteen men and fifteen women who were classed as Queen's Students who were able to compete for studentships to help cover their costs. All who earned one needed to "sign a declaration" promising to work for at least two years in an elementary school following graduation.[91]

At Indiana a Department of Physical Training was added to the *Catalogue* in 1892. The subject was not added as a result of women's admission, but when instituted, special care had to be taken to provide gender-specific instruction. The curriculum for the men included military marching (as required by the Morrill Act) and activities like "work with dumb-bells." The equipment in both the Men's and Women's Gymnasiums was described as being the newest and most advanced that was on the market at the time, and emphasis was also placed on heating, lighting, and ventilation of the spaces themselves, with every known aspect of a person's health being taken into account.[92] By 1897 the course work was specified even further, with days and times of classes noted and prerequisites indicated as well because a set sequence of courses had been established.

The courses for women had three instructors: Maud A. Davis, Juliette Maxwell, and Rebecca Rogers. Maxwell was the "Director of the Women's Gymnasium" and had earned her A.B. at Indiana, followed by study at Sargent's Normal School of Physical Training in Cambridge, Massachusetts.[93] There were six courses in physical training for the female students, including one on "Theory and Practice" which was intended for those who would be teachers and one on "Physiology and Hygiene" which added more science to the curriculum. Interestingly there were only two courses listed for men at the time. Along with exercise "class work" the men had their own "Physiology and Hygiene" course that was taught by the director of the Men's Gymnasium, Mr. Madison G. Gonterman.[94]

A final new, and unique, course that was offered at King's College, London was entitled "Ambulance Lectures." An assistant surgeon at King's College Hospital taught first aid and the basics of nursing for various ailments, which could be followed up by the female students with course work specifically in nursing. From a twenty-first century perspective the course materials, like how to make and apply bread poultices to wounds, seem archaic, but they were standard practice for much of the nineteenth century.[95] After 1900 formal nursing courses would be added to many university curriculums, much as domestic economy and domestic science were in the late 1800s. In these and other fields the application of scientific approaches to all aspects of society meant that a more structured education was thought to be needed. There was also a continuing belief in professionalization on both sides of the Atlantic, which meant that degrees would be required for people wishing to work in specialized jobs as the twentieth century progressed.[96]

Female Faculty Members and Staff

Since one of the prime goals of educating women at the university level was to make them teachers, the need for role models was important and the lack of female leadership may have limited the aspirations of female collegians.

As Angie Warren Perkins, acting dean of the Woman's Department at the University of Tennessee put it: "The idea of the higher education for women as demanded by the present age is a scheme of study which shall best prepare her for the responsible work of life."[97] The "scheme of study" included both time spent in the classroom and outside of it, as women who attended university were guided into their adult lives with great care. The opportunities for women to become members of the faculty at coeducational universities were more limited than at women's colleges. Indeed, M. Carey Thomas commented in a speech at the seventy-fifth anniversary of Mount Holyoke College that women's colleges were the only place females could freely assert their academic abilities.[98] She argued that single-sex colleges for women were the only place female professors were able to compete with men for the same positions and pay. While there were some exceptions to this statement, it remains largely correct.

At the University of Wisconsin this issue was manifested in a fairly typical way. Although there was no outright discrimination against women faculty, they were only found in the lower-paying positions within the institution. Law professors at Wisconsin, who were usually members of the State Supreme Court, received $2,000 a year, while the top position for a woman at that time, preceptress (or head) of the Female College, received only $700 a year. The discussion of female salaries also took on a significantly different form at the Board of Regents meetings. While the elections to professorships were done on an individual basis, positions held by women were considered collectively. The board resolved that "the Executive Committee be and are hereby authorized to increase the salaries of the Professors and Lady Teachers, whenever in their judgment such salaries ought to be increased, and whenever the income of the University will admit," so that the committee were left sufficient room to treat women faculty differently from the men, if they chose to do so.[99] It becomes difficult, if not impossible, to discern whether the motivation for the Regents' procedures developed out of a general belief that certain jobs were of more importance because of the nature of the job or because of the nature of those holding the job. This may seem a trivial point, but it does help to illustrate a "separate spheres" notion within higher education. Women had different roles to play in the institution, roles which were of less value to the Regents and therefore deserved significantly less consideration.

Many of the curricular options for the female students were suggested by members of the faculty to the Board of Regents. As a result of this the Regents often gave the faculty the right to make the final decision on coeducation. In June 1870 the Regents recommended "to the Faculty the adoption of uniform textbooks for the male and female students, so far as the same is or may be made practicable."[100] As this statement was simply a recommendation, there was a great deal of leeway in the application of the reform. In his report to the Board of Regents in 1877, President Bascom conceded that the faculty "in the outset opposed to co-education."[101] Not surprisingly, the faculty at Wisconsin was predominantly male prior to 1909. Figures from its founding until 1900 list sixty-eight professors, four of whom were women. The first female professor hired, and the only one who taught in the classroom, was Almah J. Frisby who was on the faculty from 1889 to 1895. The other three women classified as "professors" were all, at some point, in charge of Ladies' Hall. As a result, their influence would have been as social authorities, not academic ones. Women were looked to as representatives of devotion and hard work. The young ladies in their charge were thus provided with a good "Christian example" through this contact in Ladies' Hall. The other group of classroom teachers were the "instructors and assistants." There was a great deal of turnover within this instructional force. The longest anyone stayed in his or her position was six years; the average length of stay (for both men and women) was about two years. So, although there were a significant number of women in the instructional force (of the 226 instructors and assistants in the 1800s, 179 were women), they did not remain long enough to become mentors to the female students at Wisconsin. There was also one notable difference between the men and women in these positions. For men they were often the first step toward a professorship. This was clearly not the case for women, who presumably held the positions prior to marriage, but not after.[102]

A relatively common occurrence was for female graduates of an institution to return or stay on to teach there; they would guide new generations of students through the same process they had experienced. This was particularly helpful for women who had fewer classmates of their own sex to support them during their time at university. Almah J. Frisby, mentioned above, was a graduate of the University of Wisconsin who eventually headed the Department of Hygiene in the 1890s.[103] Another example of this career path was Blanche P. Miller who graduated from the Pennsylvania State College in 1885. In 1898, she started a women's basketball team and oversaw the women's physical culture on campus. According to the historian Carol Sonenklar, the "main emphasis was on gymnastics, but tennis, horseback riding, sleigh riding, sledding, and dancing were also included."[104] The introduction of women faculty at the Pennsylvania State College had been "concurrent with the admission of women students" because the administration acknowledged the need for female role models.[105] The first female instructors, Mary E. Butterfield in German and Sarah E. Robinson in piano music, like Almah Frisby in hygiene and Blanche Miller in physical culture, did nothing to challenge male domains in higher education, all teaching subjects that were intended for female students or were generally thought to be suitable to them.[106]

Women on the faculty were often limited to positions as instructors or assistants, generally because they had not had a chance yet to reach the highest degree levels needed to be lecturers or professors. This often depended on the subject in question and the minimum amount of schooling required to be able to teach at the university level. At the University of Michigan in 1895 Alice L. Hunt was an assistant in drawing, Allison W. Haidle was a demonstrator of dental mechanism, Jeanne C. Solis was an "Assistant to the Professor of Nervous Diseases in the Department of Medicine and Surgery," and Jennie Hughes, M.D. was an "Assistant to the Professor of Gynacology and Obstetrics, and to the Professor of Materia Medica and Therapeutics, in the Homœopathic Medical College."[107] The Board of Curators recommended in 1890 that the University of Missouri hire a "[l]ady instructor of young ladies in Physiology and Hygiene."[108] Later in the decade they also hired the "first professor of pedagogies in America, Miss Bibb," showing that by the end of the Victorian Era more women had reached the level of "professor" in the country and were being recognized for their achievements. Still, the study of education, in this case, was safely within a woman's intellectual domain and would not have threatened the male faculty at Missouri.[109]

One of the few female faculty members at the University of Mississippi in the 1890s was Miss Sallie McGee Isom, daughter of one of the state's first white settlers, who worked as an instructor in the School of Elocution.[110] Courses she offered were intended for male and female students alike, with the stated goals of the school being the production of "effective readers and speakers... in the reading circle, the college, the pulpit, on the platform and the stage."[111] The inclusion of "the stage" in this list shows that the some of the curriculum was bordering on theatrical studies, and the outline provided for the school included topics like "Gestures" and "Physical Training" that did extend beyond what one might expect from a course on elocution. This is attributable to Miss Isom's own theatrical tendencies, as many people who knew her thought she could have had a great career as an actress, should she have chosen to take that path in life.[112] One of the additional selling points of the courses taught by Miss Isom was that they were "eminently conducive to bodily health" and would help students not only in their careers but also in their personal lives.[113]

Female faculty members were limited in universities in the United Kingdom as well, and as at U.S. institutions they were often hired in "acceptably female" areas of study. St. Andrews listed only one woman among its lecturers during the nineteenth century, Alice Marion Umpherston, who lectured in physiology from 1896 to 1897.[114] Miss Edith C. Wilson served as assistant secretary and tutor of women students at Manchester from 1883 to 1904.[115] At Aberystwyth Miss Anna Rowlands was a member of staff in the Normal Department for teacher training. This department provided a two-year course to prepare primary teachers and in the 1890s admitted thirty men and thirty women to study. They also had a course for secondary school teachers that led students to take the exam for the London and Cambridge Teachers' Certificate. Despite the popularity of the agriculture courses in Wales, the Normal Department served the largest proportion of students at the institution at the end of the century.[116] In 1896 W. J. Wallis-Jones noted that women came from great distances to study at the Welsh college, including some from America, India, and Switzerland, along with other nations in the British Isles. Many of the women were already employed as teachers, head mistresses, and other academic positions and felt that further study would be beneficial to their careers.[117]

The only other woman in a position of authority at Aberystwyth was the lady principal of the women's residence hall, commonly the strongest female presence at an institution on either side of the Atlantic.[118] University College, London also had special supervision for their female students in the form of their principal, Miss Grove, and their lady superintendent, Miss Morison. Morison, along with day-to-day supervision of students, even had the final say on whether a woman was admitted to study, after students provided her with character references from former instructors.[119] At St. Andrews Miss Louisa Innes Lumsden, a graduate of Girton College, Cambridge, took on the equivalent position as warden of University Hall. Miss Lumsden believed that women students should focus on study and solitude and that living at home with their families would make both difficult. According to Lumsden students "*must* make time for methodical reading and thinking—historical, metaphysical, recreative." By residing on campus with other students who had the same goals of "self-development," a woman had a better chance of meeting her full potential while at university.[120] As a person who had long been at the forefront of women's admission to higher education in Britain, she knew what attributes her charges would need to develop in order to succeed in their studies and lives after graduation. More time will be devoted to the extracurricular aspects of university housing in the next chapter, but it is important at this point to note that many of the women who oversaw this aspect of university life, like Lumsden, believed that they had a responsibility to direct students in their academic lives too.

In the United States at the end of the nineteenth century the position of a hall director, or "preceptress," was taken to a higher level administratively with the introduction of lady principals or deans of women on many university campuses. These women were counterparts to the dean of the college, though slightly lower than the men in the administrative hierarchy. In some instances they reported directly to the board of trustees or its equivalent. One of the earliest to do so was Harriet A. McElwain, the principal of the Ladies' Department at the Pennsylvania State College from the 1880s.[121] Like women in corresponding positions at other colleges and universities, McElwain's responsibilities were manifold and covered every aspect of a woman student's experience at the institution. In 1897 the University of Tennessee set up the position of "Dean of the Woman's Department."[122] The job description stated that the woman hired would "have charge of all the interests of the women students; will overlook the Woman's building and direct its superintendent; and will advise the President and the Dean of the College with regard to all matters touching women students." Additionally she would be an adviser to the various women's organizations on campus. including the Barbara Blount Literary Society, the Woman's League, and the Young Women's Christian Association, and was to be generally available to the students if they had any questions or concerns.[123] The students at Tennessee took a lighthearted look at the relations between the male administrators and the new dean of women in 1901 with the following limerick:

> There was a young man called "The Bursar,"
> Whose manners grew worser and worser,
> When the beautiful Dean
> Brought a cat on the scene
> He immediately jumped up to curse her.[124]

There is an illustration of the dean holding her cat by its tail after it had made a mess of the Bursar's paperwork. Whether this incident had actually taken place, or the students were simply speculating on the reception male officials had to female ones is not known, but it is clear that the increasing presence of women on campus was sure to alter traditional university life.

A final area of university academics that women were hired in was the library. Indeed, at many institutions the only female presence among the faculty or staff was in the library. For instance, at Alabama Amelia Gayle Gorgas worked as the university's librarian and sometime nurse and was the wife of the university president.[125] She became librarian in 1883 and stayed in the position until 1907, and until a supervisor, Miss Sallie J. Avery, was hired for the Julia S. Tutwiler Annex in 1898, she was the only woman in a position of authority on campus.[126] The male students were clearly quite fond of Mrs. Gorgas, dedicating the 1896 issue of *The Corolla* to her and describing her in glowing terms as a true and noble person of "the highest type... who has been to us always tender, thoughtful, unselfish—a mother."[127] The maternal side of her personality also extended to her second duty on campus, overseeing treatment of any ill or injured students in the university "hospital" located in the Gorgas' home initially (apparently injuries were sometimes the result of hazing).[128] Other institutions also entrusted their libraries to women, as this was widely seen as an acceptable profession for women to enter. The University of Mississippi owes its transition to the Dewey decimal system to a female librarian, Miss Alice Beynes, who began the process of cataloguing their collection after she was hired in 1887.[129] And at West Virginia University Eliza J. Skinner also introduced the Dewey decimal system and devoted time to teaching the students how to use the library's resources for research purposes.[130] This extension of her authority into an instructional capacity was done of her own choosing because she felt the library's resources were underused. The assertiveness on her part was seen positively by the faculty and students, and there is no indication that they felt concerned by her desire to enlarge her sphere of influence on campus.

Life in the Classroom

Day-to-day classroom encounters are not always recorded in official histories, but when women first entered the classroom, it often led to memorable comments. Many college and university officials felt that the presence of women as students caused their male counterparts to behave better out of deference and respect. The president at Queen's College, Cork thought that women would be a calming influence on the men.[131] This is somewhat surprising since Cork had a history of being the calmest and least disruptive student body in comparison to the colleges at Belfast and Galway. In 1864 an article about "Queen's College Morality" praised Cork for having a higher moral standard than her "sister" colleges that had "no inconsiderable share of scandal."[132] At Michigan female students were credited with an atmosphere that was more quiet and orderly "because the men... exercise the courtesies shown in this country by all well-bred men to the other sex."[133] President Hutchins, a professor at Michigan at the time Madelon Stockwell was admitted, recalled later her first day as a student "and the curiosity that she excited among the undergraduates."[134] The responses from the male faculty and students to women's presence in the classroom were more pronounced than in other areas of university life and were easily relayed to the public at large. Because the proximity of male and female students was always of concern, where students sat in the classroom and how they behaved while there were continually noteworthy.

Photographic evidence at the University of Missouri shows that the women students there tended to sit together in the classroom, though there was clearly no preference for the front or back of the classroom as there was at some institutions. In Professor Edward A. Allen's English class, pictured at the start of this chapter, they can be seen sitting in the back and center rows on either side of the room, and in Professor Isidor Loeb's history class they can be seen sitting in the front of the room.[135] In Aberystwyth, Belfast, and Glasgow the women sat in the front rows of the classroom, either as a preference or as a sign of respect from the male students.[136] Students at Alabama were seated alphabetically in their classes and were "questioned in recitations in order" to maintain order and prevent any favoritism (or discrimination) toward the female students.[137] And in London female students at University College "attend the professors' classes along with men, 'sitting cheek by jowl,' as opponents of co-education phrase it."[138]

The suitability of men and women sitting near each other was only part of the worry expressed by those who opposed mixed classes. The idea of students flirting or communicating over a distance was raised by officials and students, as well as the public, at many institutions. At the University of Durham when the proposal was first introduced to admit women to degrees, Thomas Saunders Evans, a professor of Greek, "predicted that there would be 'ocular telegraphy'" between the students, regardless of what disciplinary rules were instituted by the administration.[139] Steps were sometimes taken to divide the students to minimize the distraction caused by being close to each other. For instance, Helen Nimmo, an early female student at Glasgow, recalled the care taken with the female students at her first class at the university: "[W]ith what courteous precaution we were slipped in by the side door of the Humanity class-room to an unoccupied bench."[140] At Aberdeen they were seated separately in the university chapel on either side of the center aisle.[141]

This space between the male and female students was commented on in a poem that appeared in the Aberdeen student magazine *Alma Mater.* In the poem, entitled "In Chapel," the author referred to the seats that were "Occupied by maidens, stealing/Glances (words unspoken)."[142] At Michigan a story entitled "The Romance of a Freshman" discussed "mental telepathy" between the narrator and a young lady classmate who he hoped would notice him.[143] And at Missouri, in a poem entitled "Edwin Hammett's Sigh for Telepathy" that was included in the 1895 yearbook, *The Savitar,* the narrator longed for "subtle sweet telepathy" between himself and the lady he was in love with.[144] Unlike the references from Aberdeen and Michigan, neither the narrator nor his beloved in this case were explicitly said to be students. Regardless, the universal desire for young men and women to wish they could read each other's thoughts is both timeless and unmistakable.

A particularly illustrative story about women's inclusion in the classrooms of the University of Tennessee was included in the first issue of *The Volunteer,* the student yearbook. "The Vacant Chair" shows two male students who were looking at three empty chairs in a seminary room, one of which had a lady's belongings on it. They decide to sit on either side of the woman, expecting to romance her when she returned to her seat. The female student apparently had the same idea when she saw the men waiting for her, as the narration stated that she "sat down and with anticipation [preparing] to engage in a double flirtation." Unfortunately for all involved, the "maiden" was unattractive and the men were instantly turned off by her appearance.[145]

A more dramatic decision was made to keep students apart at Wisconsin. There, while women "had the privilege of attending university lectures," their recitations and tutorials were kept distinct from the male students in the university. In 1870 the Regents began to feel that holding separate recitations was an unnecessary drain on both faculty and funds, because there were not enough professors and instructors to conduct separate classes full-time. The administration's solution to this problem was the construction of the Female College, later to be renamed Ladies' Hall, in 1871. This building functioned as both a student residence and as an academic building for the women of the university. The Regents expressed a desire to "do all in their power to provide for ladies the same facilities for college education enjoyed by gentlemen."[146] They were, therefore, continuing the trend of having duplicate accommodations for female students, which enabled the university to keep the sexes separate.[147] Sophie Schmedeman Krueger, one of the early women at Wisconsin, recalled that women were also given separate entrances to the few coeducational classrooms they used at this point, and if they entered a wrong door in the building, controversy would ensue.[148]

The time between classes was also a chance for student flirtation, should they choose to do so. W. J. Wallis-Jones described the ten-minute break time between lectures at Aberystwyth as being a time "when the splendid quadrangle is filled with an heterogeneous mass of students intermingling in a most picturesque manner."[149] At Missouri the students crossed the distance between the main academic buildings and the Normal College frequently enough that one pathway from the northern to southern ends of campus, via Lake Saint Mary, was named "Flirtation Walk."[150] It is not surprising that having young men and women sharing the same spaces would lead to pleasant "intermingling," and the administrators did want a certain amount of this sort of interaction to take place since the women students were supposed to be on their way to marriage. And as long as women remained a small group on campus, they were not expected to be a great threat to the status quo.

Competition between the Sexes

A constant theme found in many of the colleges and universities that chose to begin admitting women was that their female students often fared better academically than the male ones. There are several ways to assess this conclusion including graduation rates, scholarships awarded, and class rank, though every institution did not deal with the gathering of these statistics in the same way or with the same standards, and this makes exact comparisons impossible.[151] Anecdotal evidence is also prolific, though it can result from inaccurate conclusions. For instance, it would be wrong to assume that the women who left university before completing a degree did so because they could not handle the academic workload. The first woman to enroll at South Carolina College, Frances Gibbs, never completed her degree because she chose to only take courses that she felt would aid her in her chosen profession as a writer.[152] Caution is also needed when looking at the statements made by supporters of women's higher education because they felt it was important to highlight female successes to show that the correct decision was made by officials to shift to coeducation. The contemporary historian Andrew Cunningham McLaughlin remarked on this tactic saying, "Certainly the young man would not be frightened by a statistical 'spook' from entering into competition with the women."[153] The fear over competition between men and women, that it would damage gender relations or that women would deplete resources for men, was often the final obstacle to women's inclusion in the academic life on campus.

Returning to the situation at South Carolina College, some accounts of the admission of women indicate a good deal of resistance to their presence from the male students. Initially some of the men were rude to their new classmates or even "downright hostile" at times.[154] This less than gentlemanly attitude was not held by all, as some chose instead to simply ignore the women. Even when women excelled in their studies, these accomplishments were not acknowledged by some of their male counterparts.[155] The Class of 1899's history, provided in the student yearbook of that year, dispels these assertions somewhat, saying that if future female students "follow in the footsteps of our one, there will never be cause to regret the establishment of co-education."[156] It was not a glowing description of women's presence on campus, but it was positive. The Class of 1901 took their praise a step further, saying that the three women in their class "by their achievements, prove conclusively that woman is intellectually the equal of man."[157] This comment may have been a delayed reaction to the debates in the 1880s about women's mental capacity, but it should be remembered that those arguments survived well into the twentieth century and still needed to be proved "wrong" by the women at coeducational universities and elsewhere.

As noted in the previous chapter, at the first lecture of what was to become the Glasgow Association for the Higher Education of Women in 1868, Professor Nichol, as a supporter of women's higher education, expressed his concern about "an over-stimulus in the direction of competition" for female students.[158] His reservations were shared by many supporters of women's higher education who thought that in principle it was a good thing, but in practice it might be altogether too strenuous for "ladies" to handle. Two decades later President Angell at the University of Michigan reported that, in his experience, women were equally capable in their academic abilities to the men. He did acknowledge that men seemed to be slightly better at "extemporaneous discussions" than women, but did not feel that the women were at a significant disadvantage because of this. Women studied in all fields: "the most abstract and difficult studies as well as in those which tax the mind less."[159] This constant back and forth about women's ability to handle higher education, and in particular higher education that was in competition with men, was one of the most enduring scientific questions of the day, even after concerns about women's intellectual growth causing infertility were dispelled.

At some institutions fears about women's competition with men caused them to accept women tentatively, or with serious restrictions on their academic standing within the university. Initially women were not admitted to the University of Alabama until they were at the sophomore level. This was intended as a means of working with the existing female colleges in the state where women could complete preparatory work in order to pass entrance examinations at the university itself.[160] This situation was commented on by the students in *The Corolla*. In 1894 the "Sole Ambition" of the female students was "To become a Freshman belle."[161]

The admission standards were changed for the 1897–1898 academic year permitting women to enter in any class in the university.[162] This change in policy proved successful from the outset, with women doing well in all of their studies. In a "Welcome Address" in 1906, President John William Abercrombie remarked: "The professors report that [the female students] perform their duties in a satisfactory manner, and the records show that they not infrequently win the highest honors."[163] As with the comments noted in South Carolina this was not a glowing appraisal of women's university achievements, but it was an unsolicited and positive statement on their continued presence at Alabama.

Similar comments and assessments of women's successes in coeducational universities abounded during the Victorian Era in both the United States and the United Kingdom. The first woman to graduate from West Virginia University, Harriet E. Lyon, was also "the only student to achieve perfect marks in all her classes," ranking first in the Class of 1891.[164] At Tennessee women won three of the four top academic prizes awarded by the faculty in their first year as students.[165] The women at the University of Missouri fared extremely well when it came to competing for the various honors and prizes awarded by the institution. In 1871 Eliza Gentry, as a graduate on the Normal Course, was inducted into Phi Beta Kappa, the national honorary society.[166] In 1872 Anna Ware won the "Philosophy Honor and took the Law Prize."[167] In 1874 women "carried off more than fifty per cent above their proper share" of honors, leading to questions of jealousy or feelings of inferiority by the men they beat out.[168] In Wales, the Aberystwyth women were said to "generally head" the lists of academic achievement.[169] To fully assess the success of the lady students at Aberystwyth, officials looked to their results in the examinations taken at the University of London where many chose to take additional degrees. In 1894 one of the Welsh lady students obtained top marks in the M.A. examinations in classics, English, and French, and another did the same in the exam for a B.A. in mental and moral science. Some of the Welsh women found success overseas too. One earned a scholarship and staff position at Bryn Mawr College in Pennsylvania, and another earned a bursary to study at McGill University in Canada. This showed that the university education of women was not only an international phenomenon; it was an interconnected one.[170]

In 1896 one of the three sophomore "Chancellor's Prizes in English Literature" at the Western University of Pennsylvania was awarded to Margaret Lydia Stein.[171] And two years later, when only two were awarded, one went to Anna Mary McKirdy.[172] Margaret Stein and her sister Stella Mathilda Stein both received their bachelor of arts degrees in 1898, with the latter Miss Stein making the valedictory address.[173] Another two years later, Anna McKirdy was the valedictorian of her graduating class.[174] Other early female graduates of the university included Mary E. Hamilton who graduated in 1898 from the affiliated School of Pharmacy and Mary L. Glenn who graduated in 1900 from the School of Denistry.[175] The impact of women on the Western University of Pennsylvania was not terribly great, though, since there were so few women who decided to pursue their studies there. In Chancellor McCormick's annual report in June of 1907 the system of coeducation was discussed for the first time because, for the first time, women were starting to make their presence felt with increased enrollment numbers.[176]

Also small in number, the women at Queen's College, Galway fared very well in their studies, whether they ended up graduating or not. As a first-year student Emily D. M. Daly held a junior scholarship in the 1898–1899 academic year, and in 1900–1901 Rosalind Clarke earned scholarships in both the Literary and Science Divisions. Students were "ineligible" to hold two scholarships, so she held the Science Scholarship.[177] Another set of successful female students at Queen's College, Galway were the Perry sisters who were the daughters of the Galway County Surveyor. Both Agnes and Margaret Perry held scholarships in 1900–1901, Janet was awarded a Gold Medal from the Royal University in 1906, and Alice became the first woman to get an engineering degree in Ireland in 1906.[178] All of the Queen's Colleges had "low fees and generous scholarships" although initially all prizes and scholarships were not open to women.[179] By the 1896–1897 academic year this had changed with a "recent alteration in the statutes" of the Royal University of Ireland (which awarded the degrees): "All Degrees, Honours, Exhibitions, Prizes, Scholarships, Studentships, and Junior Fellowships in this University shall be open to Students of either sex."[180] Only one such award, the "Dr. and Mrs. W. A. Browne Scholarship" for studies in French and German, was specifically described as being open to men and women, with the only stipulation being that an applicant be "a natural born subject of Her Majesty."[181] This included all citizens of the British Empire, and as one commentator noted, language scholarships or studentships that were open to all imperial candidates would be "tempting and easy prey to French-Canadians and Mauritians" or others whose first language was not necessarily English.[182]

As soon as women were permitted to take the examinations at the University of London, they were able to compete for scholarships. In many cases the awards provided were designated for men or women alone, but gradually more were open to competition for both sexes.[183] Women at Durham also did well academically, "taking a high position in the University, and have carried off some of the chief prizes."[184] Following the opposition noted earlier in the chapter that university prizes might be taken away from the men to be given to the women, there was one "scholarship and two exhibitions [that were] tenable by women students only."[185] The decision by some benefactors to earmark resources for the education of women became something of a trend toward the end of the nineteenth century. While it might have been more prestigious to endow a professorship or fund the construction of a building, having one's name on an award had the possibility of being just as enduring and noteworthy.

In Scotland, one such donation was made by Sir William Taylour Thomson who wished to aid all students, but particularly women who wished to enter the medical profession. "The Taylour Thomson Bursaries for Women" were "open to all Women Students who sign a declaration that they will enter on, and follow out, a course of Medical Study at St Andrews."[186] Other requirements included passing the "Medical Preliminary Examination" and being at least nineteen years of age at the time of application for the bursary. Records indicate that the numbers of bursaries distributed by the University of Glasgow were divided relatively evenly following the enactment of Ordinance No. 58 by Parliament in 1898.[187] The continued growth of the number of bursaries available in the early twentieth century works to further illustrate the traditional concept of the Scottish "lad o'pairts" (or indeed "lass o'pairts") who could achieve greatness through industriousness.[188] If students excelled in their academic work, they were more likely to be able to afford a university education through the aid of a bursary. Bursary competitions were established by the university itself, a private person (often in honor of a relative), or communities who wished to send their own young men and women to university. The awards sometimes consisted of a monetary amount, but often books or instruments were given, because officials thought them to be "of far greater value to the students."[189] And, as the expenses of university life were considerable, students would have been grateful for any assistance in covering their costs.

More often than not students' daily workload and grades on individual assignments have been lost to time. Margaret Boyd provides some insight about what course work was like at Ohio during her time there. Her studying included preparing for debates, examinations, recitations, and writing essays. She commented on the lack of variety in this regimen on Tuesday, February 4, 1873: "Study and recite, Study and recite what monotony! Sometimes I get tired."[190] Relations between Margaret Boyd and her male classmates seem to have been positive. At one point she said "They are good & I like them."[191] At other points, however, she did note that the "boys laugh when I read" although she did not indicate if that was a result of the topic or if it had something to do with her pronunciation or speech patterns (or something wholly unrelated to the work at hand).[192] All of her instructors dealt with her and her male classmates in a completely equitable fashion, at least when it came to assigning recitations or examination subjects, which was done by drawing "slips" of paper that listed the chapter the student would work on.[193] One professor of Greek and Latin at University College, London, Alfred Goodwin, was similarly fair minded in teaching his male and female students, though not in a positive way. He earned a reputation for criticizing student work in front of the entire class. In his comments he spared "neither age nor sex, but he never offended by a severity which was always richly deserved."[194] In this case Victorian propriety might have dictated taking a gentler approach with women, but

Goodwin's choice of equality was what the truest supporters of women's higher education desired.

Conclusions

The experimentation and variety of solutions taken in the area of academic provision for women students leads to several conclusions. There was often little opposition to women receiving higher education by the late nineteenth century, as they were felt to be capable of handling their academic studies. The Victorian appeals that women's brains were too small to acquire the necessary knowledge to graduate from universities had been sufficiently contradicted by experience. The only exceptions to this were the optional course substitutions offered by some institutions. The provision of such choices indicates the administration's desire to provide suitably feminine courses of study for those women (or their families) who preferred them. The debates over the issue of separate or mixed classes fit into a similar line of argument. Increasingly, women's educational associations in Britain had adopted the idea that a separate women's college or department would best serve the interests of female students by giving them an education *equivalent* to that of men. In addition the view that women and men should occupy different spheres within society made the development of women's colleges more logical in the Victorian Era. Emily Davies also expressed concern over the "hasty assimilation of the education of women to that of men" as one of the great perils of mixed universities.[195] Considering the fact that students had no say in the actions of the administration on their behalf, their thoughts on the practical issues of coeducation in the classroom were largely unspoken at the time. Decisions of whether men and women, who were learning the same information, should do so separately, in line with conventions of the day, or collectively so that they might learn about each other in the process, remained in the hands of those it effected least—the faculty and administration.

In examining university records and social commentary, it is clear that administrators and faculty alike knew from the outset of women's admission that the future of higher education was coeducational. The implementation of mixed classes and the integration of facilities and extracurricular activities bring out the differences between the universities, which make a comparison between them valuable. In Britain, ancient universities like Aberdeen and St. Andrews had centuries of tradition to revise when it came to the inclusion of women, while some American universities only had two decades. Ultimately the question of university coeducation came down, not to overturning history but to adjusting public perception. If women were seen to be able to compete in an appropriate way with men and maintain their femininity and still marry and raise a family at the end of their studies, then the community would accept the inclusion of women at their university. The fact that other modifications were taking place concurrent with coeducation, like the introduction of elective courses of study and specialties, meant that more than just coeducation needed to be adapted to. In 1899 *The Southern Educational Journal*, discussing the West Virginia University, remarked:

> We should add that distinctive features of the university [include] . . . the elective system, which allows the student the largest liberty in the selection of their studies. . . . It need hardly be added that a university which has shown itself in all respects so progressive and so thoroughly in touch with the modern spirit is coeducational.[196]

For institutions to remain competitive, they would need to embrace the presence of women, and the more successfully they were able to integrate women into their campus communities, the better they would fare with prospective students, benefactors, and, to some extent, government officials who controlled funds. As the century came to an end, it was not entirely left to university officials to make the case for coeducation because all institutions had students and alumnae who could make the argument more effectively than any administrator. According to one commentator in the early twentieth century, "[T]he college girl herself is no mean press agent. She can easily cajole unwilling parents into allowing their daughters to matriculate in her school, since she herself so amply demonstrates the practical worth of co-education."[197] Questions would remain about women's lives on campus, but in academic study, at least, there was no doubt that university coeducation was a valuable and permanent development.[198]

CHAPTER FIVE

Facilitating Coeducation

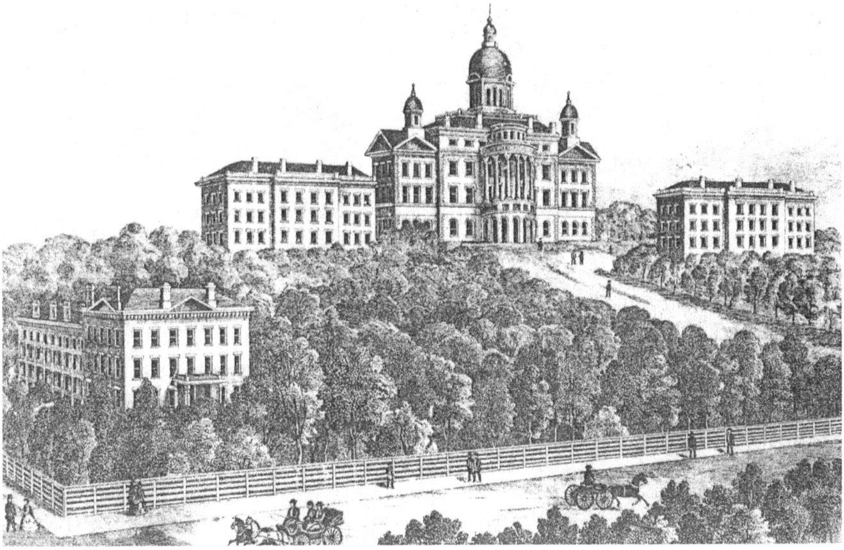

Image 5 The campus of the University of Wisconsin in Madison, c. 1873.
Credit: University of Wisconsin Archives, Madison, Wisconsin.

All colleges and universities had a limited amount of space available to students for study and other activities, and they were also limited in the amount of funds needed to expand in any of these areas. The larger the enrollment, the more potential problems with discipline might arise, and the more concern there was about the ability of faculty and staff to supervise the interactions of students on campus. The physical proximity of men and women on campus was not limited to seating in classrooms. If the administration really wanted to keep them apart, they could make sure that the buildings they used were far apart on the campus grounds. As seen in the drawing of the campus of the University of Wisconsin in 1873 above, the women's residence (seen in the lower left-hand corner) was not located near the three existing campus buildings when it was constructed.[1] The climb made necessary by the hill would not have been desirable for the delicate nature of the female students (particularly in winter), making it easier to justify holding separate women's courses within the residence.[2]

In both the United States and the United Kingdom, there was much debate over the need for "corporate life" at colleges and universities and whether an institution that did not provide accommodations for its students was truly providing "higher" education.[3] A common debate during the Victorian Era was how to extend the benefits and privileges of university education to more people, both in terms of class and gender (and eventually race as well) when funds to do so were limited. Government assistance made inclusion more possible, but institutions were still faced with a choice of how to spend the money they had. One of the easy ways to save money was by not providing housing and having to staff residence halls, so that more money could be spent on classroom facilities and faculty. Alternatively they could redistribute funds that had previously been set aside for male housing, but this would lead to tensions within the student body.[4]

At most institutions, early in their admission of women, the students in attendance were from the local area. There were numerous reasons for this, all of which dealt with concerns over expense and safety. In some cases the safety of the city was questioned, and at other times it was the safety of a young lady's virtue if she were to live in proximity of numerous young men. Universities, therefore, had to maintain an awareness of fears that "courtships will abound; scandals will arise; no prudent parent will permit a daughter to thus associate with young men when away from home."[5] In locations where there was little private accommodation available, colleges and universities felt obligated to provide at least some housing for their students. In urban areas this was not as crucial, so if supporters of women's higher education wanted accommodations to be provided, they needed to find alternative arguments to convince administrators. Even when housing was not available on campus, other facilities, like libraries or reading rooms, were provided for students to use between classes, all of which needed their share of the yearly budget for maintenance.

The question of housing took on two distinct forms. Some universities were not residential at all when women gained admission, while others did have housing provisions. This initial difference between institutions led to remarkably similar outcomes, due in large part to similar concerns amongst administrators. Those students who lived at home with family remained under their supervision when not in class, exempting colleges and universities from this responsibility. Once official housing was established on campus, the institutions took on an additional set of roles and standards of activity that would further revolutionize campus life. The question of whether to shift to the position of a residential institution faced every college and university at some point in their history. The types and extent of housing provision at each institution will be the primary focus in this chapter. The wider implications of the choices made at universities will also be considered, as housing, or the lack thereof, was seen to affect both the academic and social development of students.

In Loco Parentis

Those in charge of university students and their behavior felt the responsibility to regulate interactions between male and female students in accordance with social practices of the day. Emily Davies, at Girton College, Cambridge, argued that discipline, both inside and outside the classroom, led to increased benefit from education. Order was not only advantageous to the students trying to complete their academic careers, it also helped to regulate society in general. An idea used often in all levels of education is that of *in loco parentis*, or the authorities functioning in place of the parents. This was most apparent while the young women were living in a campus situation, as men were not thought to need as much protection.[6]

In Wales, the parents of some of the women students requested that "their daughters…be placed under the special care of a lady appointed by the Council," so the university officials rented space on Victoria Terrace that was first overseen by Mrs. Powell.[7] Later space was rented in the Queen's Hotel before a purpose-built Hall of Residence could be constructed. The accommodations at the Queen's Hotel were similar to those maintained by other colleges and universities, complete with a lady principal, Miss E. A. Carpenter, who guided and watched over the students while they were not in the classroom.[8] Miss Carpenter was described as having "energy, tact, organising power, and versatility" that set an excellent example for the students that they should aspire to in their own lives.[9] Women who attended Aberystwyth were in fact required to live in university housing if they did not have family to live with in the town. If students did not attend evening lectures, they had a curfew that had to be adhered to unless permission was given by officials to stay out late. During the week, women had to be back to Alexandra Hall by 7:30 p.m., during the winter months and at 8:30 p.m. on Sundays, and during the summer. Reports indicate that these rules were not broken often because the weather and lack of sunlight that determined the times also prevented students from wanting to be out of doors any later.[10] The male students were also required to live in "registered lodging houses" in the 1890s as the college officials wanted to maintain a certain amount of control over all of their students, as an assurance to their parents that everyone would remain safe and well supervised.[11]

The theory of *in loco parentis* surfaced at Wisconsin in the 1880s, after the Female College Building had been renamed Ladies' Hall (it would be renamed again in 1901 Chadbourne Hall). The Board of Visitors made increasingly moralistic statements at the end of the nineteenth century, referring frequently to this new role being taken on by the university: "No home can be presumed intact in all its purity without parental restraint, and no college can keep its members on a uniform moral level; such is the condition of youthful human nature without the proper regulation of the hours and habits, and moral tone of its people."[12] They acknowledged that providing quality, on-campus housing would make the school more attractive to prospective students. If Ladies' Hall in particular was kept at a standard higher than that of the average boarding house in downtown Madison, the parents of the female students would feel comfortable sending their daughters to Wisconsin.

The historian Helen Lefkowitz Horowitz points out that many early women's colleges in the United States based their housing on a similar principle. As female students were most likely headed for lives as wives and mothers, having them live in a structured home life was ideal. Additionally, such "buildings were designed to protect young women from risks."[13] It was this latter point that proved to be the greatest issue at Wisconsin, as the administration worried greatly about the respectability and security of its women students.[14] A female preceptress (Miss Delia E. Carson first held the position in Ladies' Hall) and an assistant were hired. These women lived and worked in the Hall, helping the students with their recitations and other study. There was also a "judicious matron" who directed the "Department of Boarding" in the Hall. These women held the students to strict standards of conduct, and regulated moral training, such as attending religious services on Sundays.[15]

The residential life at universities that provided housing for their students was seen as appealing by many prospective students. University officials and supporters of women's higher education often highlighted the many benefits of living in university accommodations and emphasized the chance to build a community feeling among the students. At the Central Conference on Women Workers held in Glasgow in October 1894, Agnes Maitland of Somerville College, Oxford, gave a talk about this subject, making reference to the residence halls at Aberystwyth, Edinburgh, Glasgow, and London, among others. She said that the women in these situations benefited from "combining home life with hard study" and "learnt how selfishness belittled womanly influence."[16] This moralistic tone was often used in the area of student accommodations as it applied to character building. At Queen Margaret College in Glasgow, supervisors and lecturers took seriously the fact that they were "responsible to the parents who send their daughters" to the institution.[17] From Miss Galloway, and the Mistresses of Queen Margaret College, the students received "lessons in tact, patience, and kindly consideration" from those around them. Most of this schooling was given indirectly, through osmosis, but the influence did not go unnoticed.[18] Several students noted the advice given to them by Miss Galloway as the most valuable instruction they had while at college.

Libraries, Reading Rooms, and Waiting Rooms

Most universities that did not provide housing on campus for their newly admitted women students provided some other sort of room for their use during the day. One of the more interesting spaces handed over to female students was at the Western University of Pennsylvania where the chancellor himself "turned over his large office in Main Hall" for the ladies to use (he had a second office downtown).[19] Some accounts of this gesture also mention that he "decorated it with some of his own sketches and paintings" in an effort to make the ladies feel more at home.[20] In future years the *Catalogue* referred to women having a "special room for rest and study" without identifying its location.[21]

Unlike the effort made in Pittsburgh, the women at West Virginia were not provided with a spacious room to use between their classes during the day. Members of the WVU Women's Centenary Project, in their *WVU Women: The First Century*, included among the "Famous Firsts for WVU Women" a "boycott" of the "inadequate" cloakroom that women were given to use in 1890.[22] The students may have been displeased with the austere nature of the room, or its location in the basement of a building, Martin Hall, or the fact that it was so far below the standards of what was provided for their counterparts at other institutions.

In most cases rooms like these that were specially provided for the use of female students served multiple purposes during the day. Women at the University of Missouri, who lived off campus until Read Hall was built in 1903, had a "Ladies' waiting room and hall" which they could use between classes if they needed a place to rest and relax.[23] The Waiting Room, later known as the Ladies' Parlor, was described as being "comfortable, almost luxurious" at the dedication of the reconstructed Academic Hall and several other new buildings in 1895.[24] Similarly, at the Ladies' Department of King's College, London there was a "Reading and Waiting-room for ladies who wish to remain during the intervals between Lectures, and a simple Lunch can be obtained from the housekeeper."[25] Male students also needed a place to get something to eat if they were taking multiple courses. At University College, London there was a "large refreshment room where meals can be obtained during the day" that could be used by either male or female students.[26] In Ireland a Ladies' Room was made available to women at Galway and the female students at Cork had a cloakroom that "was secured in 1885."[27] After women were admitted to Tennessee "a small building on the hill was set apart for their lunch room and study hall."[28] And advertisements for the University of Alabama also promised prospective young women that there were "rooms for study during the study-hours of the day" that they could use on campus, limiting any inconvenience they might find living off campus.[29]

Preexisting facilities, like libraries and reading rooms, also had to be integrated to some extent, even if only for a short time before funds were available for new spaces to be created. The Reading Room at the University of Mississippi, even before the admission of women, included copies of *Godey's Lady's Book* among its numerous journals and periodicals.[30] The Reading Room was located in the Lyceum, one of the six original buildings on campus, which after women's admission included a "Study Room for Female Students" to use during the day, as well as the chancellor's office, lecture rooms, recitation rooms, and other academic facilities.[31] Later the women got two rooms in the Jefferson Building and the women's study room in the Lyceum was eliminated.[32]

In Glasgow there was great concern that the intermingling of male and female students would lead to much socializing and very little studying. The staff of their General Library and Reading Room decided that women were equally entitled to use the library facilities on campus, so they would reserve a special part of the Reading Room for them. The committee also requested that Miss Galloway or another member of the college staff make occasional surprise visits to the library to ensure that the female students were not "talking in a disturbing way." The male librarian would perform the same service for the male students, but as he was "rather shy about talking to the women students himself," he could not provide total supervision of the entire library.[33] At Edinburgh the libraries and reading rooms were "fully taken advantage of by students" during the term at Edinburgh.[34] Although most were studious, there were also reports of "tittering and joking" and other unscholarly behavior from time to time. A third Scottish library, at King's College, Aberdeen, circumvented this potential problem by establishing entirely separate reading rooms for men and women.[35]

Because space on campus was limited, some buildings had to serve more than one purpose, both academic and social. For example, the Old College Library at Aberystwyth was often decorated and used for soirees. These gatherings were usually dramatic, literary, or musical evenings and were a chance for the men and women to socialize and work together just as they would be expected to do in life after graduation.[36] At Indiana University fires in both 1854 and 1883 destroyed much of their campus infrastructure. Two of the new buildings erected post-1883, Owen and Wylie Halls, were described as being made of brick with concrete floors, an iron frame, slate roofs, and "limestone trimmings."[37] The main building on Indiana's campus housed a room or rooms for "young ladies" before the fire.[38] Afterward the room was located in Maxwell Hall, "a wooden structure," along with six recitation rooms and the chapel.[39] After Dr. Mary Bidwell Breed became the dean of women in 1901, she welcomed all women students to campus personally and made sure that they had found suitable accommodations in town.[40] As the number of students on all of these campuses grew and it became more difficult to feel a sense of cohesion amongst the student body, there were increasing demands for university housing and a new expectation that institutions provide both an academic and social education.

Living with Private Families

Before residence halls were put into use, the administrators helped students to find suitable housing in the community. There were a number of benefits to this sort of arrangement, and also many drawbacks. In larger cities more accommodations were available, but there could also be less certainty that the boarding houses or other lodgings were of an appropriate standard of both cleanliness and morality. In more urban locations there was also great concern about the safety of women students and the potential corrupting influences they might encounter. This concern was not limited to female students either, some men were questioned about whether they "might escape unscathed from the temptations and snares of a city life" when attending the University of Edinburgh.[41] As noted earlier the proximity of men and women, whether students or not, was also of great concern to administrators and parents alike. Logistically living at a distance from one's classes might lead to absences that were detrimental to education. And finally, the lack of housing on campus often caused students to choose to attend a different institution altogether where such a living situation was possible.

Male students at St. Andrews did not have university lodgings and lived in town. The "Senatus and Students' Representative Council" kept a list of "suitable" places for the students to live in town.[42] Female students at St. Andrews who lived in town established the Town Students' Association in 1897 "for the benefit of all Matriculated Women Students who live in lodgings." The group organized activities or "entertainments" for members who wished to have more camaraderie despite not living on campus.[43] When women were first admitted to Manchester there were no accommodations for them or for male students either. The registrar kept a list of faculty members and other officers of the university who were willing to "receive students to board with them at their own houses."[44] To try to ensure quality accommodations within the surrounding area, the university established a set of guidelines that boarding houses or other lodgings in the city could conform to, in the area of discipline, and they would then be added to a list that would also be provided by the registrar when asked by the students. Officials disavowed responsibility for the enforcement of these rules, but they promised to do their best to see that they were followed.[45]

Queen Margaret College in Glasgow provided a similar list of houses where women might board, all of which were the homes of women who worked in association with the college. As of 1888, however, only one request for use of this housing had been made. In 1948 Marion Gilchrist, one of the original Queen Margaret students, noted the great difficulty in finding rooms in the local area, with many only providing rooms for "single gentlemen."[46] The University of Michigan had similar problems in Ann Arbor. There was usually enough housing available, but it might not be for women only, a characteristic that many female students and their parents required. When possible, officials tried to help students find single-sex accommodations, though President Angell admitted that sometimes men and women wound up having to take "rooms in the same house and take their meals at the same table" as each other.[47]

Elsewhere in the Midwest, early female students at both Indiana and Ohio Universities also lived with their families or friends in Bloomington or Athens. Because both towns were still relatively small and rural, it was not usually difficult for students to make their way to their classes.[48] Margaret Boyd did note one drawback to living with her family in her diary on Thursday, January 23, 1876: "I went to college this morning through a big snow-storm."[49] Inclement weather would be less of a problem for the women after 1896 when female students at Ohio could reside in Women's Hall, a private building located "on the corner of South College and East Union streets" near campus.[50] Advertisements for the university in the early 1900s, even though the hall was not affiliated with the institution at the time, promised prospective students that the rooms in Women's Hall were "well appointed and under efficient management."[51] An endorsement like this would have been beneficial to the management of the hall and would show parents of young ladies that the university was looking out for them even when they were not on campus.

In the southern United States the biggest concern was the proximity to male students. In South Carolina women were directed to find board and lodging in the surrounding community.[52] Female students at Mississippi were not permitted to live on campus "except in the homes of members of the Faculty."[53] And at Alabama, rather than finding a boarding house or similar group accommodation in Tuscaloosa, the university actually required them to live with "private families" during their time as students.[54] These decisions were seen by some as intentional failures to "induce" the women to attend, but the decisions were also based on financial considerations.[55] It was not uncommon for colleges and universities to wait to either reallocate space for housing or build women's residences until there was sufficient demand for it. Officials at Mississippi later described the growth of women's presence on their campus as having been "retarded by the fact that no home under supervision of the University was provided," and female enrollments certainly did increase once Ricks Hall was completed in 1903.[56]

The lack of residence halls at the Queen's Colleges meant that only students who lived locally or were able to find accommodations in Belfast, Cork, or Galway were able to attend courses there, limiting the possible number of students who could matriculate. Requests were made that residence halls be built even before women were admitted, with one proposal in 1875 receiving applause from the crowd gathered for the Queen's University degree ceremony at St. Patrick's Hall in Dublin Castle.[57] Part of the reason residence halls were not constructed had to do with the continuing religious conflict over the secular nature of the colleges themselves. In 1880 the president of Queen's College, Cork,

Dr. W. K. Sullivan, suggested that "denominational halls for residence and religious instruction" be erected as a possible means of settling the dispute, but his remark actually added yet another topic for debate.[58] The concern about the secular or sectarian nature of campus accommodations would delay the decision to approve any residence halls until the twentieth century.

Perhaps the most unique suggestion for the housing of women students came from Morgantown, West Virginia. As at other institutions students at West Virginia University were expected to live with their families locally or with family friends who lived in town because some in West Virginia were concerned that Morgantown was not a safe place for young women to live without parental supervision.[59] On hearing this, local newspapers began running advertisements for lots for sale in the community where parents could build homes to house their student children which could then be sold on after their graduation.[60] Although it is unclear how many families followed this suggestion, it did add a new alternative for providing students a home near campus.

Deciding to Open Women's Residence Halls

On many campuses there was a lengthy debate, after the admission of women, about the nature of coeducation on campus. Many issues of equality arose as women were often discriminated against, leaving officials with a need to establish formal policies regarding coeducation. In Scotland arguments arose about tradition and that the "system of the Scottish Universities" did not include on-campus residences. In 1888 a debate emerged in a local Glasgow newspaper about the possibility of women living on campus. The main argument in opposition of the idea was tradition. Along with this, the fact that there was no residence for male students made observers doubt "if a residential college is needed for Scotch women any more than for Scotch men."[61] Some also believed that providing a residence hall would prove to be restrictive to the growth of the college if students were required to live on campus. The desire to increase enrollment figures in size and, more importantly, in scope, usually led to residence halls being opened.

The Board of Curators of the University of Missouri noted in their Biennial Report to the Board of Regents that there was undue discrimination against the female students with respect to housing accommodations. They found fault with the fact that the men were "supplied with a club house in which they can live comfortably and well at a cost of $1.75 a week" but the women had to live with private families at an average cost of $4.00 per week. Mentioning this discrepancy was part of a formal request for an appropriation of $20,000 from the state legislature to fund the construction of a residence hall for women on campus.[62] The students also hoped for better accommodations. Just before the second men's dormitory, Lathrop Hall, was built at the University of Missouri, the editors of *The Savitar* in 1898 referred to boarding houses as "One of the most horrible instruments of torture ever contrived by man."[63] Certainly this was an extreme, and possibly tongue-in-cheek, description, but if the boarding houses the men were living in were wonderful places, there would have been no reason to make such a statement.

Student agitation for a dormitory at the Western University of Pennsylvania predated the introduction of coeducation. Unlike other institutions the distances involved for students to traverse on their way to classes could be quite large in a burgeoning city like Pittsburgh, and there could be real dangers along the way as well. At the time of the students' complaints in the 1880s and 1890s the campus was located in Allegheny, having moved there when the earlier buildings were sold to the county after the Court House burned in 1882.[64] The students argued that their commuting was wasting time that they could be studying. They felt this was "conducive neither to the highest welfare of the student nor the University."[65] In addition they believed that if they had more time to spend on campus, more students would take part in university activities, like athletics and debating societies, and would be able to make better use of the library. Students would also have the chance of camaraderie with other students that they did not take classes with because they were in different departments of the university. This increased kinship between the students would translate into greater loyalty to the university itself, if for no other reason than because the institution would be their home, not just a place of study.[66]

Concerns over student safety were common in both large and small cities. As the state capital, Madison, Wisconsin was long considered to be "an expensive and unsafe locality for an educational Institution."[67] The admittance of women duly magnified this concern as they were felt to be more naive in the ways of the world. The dangers of life in another capital city, London, were apparent for all students, not just women. College Hall, located on Gordon Square, opened as a hall of residence for women who studied at University College, and also at the London School of Medicine for Women. Because there was not enough space to house all the women who attended these institutions, there was also a list of families in the area who were willing to take in women boarders that was kept at University College.[68] The hall was described as offering "a bright and cultivated home to its inmates" that was an enhancement to their students and would "safeguard" them while they were students.[69]

At Tennessee, pressure came from parents to provide a dormitory for women, with the implication that it was the duty of a state university to take care of their students in every way.[70] In particular students who traveled from a distance had greater difficulty in finding local accommodations if they had to wait until they arrived to do so. There was enough uncertainty about sending daughters to study at a coeducational university without also leaving their residence while there to chance. Once a decision was made by a university to provide housing for women, the next question became if that housing could be located in an already existing building or one nearby or if an altogether new structure was needed.

Renting or Purchasing Residences

In England and Scotland the universities included in this study were all in established cities; this left them little room for expansion in the form of new construction. When new facilities were needed, they often had to rent or purchase nearby buildings and refit them for a new purpose. Providing housing for students often took this form. It was good that housing was made available, but it could also lead to its own set of challenges. For example, in St. Andrews, before University Hall was built in 1896, the university provided temporary accommodations for female students at a house on North Street for the 1892–1893 academic year, and at "Argyle Lodge for 1895–96."[71] There was no official lodging provided from 1893–1895; this left the women students to find their own accommodations just as their male classmates had to do.[72]

On June 27, 1891, an organizational meeting of the Queen Margaret Hall Company began efforts to open a residence for its students in Glasgow. By July they had received 1,500 subscriptions and had taken a lease on Lilybank House for ten years, at an annual rent of £130. Lilybank House was designed by renowned local architect Alexander "Greek" Thomson for the late Glasgow provost John Blackie. It was located close to the university and, following incorporation with the university, this was a particularly convenient location. After refurbishment and refurnishing, the newly named Queen Margaret Hall opened in October 1894 housing eighteen residents.[73] The building enhanced social interaction of the students at Queen Margaret College, and notices printed in the local papers worked to promote the new living option and its rates. Rooms cost either £30 or £37.50 a year and were comparable to those provided at other universities at the same time.[74] The cost was somewhat prohibitive to many Queen Margaret students, who continued to live in private homes or with family who lived in Glasgow. The division between those students who lived in Queen Margaret Hall and those who did not was most noticeable in terms of social activities. Those who had to commute to campus were often unable to remain after classes for social gatherings or club meetings, thus limiting the amount of interaction they were able to have with their classmates.[75]

After women were admitted to the University of Durham under their Supplemental Charter in 1895, they were considered to be "unattached students" or "home students" because there was no housing provided for them at the university itself.[76] Lodgings were eventually provided for them in town beginning in 1899. The Women's Hostel was first located in Claypath, but was moved in 1901 to the "more convenient premises on the Palace Green."[77] Women were subject to the same discipline as the male students and were looked after by the Principal of the Women's Hostel. Laura Roberts first held this position from April 1899 until July 1900, and she was succeeded by Elizabeth Robinson.[78] Women were also permitted to live with family or friends in town, as long as they had been "approved by the Council of the Durham Colleges."[79]

Masson Hall, a women's residence at the University of Edinburgh, opened in 1897 and functioned as the "central meeting place for women students attending the University" whether they lived there or not.[80] Located at 31 George Square, the hall was convenient to lectures and also to other attractions in the city. Isabel Maddison, in her *Supplement for 1897*, included the fees for board and lodging at Masson Hall in both British pounds sterling and U.S. dollars, indicating that she anticipated an interest among women in the United States to attend the Scottish university. The rates for the full winter session ranged from "£25 ($125) to £30 ($150)."[81] Students who did not reside in Masson Hall were able to take meals there for a "moderate" rate, thus encouraging strong bonds between all the female students, whether they lived in university accommodations or not.[82]

A second residence at 12 George Square, Muir Hall, was soon opened for women studying medicine. Selling features of this building were "two Common Rooms...and Bath Rooms on each floor." The range in prices here, as in other locations, had much to do with the amount of space or privacy a resident had, though the *University Calendar* also noted that the "outlook" or view that each room had out the window was also a consideration in the fee charged.[83] In 1900 it provided accommodations for twenty-three women.[84] And a third hall, known as Crudelius University Hall, at Burns House (457 Lawnmarket), was used for a few years by nine "women engaged in professional work."[85]

King's College, London made similar provisions for their Ladies' Department by acquiring two "freehold houses" in the "wealthy London suburb" of Kensington.[86] Located at 13 Kensington Square and 28 Kensington Square, the buildings were "fitted up...for the use of the Ladies' Department, at a cost of over 9,000*l*."[87] The patroness of the Department for Ladies of King's College, London was the Princess of Wales, and there was a committee of "Lady Visitors" who made sure that the department was functioning as it should. There was also a "Committee of Management" that was made up of men and women and was led by the principal of King's College, and included the dean of Westminster.[88] Mrs. Cornelia Gertrude Wace worked as the first lady superintendent from 1885 to 1890, at which point the position was given the new title of vice principal. She also functioned as the department's secretary, handling all correspondence.[89]

Ashburne House, a hostel for women students in Manchester, opened in October of 1899.[90] Miss Helen Stephen was the first warden of the residence hall, which could house sixteen students. The cost of living was kept as low as possible, but there were "scholarships and bursaries for necessitous students" that could not afford to live in the house.[91] The grounds around the building were extensive, including gardens and a tennis court that the women could enjoy when they were not studying. The walk to campus took about fifteen minutes, according to contemporary estimates, and the house was seen to be highly convenient to the university. The building contained a common room, drawing room, and reading rooms, as well as the study bedrooms that the women usually shared with another student. Advertisements also noted that the building was "lighted by electricity," showing they were keeping up with the newest technologies of the day.[92] Two additional residences, The Oaks and the Victoria Church Hostel, would be added for the benefit of women studying in Manchester in the early twentieth century. Ashburne House and The Oaks were described as being located "in open, healthy neighbourhoods," reminding people that there were still worries about the effects of higher education on women and that housing provisions needed to take this into consideration.[93]

Building Residences

Purpose-built university housing, even more than rented or purchased facilities, served their university communities in multiple ways. Because officials had to approve the plans drawn up before funding was approved, it was possible for them to include rooms for meetings, study, or even teaching underneath one roof in the name of convenience and fiscal responsibility. These buildings also provided them with further chances to organize student life and interactions, making sure that propriety was observed at all times. Even the position of a women's residence hall on campus could reinforce the institutional hierarchy. As argued by Annabel Wharton in her article "Gender, Architecture, and Institutional Preservation: The Case of Duke University," the importance of building arrangement in displaying hierarchical order cannot be underestimated. Campus architecture can be used to control social interactions, and "is a purveyor of status and authority…representing a set of values."[94] This is especially apparent at the University of Wisconsin, as seen in the image at the start of this chapter, where the main administration building was placed at the top of university (later Bascom) hill, with subsidiary academic buildings farther down the hill, and Ladies' Hall tacked on at the farthest reaches of what then made up the campus. The threat posed by the possibility of women living in close proximity to their fellow male students was also considered. In the end, Ladies' Hall was built at a safe distance from the men's residence, North Hall.

Initially women at the Pennsylvania State College lived in the main building on campus, on the top floors, "segregated as much as possible from the men."[95] In 1888 the college officials decided to provide meals for students who lived off campus at a "private table," but they hoped for a more permanent space for subsequent years. There was a greater concern for the welfare of the female than male students because it was believed that "a refined family life plays no small part in the liberal education of young women."[96] A push was therefore made for a residence hall to be built on campus for the women students, with a proposed capacity of forty. This building could also be used for courses in domestic economy, with the kitchen being used "as a laboratory."[97] In 1890 the student publication, *The Free Lance*, reported on the opening of the "co-ed cottage" on campus as a residence for the female students. They described it as both "neat and enchanting" and felt sure that the women who would live in it would find it pleasant.[98] The building boasted electric lights and steam heating, as well as a gymnasium for the use of the residents. It is clear, looking at images from the Ladies' Cottage at Penn State in the 1890s, that it was decorated as any middle- or upper-middle-class home would have been at the time. Furniture, lighting fixtures, and carpets would have shown the women who lived there what they should look to purchase for their own homes once they were married.[99]

University Hall for Women Students at St. Andrews, opened in 1896, had as its warden Miss Louisa Innes Lumsden.[100] Unlike some universities, women were not required to live at University Hall and could instead stay with family or friends in the area. Although staying with family was common, the inclusion of "friends" meant a wider selection of housing options in the relatively small and remote community of St. Andrews.[101] The provision of permanent housing for women made the university more competitive with other institutions that increasingly sought to recruit both male and female students. Having a woman with respected credentials in charge of the residence increased the drawing power of the institution as well. Lumsden's success in this respect was "proved by the fact that the students of University Hall have been drawn in fairly equal numbers from both sides of the Border" during her tenure.[102] When she resigned her position at University Hall in 1900, *The Journal of Education* described her work there as combining the residential format of the English colleges at Oxford and Cambridge with the "freedom of the Scottish life."[103]

Contemporary sources reveal that the compulsory nature of the housing provisions in Aberystwyth "seemed to attract students" in subsequent years; this caused officials to make plans to build a larger women's residence hall, as well as residences for the male students.[104] In 1896 Alexandra Hall was opened by its namesake, the Princess of Wales, wife of the future King Edward VII (who was installed as chancellor of the university on the same visit), for the use of the women students. The building cost more than £30,000 and was designed by C. J. Ferguson. Its location was at the northernmost spot on the Marine Parade, or Marine Terrace; this made it quite a distance from the academic buildings of the college, or as one modern historian at the university put it "as far away from the Old

College as possible."[105] By its tenth birthday over 200 female students lived in the building.[106]

Alexandra Hall is five stories tall (plus a basement) with a courtyard in the center and was built with local grey stone. Students could attend prayer in the dining hall each morning if they wanted to before breakfast and also ate dinner and tea (lunch and dinner to Americans) there daily as well. There was also an optional supper offered at 9:00 p.m. Other rooms in the Hall included a library, "a suite of drawing-rooms...a large recreation room, a matron's room, a servants' hall, and a commodious kitchen."[107] Students also had a choice in the type of bedroom they had, depending on how much they wished to pay for their accommodations. The price range in 1900 was between 27 and 42 guineas.[108] They could have a bedroom alone or a study bedroom that included both sleeping space and a sitting room they could use for studying.

A Woman's Building, eventually renamed Barbara Blount Hall, was constructed at the University of Tennessee in 1898. It housed fifty students and had an office for the acting dean of the Woman's Department, Angie Warren Perkins.[109] The building was described by T. W. Jordan, the dean of the college, as "elegant" and "the envy of all."[110] Whether he felt the male students were envious or other universities were envious is not clear. The increasing competition between colleges and universities at the time makes it possible that he was referring to both. The intended purpose of Barbara Blount Hall, aside from providing lodging, was to help the young women who attended the University of Tennessee to learn proper household management. Their new dean, Florence Skeffington, explained that she tried to recreate "as far as possible...the moulding and restraining influence of a well regulated private family" in the Hall. The rules and guidelines established for the running of the Hall were adhered to "cheerfully" by the students who knew what was expected of them and did not wish to create problems. Furthermore, Dean Skeffington felt that the presence of women on campus helped the male students' behavior as well. She stated simply that coeducation made "the boys more manly and the girls more womanly."[111]

Alternatives to Residence Halls

Another housing option for women in the United States that started in the late nineteenth century, and one that ensured the observation of proper gender roles, was the sorority. Initially formed as secret societies similar to literary societies, these groups were designed to be more social and less academic in nature.[112] Fraternities and sororities (the male and female counterparts) came into great prominence on university campuses across the United States by the turn of the century. The role of these groups is often downplayed by historians due to their exclusive nature, making them unrepresentative of the wider university student's experience. Their significance, however, in the evolution of student's activities on campus cannot be discounted. In particular, the role of sororities as university-sanctioned, on- or off-campus accommodations, which may or may not be strictly monitored by campus officials, was crucial to the development of other university housing. At some institutions, such as the University of California at Berkeley, sororities offered the only housing on-campus for women.[113] Where sororities existed alongside university residence halls, the liberal attitude held at sorority houses put pressure on officials to have similar standards in their residence halls or risk losing the income from students who preferred the option of sorority houses.

One national sorority, Kappa Kappa Gamma, had chapters at several midwestern universities, including Indiana, Missouri, and Wisconsin. The Indiana chapter was founded in 1873 and was their second sorority after their Kappa Alpha Theta chapter was founded in 1870.[114] The Missouri and Wisconsin chapters of Kappa Kappa Gamma both began in 1875 and were the first sorority on each campus.[115] At Missouri the life in the sorority chapter house was self-governing, while at Wisconsin officials argued in 1899 that the system of regulation over on-campus housing "be extended to sorority houses as well," indicating that it had not been previously.[116] And at Indiana there was no housing provided for men or women on campus until 1915 when sorority and league houses were built by various student organizations on university grounds.[117]

In time university officials decided that having their women students live in one of the approved houses, rather than in a questionable rooming house in the city, was preferable, if not ideal. Sororities were increasingly used by officials as an acceptable alternative to living in residence halls as well. The "Greek" groups could at least be held liable to the regulation of student government, and the fact that fraternities and sororities had selective "pledging" practices meant that their intake of new members was limited. This was often due to the size of the house they occupied, and it enabled them to serve only a small portion of the female student body. Lynn D. Gordon notes that the deans of women at the University of Chicago "worried that the formation of national sorority chapters . . . would exacerbate social-class divisions among women students."[118] Although some universities forbade the formation of Greek letter societies, most in the country permitted them, paving the way for the establishment of both social and housing hierarchies on campus. According to David Sansing, 56 percent of the students at Mississippi in 1900 were in a fraternity or sorority and they "dominated every aspect of campus life."[119] The social "pairing" of different fraternities and sororities also furthered the relationships between students who joined them. Ideally the men and women in these groups would come from similar societal backgrounds, making the relationships which formed in university ideally suited to continue into adult life.

Discipline for Students Living on Campus

Along with developing codes for academic discipline, universities that provided room and board for their students needed to establish codes of conduct for students when they were outside the classroom. In some cases student discipline was in the hands of the faculty who would decide on "rules and regulations...for good and orderly government" of the campus.[120] In other cases the discipline was military, as it was at the University of Alabama. According to their *Catalogue* in 1880, "this mode of discipline is the most efficient means of maintaining good deportment, and of promoting good morals and studious habits; while it enforces just so much manly and invigorating exercise as is conducive to healthful development."[121] Clearly the same angle could not be taken on disciplining women students, but at most colleges and universities it was assumed that the male students were the ones who were likely to act up. Regulations at the University of Michigan were not as extensive, or as intrusive, as they were at some other institutions. There was no curfew for students, and there was no prescribed conduct for time spent outside the classroom or off campus. Officials hoped simply that students would act in a "becoming" manner and not bring negative attention to the university.[122]

Soon after John Bascom arrived at the University of Wisconsin, the Board of Regents resolved that the president would have sole responsibility for instituting regulations for the student body in the name of "good order." Specifically the Regents

> [r]esolved, that the following rules for the maintenance of good order and discipline be hereby and henceforth established subject to such deviation in special cases as the President of the University may deem proper....That the doors to Ladies Hall be closed at Eleven o'clock P.M. and all students there occupying rooms, be required to be within doors before that hour.[123]

One of the more intriguing regulations added by the president after this ruling actually specified the men allowed in Ladies' Hall. Although it would seem "proper" to forbid all men from spending time in the women's residence or spending time there unchaperoned, the only specification made was that the men who visited be family or Wisconsin students.[124] This supports the historian Ronald Hogeland's conclusions in his article, "Coeducation of the Sexes at Oberlin College: A Study of Social Ideas in Mid-Nineteenth Century America," in which he argues that administrators felt it was part of their duty to provide their graduates with suitably educated wives.[125] From this standpoint "the presence of young maidens at the school was essential for the well-being of their male student's sanity."[126] His further claim, that this was the only reason the women were admitted as early as they were, may be a bit too harsh, but the encouragement of "friendships" between male and female students cannot be denied.

Activities at the Ladies' Cottage at the Pennsylvania State College were of great interest to the students who wrote and edited the yearbook, *La Vie*. In 1892 an illustration is shown with a male and female student having a conversation, with an unattractive-looking matron sitting on a chair between them chaperoning the scene. The caption reads "A supposed case at the Cottage" for this type of direct supervision was not, in fact, the norm at the women's residence. The women did need to get permission to have male visitors, but that did not equate to a chaperon listening in on their conversations.[127] *The Free Lance* also reported regularly on the situation at the cottage, which quickly became a social center on campus. Even before it was built the male students anticipated "big porches, cozy rooms, and lots of hammocks" which they hoped would get plenty of use.[128] The Ladies' Cottage was also the scene of some shenanigans among the male students at Penn State. One of the relatively harmless forms of hazing of incoming students was to tell them that if they needed a blanket, they should ask for it at the cottage. According to one description of it in the 1898 issue of *La Vie,* the custom developed as a way "to relieve the monotony." At other times, rather than collecting a blanket, the errand set out was taking laundry to the building to have the ladies wash it for them.[129] These activities may have been harmless, but they would lead to further restrictions being placed on student residential life in the twentieth century.

In a continuing effort to align policies in Glasgow with those of other institutions in Scotland, Mrs. Riddoch, the superintendent of Queen Margaret Hall sent enquiries to the Universities of Edinburgh, St. Andrews, and Dundee in 1902 to determine their regulations for visitors. The first response she received was from Frances H. Simson at Masson Hall in Edinburgh. Miss Simson reported that there were no formally printed rules at the hall, but that there was "an understanding that none but fathers and brothers may go to the students' rooms."[130] If the residents desired to host a guest, they could arrange for "undisturbed possession" of a public room for the visit. The possibility of violating the accepted social code of relations between the sexes was an often unspoken, yet undeniable apprehension about coeducation. St. Andrews took a similar approach, with the rules simply passed "from one generation of students to the next." Their warden, Frances Melville (who would later hold the equivalent position at Glasgow), considered individual requests to allow visitors, with tea provided for guests in a sitting room if desired.[131]

Mrs. Riddoch wrote back to Miss Simson in Edinburgh to follow up on an anecdote she mentioned concerning disciplinary problems with a student who disregarded the implicit policy on visitors. Apparently the student in question had invited two brothers and a friend to her room for tea following a funeral. Due to the circumstances the university overlooked this indiscretion, but in most cases any resident who was "not content with the permission to entertain her friends in one of the public rooms, I should think she was rather too 'emancipated' a young woman to be a desirable inmate of a Hall."[132] Miss Simson added that she should welcome the departure of such a student, indicating the extremely serious nature of the offence at the time. The maintenance of Victorian standards of behavior was a constant focus at the university, and they felt particular resonance in respect to female students. Relations between male and female students would continue to be approached with considerable reserve, both in the initial stages of coeducation and well into the twentieth century.

The Health of Students on Campus

Despite the success of women in the classroom, there was a continued belief that women's health would be damaged by university life that was not easily overcome. Ohio's first female graduate, Margaret Boyd, commented on her own physical abilities as a student in her diary. On Saturday, January 11, 1873, she remarked that she "Did not write any and studied very little. If I just could study all the time, but I find I can-not. The body I think has a great influence on the mind."[133] Whether Boyd's views were purely her own or were reflections of the concerns society had about the physical strain higher education caused women cannot be determined from her statement, although it is likely that she was influenced by the views of those around her. And certainly women's fashions of the day, corseted or not, would add warmth and weight that students would need to bear during the day. The ultimate question remained, however. Could women physically handle a university education at the same level as men?

The first evaluation of the effects of higher education on women at the University of Wisconsin came in 1877, following the publication of the Board of Visitors' Report of that year. After several years of support for the school's coeducational policies, the Visitors devoted one-fourth of their report to the ill effects of the system on women. The Visitors observed that the female students appeared sick with "sallow features, the pearly whiteness of the eye, the lack of color... and an absolute expression of anæmia." The Visitors considered Ladies' Hall to be clean and hygienic, so they felt the only remaining explanation for the poor health conditions to be the demands of the university.[134] The report continued by pointing out the belief that "nature makes a great demand upon the energies of early womanhood" and that the hard work done at the university was just too much to handle. They moved on to apply the theory of Republican Motherhood to this as well. While the Visitors realized education was important, they believed "it is better that the future matrons of the state should be without a University training than that it should be procured at the fearful expense of ruined health." Their recommendation to the Board of Regents was to alter the academic requirements so that "each sex should be enabled to secure that form of education best fitted for his or her respective sphere."[135]

The university's president at the time, John Bascom, adamantly contradicted the Visitors' Report. He said that they might have "inadvertently" made some "mistakes" while drawing their conclusions, arguing that the female students, who made up about one-fourth of the student body, were extremely fit and capable of maintaining their course work. He went so far as to say that they "do their work with less rather than with greater labor than the young men" and "that a young woman who withdraws herself from society and gives herself judiciously to a college course, is far better circumstanced in reference to health than the great majority of her sex."[136] As a result of President Bascom's impassioned comments, the Board of Visitors looked at the situation differently the following year when they "observed with pleasure the robust appearance of many of the students" and that they did "not concur in the criticisms made by some upon the system of co-education."[137] Also worthy of note is a fact that the Visitors failed to recognize in 1877, regarding the health of their female students, which was located on the first page of the Regents' Report to the Governor. In discussing the many "improvements" to buildings on campus they listed the new use of gas as a "healthful and necessary convenience of Ladies' Hall" during the year. So, although the Visitors did not realize it at the time, the ill health they reported was due, apparently, to the gas fumes introduced to Ladies' Hall.[138]

The 1870s and 1880s were a time of reflection and assessment for the universities that were among the first to open their doors to women. In University of Michigan president Angell's Annual Report to the Board of Regents in 1879 he made a decisive statement about the success of coeducation at the university:

> After our nine years' experience in co-education, we have become so accustomed to see women take up any kind of University work, carry it on successfully, graduate in good health, cause no embarrassment in the administration of the Institution, and awaken no special solicitude in the minds of their friends or of their teachers, that many of the theoretical discussions of co-education by those who have not had opportunities to examine it carefully read strangely to us here on the ground.[139]

A few years later he was one of the respondents to a survey done by W. Le Conte Stevens of several coeducational colleges and universities, on behalf of the Association for Promoting the Higher Education of Women in New York. The purpose of the study was to see what the status and health was of women attending both public and private institutions. President Angell reported that as long as a woman was healthy when she began her university career, she would still be healthy at the end of it, presuming she "exercises a fair degree of prudence" in her behavior. He concluded that higher education was no more demanding than the expectations of women in "society."[140]

Questions of a woman's health in Britain were made somewhat later, since in many cases there were not significant calls for concern until the numbers of women in attendance were great enough to warrant it. Understandably, the health of the women students living in halls was a constant concern for the university officials who were responsible for them, and for their parents. Miss Maitland addressed this topic in her speech at the Central Conference on Women Workers in Glasgow in 1894. She felt that the health of women students on campus was improved because they were well supervised. By eating meals together, a proper diet could be maintained, and having regulations that set times for them to be inside for the evening greatly reduced the danger to their health posed by bad weather, tiredness, or any inappropriate activities.[141] Miss Maitland's experience at Oxford colored her view of the situation because all colleges there were residential. The increase in the number of other universities providing housing for their women students made the experiences she described transferable to many other locations.

Aberystwyth was one university that made extensive provisions for health care, going so far as to provide an infirmary for its students, located near the other buildings of the campus. Apparently this building was often "empty, due to the healthy situation of the place and to the fact that only the physically fit are admitted to College."[142] The location of the college directly on the coast meant that students had the benefit of plenty of sea air, and it was a commonly held view in the Victorian Era that "invigorating breezes" helped people remain healthy.[143] Claims were even made that the students did not need to take part in organized physical activity on campus because they had so many opportunities to enjoy the "bracing climate" that was common on the Welsh coast.[144] Additionally all of the rooms in Alexandra Hall were considered to be "light and airy" by Victorian standards, and a separate "lavatory block" was attached to the building by "a narrow cross-ventilated 'bridge'."[145] This indicates that great lengths were taken in the area of sanitation and modern hygiene to protect the health of the female students.

Expenses of University Life

At many state-supported universities in the United States, tuition was free to students who were residents of the state until well into the twentieth century. Since state tax dollars were being used to fund the colleges and universities, it was seen as wrong to charge parents again for an education that they had already paid for through taxes. Or, if tuition was not free, some other aspect of university life was made more affordable to keep the costs as low as possible; this made a clear choice between a state-supported institution and an expensive private one. At West Virginia University tuition was "practically free to all students from West Virginia."[146] Similarly, advertisements for the University of Alabama noted that if you were a "*bona fide*" resident of the state you could become a student in any course of study, except law, free of charge (tuition for the Law School was $50.00 in 1895 and 1896). The expenses of college life typically came outside the classroom for room and board, which for a cadet cost $172.50 in 1895 and 1896, and residence in the barracks was required for male undergraduates.[147]

Starting in 1897 expenses at South Carolina College included both a "Term Fee" and a "Tuition Fee" that were paid by the students at the start of each academic term. The cost of tuition was the same for all students, unless it was remitted for some reason, but women were actually charged less than men for their "Term Fee."[148] This was due in large part to the fact that there were dormitories for the male students, where they could live free of charge.[149] The catalogues produced by the administration, as at other colleges and universities, gave a further rundown of potential expenses students might incur, but also suggested that there was "nothing in the customs and habits prevailing in and around the College that encourages extravagance or useless expenditure of money."[150] Many universities did their best to convince students not to have money in their pockets while on campus. This would cut down on temptation and would limit the possibility of theft.

Expenses were somewhat different at Mississippi during the 1890s. There was no tuition in the Department of Science, Literature and the Arts, but there was a $10 matriculation fee. Students also had to pay a $2.50 "fuel fee" to help cover the cost of heating the buildings, and the men who lived on campus had to pay an additional $10–$20 for their fuel use there. If students lived with private families, as some of the women did, they paid a fee for their board to the university of $10–$12 per month to cover their meals. Further expenses paid to the university included those for lights and washing, and all fees needed to be paid at the start of the academic year. Similar to modern universities, students who lived in the dormitories also had to pay a deposit of $3 "to cover any injury by him on the building" which would be refunded if he did not do any damage.[151] Again the university also advised parents and guardians who were sending young men to campus not to send them with too much money, but to send it as needed at the start of each term. Their warning was strong: "Almost any student will be ruined by having always plenty of money in his pocket, and he will be the cause of ruin to others."[152]

At Indiana students were told that the Bloomington community, because it was primarily rural, had a low cost of living and that they would be able to find affordable rooms to rent with private families. They could either take their meals with that family or eat at a boarding club to help minimize expenses.[153] Female students were often resourceful and found ways to help each other through their time at university. At Ohio University four women "formed a self-boarding club" to reduce their food costs and make life more affordable.[154] All of the efforts of students to save on expenses and the encouragement of university officials that students live frugally did not prevent some students from flaunting wealth, whether it was done consciously or unconsciously.

Class issues were apparent at Queen Margaret College from its inception. Differences in housing fees on campus separated residents in subtle ways, as some students could not afford the convenience of living on campus.[155] At the university itself housing and meals were also dependent on wealth. In 1948 Marion Gilchrist recalled that they tried to provide lunch at the college, but "the cost (2s. 6d.) was prohibitive for most of the students."[156] A similar approach was taken in Manchester where students were able to eat at the refectory that was located across the street from the campus for "a moderate charge."[157]

In 1902 a Glasgow newspaper reported: "Cheapness is a thing of great importance to the Glasgow Student, and when Professor Geddes lectured to a Glasgow undergraduate society on this subject the point was made during the debate that the Edinburgh Scheme would create an aristocracy of wealth." The Edinburgh Scheme referred to began with the foundation of University Hall in 1887.[158] Room rents varied "from 8/6 to £1 per week, according to situation and style of room." Additionally, the costs of meals, heating, cleaning, and "coaching and general supervision of work" by "resident Graduates" could vary depending on the ability of the resident to pay for these items. The acquisition of a room in University Hall in Edinburgh brought with it an amount of status, with internal status in the hall depending largely on seniority. Within this structure, financial differences would have been clearly visible as well.[159]

There is a similar sense of an "aristocracy of wealth" in the women's residence halls elsewhere in Britain. University Hall, St. Andrews, was a small residence hall in comparison with others in the United Kingdom with housing for only twenty-four women in its first year. Similar to its competitors it had a range of fees depending on the room and board combination chosen. In 1900 the fees were £30 to £50.[160] The least expensive rooms were study-bedrooms the two portions being "divided by a curtain." The most expensive rooms had a "small bedroom and private study."[161] There were different costs for rooms in Queen Margaret Hall, Glasgow, as well. So within the university housing a hierarchy formed between those who paid £30 for their room and those who paid £37.50. Early on, the administration at Glasgow attempted to prevent the further stratification amongst the students by making a conscious decision to charge a flat rate for matriculation to courses. The differences remained, as the financial constraints placed on those who needed to commute to university restricted their social life. This difficulty was perhaps best summarized by Katharine Lake in her reminiscences of life at the University of Durham when she remarked on the "pinched and straitened circumstances" many students faced when it came to being able to afford the costs of a university education.[162]

At Wisconsin there were similar divisions within the student body on the basis of family wealth. Mary Caroline Crawford noted that at Chadbourne Hall the "cost of the table accommodation is three dollars and seventy-five cents a week, the price of rooms varying from forty to ninety-five dollars a year, according to location."[163] The question of hierarchy within Chadbourne Hall should be noted at this point, with differences in status based on ability to pay. Off-campus housing was similarly divided, with more acceptable residences costing more than others. With housing options limited, those students from poorer backgrounds generally had to live with relatives in the city to economize. This situation left them on the periphery of campus activities, while those who could afford to live on campus were thought to gain a more complete university education. The administration often argued that the "men" attending university should be treated as adults and did not need constant surveillance by the faculty. Once there was sufficient competition for housing in Madison to keep the rates within reasonable limits, the university chose to cease provision of housing for their male students, and by 1887 the only campus residence hall remaining was Ladies' Hall.[164]

A final consideration in the area of expenses for students living on campus was the creation of scholarships or other funds that could be used to defray the costs of living. The Council of College Hall, London, offered two fellowships from the Pfeiffer Bequest for women residents who were pursuing degrees, one for a graduate student and one for an undergraduate.[165] In Scotland the question of wealth in determining a student's ability to undertake higher education changed significantly with the introduction of the Carnegie Trust in 1901.[166] Money available from this fund "made it possible for 'intermediate' and 'working class' groups to attend university." The historian R. D. Anderson notes that by the period 1904–1908, fifty percent of all Scottish students were aided by the trust. Since these funds were available to both men and women, the matriculation of female students rose considerably.

Increasingly individual communities also began to appreciate the value of women's higher education and therefore established bursaries to support local students who wished to attend university. For example, an article in *The Buteman* in Rothesay reported the establishment of two bursaries "for ladies taking the Glasgow University Local Examinations."[167] As the historian Catherine Mary Kendall concludes there was a fundamental shift from the early years of higher education for women when it was "more for culture than professional advancement and hence more a preserve of the wealthy" to the end of the Victorian Era when a more democratic nature was introduced, particularly following the establishment of the Carnegie Trust.[168] The women no longer had to be those from the wealthiest areas in the city that surrounded the university or from wealthy families who could cover all the costs of university life upfront.

Conclusions

The organization and regulation of residence halls led to considerable variation within universities. At the heart of this issue was the question of what role should be undertaken by an institution of higher education. The collegiate model, as developed at Oxford and Cambridge, followed the view that higher education was both a social and intellectual enterprise.[169] Along with convenience and protection, the argument was made that residence halls were a way for the students to have "companionship without which there can really be no higher education."[170] The conclusion that higher education necessarily includes a residential component is largely a result of the Oxbridge model, and there is ample evidence that students during the nineteenth century still felt camaraderie with their classmates whether housing was provided for them by their institutions or not. Ultimately the interaction of faculty, staff, and students outside the classroom was often given the same importance as that within it. Once institutions chose to take responsibility for the development of their student's character, as well as that of their academic training, they necessarily began making moral judgments about life at university. This issue was in many ways simpler when it came to university women, as all institutions acknowledged that they would need to provide an additional amount of protection and guidance for them than they had to do for men.

The collegiate system remained for women on campuses on both sides of the Atlantic, with the universities stipulating that part of the higher education received by young women was that found in university residence halls. In many ways this "education" was equivalent to that of a finishing school, with students learning proper manners and behavior from the women in charge of them. The focus placed on the establishment of a simulated home life for the students while on campus often eased their transition to universities.[171] This focus can also be seen as training for women's expected future life; that of being wives and mothers. The mother-daughter relationships formed by way of university housing was often cited by university alumnae as the most valuable part of their educational experience. This structure was also pleasing to those in society who still harbored reservations about the place of women in universities. If the individual institutions could show that they were preparing their female students to take on traditional roles in society once they left university, there was less reason to challenge their admission.

The housing practices that developed at each institution were also key in the formation of social structure on campus. The hierarchical system at the universities was visible within housing provisions, with women clearly receiving separate treatment to that of their male counterparts. This, coupled with the regulations on visitors to residence halls, helped to guide women into forming proper relationships with male students. These relationships will be investigated more fully in the following chapters on campus social life, but it is important to note the difference in experience between the male and female students that was emerging at this stage, to consider how it played out in other areas of campus life. The restrictions placed on women students, which were often contrary to the freedoms granted to male students, perpetuated the notion of separate spheres that had so fully pervaded Victorian society. Overall, the ongoing inculcation in traditional gender roles made possible through university housing resulted in a significantly different university education for men and women.

CHAPTER SIX

Extracurricular Student Life

Image 6 Women's Boating Club in the 1890s. In this photo you can see Craig Glais/Constitution Hill in the background. The newly constructed Alexandra Hall is just visible on the shoreline, the farthest building to the left.

Credit: Llyfrgell yr Hen Goleg/Old College Library, University of Wales, Aberystwyth, Ceredigion, Wales.

The extracurricular, or "extra-academic life," of students at coeducational universities provides some of the most compelling evidence that the students themselves wished to maintain separate spheres of activity despite working toward the same degrees.[1] As discussed in the previous chapter, administrators in the nineteenth century did their best to restrict male and female students to separate areas of the campus by offering separate facilities to each. These regulations were relaxed over time, and the students slowly began integrating in all areas of campus life. The amount of gender mixing remained limited, though, and as Lynn D. Gordon points out in *Gender and Higher Education in the Progressive Era,* "men's and women's student lives proceeded along separate, although parallel, paths."[2] Sometimes this separation was encouraged by the administration, sometimes by the male students, and sometimes by the women themselves who preferred to have their own groups that appealed to their own skills and plans for their future lives. A mixture of academic organizations and those which were intended to be primarily social was a standard feature on most campuses. Athletics of an informal or formal nature were also increasing in popularity during the nineteenth century, with students having a chance to participate or be spectators. This variety of groups and activities meant that students had ample opportunities to improve their minds and bodies, and some associations even enabled them to look after their souls as well.

In her book *The College Girl of America*, Mary Caroline Crawford emphasized the "many-sided" education women received while at university as being one of the most positive aspects of attending a coeducational institution. Along with athletics and literary societies, she also pointed out the formation of student self-government associations and Christian groups on campus.[3] Although she was discussing the University of Wisconsin, her glowing account of women's lives on that campus could easily have been made about most coeducational universities during the late Victorian Era. For instance, women students had become very active at the University of Missouri by the 1890s. Miss May Mansfield was on the editorial staff of *The Savitar*, was an active member of Kappa Kappa Gamma, was the only woman officer of the Athletic Association, and was the ladies' singles champion in lawn tennis.[4] One of her classmates, Miss Iva Jane Todd, was a business manager for *The Savitar*, was both a president and treasurer for the Philalethean Society, and was the president of the Y.W.C.A.[5] Students at St. Andrews had a broad range of possible student societies including the Celtic Society, the Missionary Society, the Musical Society, the Science Club, the Shakespearean and Dramatic Society, the Theological Society, the Total Abstinence Society, and the University Gymnastic Club.[6] The full range of student extracurricular activities will be the focus of this chapter, with the exception of student publications that will follow in Chapter Seven.

Academic Clubs and Organizations

The first type of student group to consider are those tied directly to their courses of study. These societies were usually formed by the faculty members in the first instance, though they were run by the students. In many cases the line between classroom activities and extracurricular ones was not entirely clear, and at times there was an implicit expectation that students studying a subject would pursue further enrichment in their free time in these groups. Along with discussing topics related to specific fields of study, many of these organizations maintained their own collections of books that members could use to save on the expenses or time needed to do research. At the University of Edinburgh "Class Museums" included "Museums in connexion with the classes of Natural Philosophy, Materia Medica, Midwifery, and Botany."[7] At St. Andrews the chemistry class, humanities class, logic class, and mathematical class all had libraries too which could be used by students in those courses.[8]

Elsewhere in Britain student and faculty curiosity often led to the formation of academic interest groups. All of the ancient universities in Britain had a long history of extracurricular activities, most of which were related fairly directly to the academic studies pursued by the students. These longstanding groups would inspire more groups to form as new disciplines were added to the university curriculum and new topics became popular in society. At Edinburgh, the youngest of the ancient institutions, organizations like the Dialectic Society (started in 1787) and the Scots Law Society (started in 1815) provided the male students with additional opportunities to hone their skills and knowledge of subjects that they would continue to use in their professional lives.[9] At Aberystwyth, like St. Andrews, there was a Celtic Society and a Scientific Society, whose remits are clear in their names.[10] One difference between the Celtic Societies in the two countries was that the Scottish one focused on Scots topics, while the Welsh one held debates in their native language.[11] The women of the King's College Ladies' Department had social activities to serve their academic interests as well. The Browning Society met weekly "in the Vice-Principal's room, to read and discuss Robert Browning's poetry."[12] Miss Lilian Mary Faithfull became the vice principal of the Women's, or Ladies' Department of King's College in 1894, after teaching English at Royal Holloway College, which led to her affinity for Mr. Browning's work and encouragement of the same love in her students.[13]

In Ireland both the Queen's Colleges in Belfast and Cork had Medical Students' Associations that promoted social interaction between students, their professors, and members of the public as well. In 1890 *The Lancet* reported on the "annual conversazione" of the Belfast Medical Students' Association that served as a "social reunion" of past and present members. The event included the demonstration of experiments from various faculty members followed by a concert and refreshments.[14] The organizations also gave the students a collective voice to ask for changes at their institutions and in the wider medical community. The Cork Medical Students' Association wrote to the Royal College of Surgeons in 1898 to ask that the examinations they took for the Royal University of Ireland be recognized as equivalent to those given in England.[15] This assertiveness was noted in both *The Lancet* and *The British Medical Journal* and showed that the students were gaining the confidence needed to succeed in their chosen profession.[16]

At Indiana University the Zoological Club was made up of instructors and students in that field. The group usually met once a week, and sometimes had guest speakers or group discussions on set readings. In 1894 they discussed "Geddes and Thomson's Evolution of Sex" and "Wallace's Darwinism" at their winter and spring meetings.[17] The first book was written by the Scottish professor Patrick Geddes and John Arthur Thomson of the Universities of St. Andrews (University College Dundee) and Aberdeen, respectively, and was first published in 1889.[18] The second book, *Darwinism: An Exposition of the Theory of Natural Selection with Some of Its Applications*, was published in the same year and was written by Alfred Russel Wallace. Both books were popular enough to go into multiple printings and editions and were often used as required textbooks in colleges and universities.[19] Other academic organizations at Indiana were the Biological Society, Classic Club, Historical Club, Language Teachers' Club, Mathematical and Physical Club, Scientific Society, Shakespeare Club, and Social Science Club.[20]

Dancing, Drama, and Musical Clubs

Artistic pursuits were popular with students at all universities, though some had more organized forms of entertainment than others. Even before coeducation the male students often sought out female companionship for the purpose of dancing and dramatic presentations that required mixed-sex participation. At South Carolina College one student, Charles Woodward Hutson, mentioned in a letter to his mother in 1857 that he had gone to supper at the home of a classmate that was "awfully rowdy...and showed most sensibly the want of female society."[21] The belief that women would cause men to behave better surfaced in many areas of discussion at universities, though in this case the desire to have women at a social engagement probably had as much to do with simply wanting to have them as companions, as having them help maintain decorum. The men of the Western University of Pennsylvania also saw the lack of women as classmates as a disadvantage and invited the women at the Pittsburgh Female College to play the female characters in their theatricals, and the young ladies "did much to add to the evening's pleasure."[22]

Several universities had single-sex dramatic groups which could perform together depending on the situation. At Tennessee the men's society was the K.K.K., which stood for Kit Kat Klub. The women's dramatic society was the Rouge and Powder Club.[23] The membership of this group also included the "present librarian of the University and the dean of the woman's department. By the constitution, no play can be presented until the dean of women has approved the selection of both play and cast."[24] Photographs included in the student yearbook, *The Volunteer*, show that the men sometimes chose to take on the female roles themselves as was done in Shakespeare's time. The students appear to have done it more for laughs than out of a sense of theatrical tradition, however. At other times they performed with their female counterparts in the Rouge and Powder Club.[25] Southern propriety did not provide an objection to this arrangement, and at South Carolina College their Carolina Dramatic Club was open to both men and women who wished to take part in campus theatricals.[26]

In Britain coeducational groups were more common than single-sex ones. At St. Andrews there was a Shakespearean and Dramatic Society that both studied and performed the playwright's works.[27] The description provided in the *Calendar* stated that they wanted to "give an opportunity for the cultivation of dramatic talent among its members, and to foster the study of good reading."[28] Furthermore they would put on "a public representation" of some plays at the end of each year. In Aberystwyth a slightly different style of organization was chosen when the students formed a Dramatic Committee to discuss plays that later became a Dramatics Club to perform them as interests evolved.[29]

Playing music and singing were also popular amusements on campus. Students at Aberdeen, Edinburgh, and St. Andrews each had a Musical Society.[30] Durham had two Choral Societies, one for men and one for women.[31] The Chapel Choir at Missouri was coeducational, as was the South Carolina College Glee Club.[32] At West Virginia there were four singing groups—the Men's Glee Club, the Women's Glee Club, the W.V.U. Choral Society (for women), and the Choral Union (for all students). The Executive Committee of the third group was made up of students who were members of "the Young People's Societies of the various churches."[33] Michigan's first Glee Club was formed in 1867, before women were admitted, and was revived in 1884 as the University Glee Club when interest in the group was renewed. In the interim the desire of students to sing was satisfied by the Choral Union, which was established in 1879.[34] The considerable love of music in Ann Arbor led to the arranging of an Annual May Festival that was first held in 1893.[35]

Sometimes the various musical groups on campus would combine their talents in joint concerts or recitals. This was a common occurrence at the Pennsylvania State College at the end of the century:

> On March 25, [1899] the Women's Glee Club gave a recital, in the Chapel, of "The Fisher Maidens." This difficult cantata was very credibly rendered under the leadership of Miss Hattie Atherton, to whom special credit is due. The Glee Club was assisted by the Mandolin Club, whose selections alternated with those of the Glee Club.... Those who staid away on account of the inclemency of the weather missed a rare treat.[36]

Less than a month later:

> The Concert given in the Chapel on April 15th, [1899] for the benefit of the Track Team, was rendered before a very appreciative audience,—a double programme having been given. It consisted of vocal and instrumental solos, piano duets, and selections by the Women's Glee Club and the Mandolin Club. The banjo solos of Chas. M. Atherton in particular received repeated encores.[37]

The added element of fundraising for the track team shows that the students were taking it on themselves to use their accomplishments to help those who needed assistance, a common motivation of university students at the time. The concerts may have been successful had only the Women's Glee Club performed, but having the men in the Mandolin Club join them meant that a larger audience was probable, regardless of the inclement weather at the first event.

One of the favorite extracurricular activities of the students, both male and female, was dancing. It was one of the more controversial pastimes of students because it required physical contact that could be considered unseemly. At the Pennsylvania State College there was a ban on women attending campus dances until 1890, at which point carefully supervised socials were held on campus at Old Main.[38] In Alabama there was also a ban against dancing in Tuscaloosa. The students found ways to get around the rules by cleverly titling dances as receptions or "German" club meetings. According to the historian James Sellers, the term *German* indicated both a dance step and a dance where women asked men to dance.[39] By using code words, the students were not in direct violation of the regulations against dancing, and university officials were apparently unable to crack down on the misbehavior.

The Junior and Senior Hops at Michigan were the social events of the academic year in the 1870s and 1880s, their names being changed in 1895 to Annual Ball.[40] Students in either the junior or senior class organized the evening themselves, taking care of decorations, invitations, and refreshments and trying in each year to outdo the ball of the previous year. The dances were held off campus, first in a local hotel and later in the armory where a larger group of students could be accommodated. Once Waterman Gymnasium was built on campus in 1893, it became the new home of such dances; this made the university officials happy as they could oversee the activities.[41] Students had the opportunity to get to know each other better in a proper social setting, with the hope that suitable matches between the male and female students would be made. Much as the university officials or the women students and their families may have wanted to use such events to encourage appropriate relationships between male and female classmates, the men themselves did not always see the classroom as a place to find romance. Toward the end of the century the trend among the male students at Michigan was to take "out-of-town" girls as guests to their dances rather than their female classmates.[42]

Literary and Debating Societies

Though the mixing of male and female students outside the classroom was often looked on by administrators and society with trepidation, the rules and regulations of debating might provide a safe atmosphere in which to attempt interaction between the genders. This proposition would bring on a new problem with the formation of literary and debating societies for women, as they challenged many societal perceptions about gender roles. There was much worry that debating was not a suitable activity for women to undertake. As an activity confined to the public, male sphere of Victorian life, debating was considered by many to be an unsuitable activity for women. Debates over women's ability to serve as orators abounded in the nineteenth century, with many people arguing that it was unseemly for them to be on display in public. In some cases the concern was about the subjects being discussed; in some it was chauvinism from men who did not want to share the spotlight with, or possibly be upstaged by, a woman. More practical arguments were made as well. Some thought the tone of women's voices did not project well from a platform, which was further exacerbated by the size of the audience. In 1867 one writer attributed this to "popular curiosity [which] compelled them to speak in the largest buildings."[43] She also stated that many men, as well as many women, did not have voices that were suitable to public speaking.

With all of these concerns surrounding women's inclusion in literary and debating societies, university students were careful to explain that the groups on campus were intended as an enhancement to their education and were in no way intended as practice for women to become public speakers after graduation (even if many of them would eventually do so). One of the literary societies at Wisconsin—Castalia—argued that their group worked toward "the improvement and discipline of mental faculties."[44] Thus exercising their brains in a more social atmosphere, it was thought, would further their education in the classroom. This emphasis on mental exercise proved enough to allow for the formation of women's literary and debating societies, though many in the faculty and administration at each institution reserved their judgment for the time being. There was a great tradition of literary and debating societies at most universities prior to the acceptance of women. One historian of the University of Wisconsin remarked that they were "very deeply rooted in the institution, and profoundly influence the intellectual tone of the University."[45] These groups typically remained all male, with women beginning their own, separate societies. Unlike dramatic and musical groups there were concerns about holding coeducational debates because the competition with men might be too much for the women to handle emotionally or physically. Not surprisingly, these concerns were unwarranted, and the women's groups proved as successful as their male counterparts, with the women often being victorious. The first winner of the Junior Exhibition at Wisconsin was Emma J. Sarles of Castalia, while the following year a man—Fred J. Turner of Adelphia—was triumphant.[46]

South Carolina College had two long-running literary societies, both founded in 1806. The Clariosophic and Euphradian Societies had rooms provided for them on campus and were described as providing "a valuable auxiliary to the educational work of the College."[47] They debated each other annually, and focused on recitations and other activities at their regular meetings or held debates as part of the "Southern Inter-State Oratorical Association."[48] An interesting twist to their work came in the formal courses on public speaking that were offered in the Department of English Language and Literature. Credit was given to students in these classes for their "work done in the public contests of the Literary Societies."[49] When literary societies were in decline on other campuses, they maintained their importance at South Carolina in large part because the faculty and administration reinforced them in this manner.

Women were not permitted to join the two existing literary societies at South Carolina College, so they began a group of their own, the Parthenian Society.[50] Beulah Gertrude Calvo and Edith Eloise Bollinger each served twice as president of this organization, whose motto was "Strive to be good and beautiful, if you cannot be both, then be beautiful."[51] The women of the college clearly had a good sense of humor about life, which was evident in the "Minutes of the Last Meeting" which they published in *The Garnet and Black* in 1901:

> Miss Evans, in her usual practical way, moved that all boys be abolished from the South Carolina College, and further that a few men be induced to come. This was carried in a most satisfactory style. Miss Bateman was much moved and wept softly at intervals. Miss Nelson was completely overcome and promptly fainted, but was revived later on. Regular order of business resumed.[52]

The description of their meeting was followed by an illustration titled the "Co-ed Sewing Society" which showed three female students doing mending for their male classmates. A sign on the wall behind them promised to repair "Football and Baseball Suits...while you wait!" And for good measure there was a warning against flirting.[53] Both the Parthenian Society minutes and the sewing society were presumably jests, but there was undoubtedly at least a grain of truth in them.

Literary and debating societies were organized in all four nations of the United Kingdom, where students held regular debates on political topics of the day.[54] The Union Society at Durham "proved very beneficial in bringing all classes of men together, promoting good feeling among them, practising in debate and in management, and providing a good supply of papers and periodicals."[55] Oftentimes the women at these universities established their own separate groups so that they could focus on subjects they wanted to debate such as "that clever men prefer unintelligent women" and "that University Women as a class have an exaggerated sense of their own importance." Topics like these indicate a great amount of introspection among the women. They were clearly concerned about their new role as university students, as they contemplated their position in the wider society.[56] The question of holding coeducational debates between the students did arise occasionally. In his *Student Life at Edinburgh University* Norman Fraser recounts a conversation he had with a fellow student named Brown. It seems Mr. Brown had become a good public speaker but he "thought it wouldn't do to be beat by a woman."[57] In the end, coeducational debates would not become common until the twentieth century.[58]

The women at two universities, Edinburgh and Glasgow, even held annual joint debates.[59] When the Edinburgh students hosted, the Inter-University Debate was held in Masson Hall, and when Glasgow hosted, it was held in Queen Margaret College's lecture hall.[60] The historian Sheila Hamilton found in her research on the Edinburgh Debating Society that the groups attempted to maintain a balance between literary and debating pursuits, with topics considered varying "from the flippant to the serious."[61] Interestingly she notes that the groups did not debate the issue of female enfranchisement at any time before 1915, with general issues on women's behavior and place in society taking precedence (such as smoking, flirting, and marriage or family obligations). Topics also included those of wide social significance as seen in the subject of their first Joint Debate in 1898: "That war has benefitted Mankind."[62]

Many of the students themselves considered the debating societies to be at the forefront of their social lives, and the joint debate was often the high point of their academic year. Helen Nimmo, a Glasgow student, noted that the women from the two campuses "feel like old friends now" and learned a lot from each other during their competitions.[63] The minutes of the QMC Literary and Debating Society described the first Joint Debate in more a more serious, though no less feminine, tone than their counterparts in South Carolina:

> Twenty-eight members of the Edinburgh University Women's Debating Society arrived in Glasgow, by the 3.45 train, and were met at the station and conducted to Queen Margaret College by Miss Hay and the Secretary. Tea was served in the drawing room after which the debate took place in the well filled lecture hall.[64]

When the joint debate became more popular, ensuring a larger attendance, the women asked the Glasgow University Dialectic Society if they would sponsor the event in the University Union. This request renewed contact between the male and female groups in Glasgow, eventually leading to debates between the male and female groups on campus.[65]

Many U.S. universities had more than one literary or debating society; this caused competition within the student body that was not necessarily male versus female in nature. In fact, it usually was not. West Virginia University had two societies—the Parthenon and Columbian—that were said to be "of great advantage to the student" by teaching them "parliamentary forms, and the acquisition of business habits."[66] Literary societies at Indiana, at various points in the Victorian Era, included the Athenian Society, the Century, and the Philomathean Society for men, the Hesperian Society for women, and the Independent and Union Societies which were both coeducational.[67] As at other universities, literary societies at Indiana were seen as a good way for students to learn "to think and act for themselves" by taking on responsibilities in a forum that was both academic and social at the same time.[68] The Western University of Pennsylvania had three literary societies, the Franklin, the Irving, and the Philomathean. The third group had its own room, described in the *Catalogue* in 1895 as "handsome." The group held their meetings every Wednesday and practiced "declamation, composition and debate."[69] The other two literary societies were also active on campus during the Victorian Era, the Franklin, focusing their energies on elocution, and the Irving, on helping to publish the university's student journals.[70]

In the southern United States these groups were spurred on by a general love of oration that many politicians in that region were known for.[71] If the students wanted to find their place in southern society, then, they would need to become masterful public speakers. At Mississippi two literary societies, Hermæan and Phi Sigma, were formed by the male students before the Civil War.[72] At Tennessee the literary societies, Chi Delta and Philomathesian, "provided students with the greatest relief from classroom studies and from the many restrictions placed on their daily lives" by offering "relaxation, amusement, and a measure of excitement."[73] Like some of the academic societies at other universities, these two groups also maintained their own library which members could use to enhance their studies.[74] The original male societies did not want to admit women to their ranks, so the women formed their own group, the Barbara Blount Literary Society, named for the first woman to take classes at the institution, when it was still Blount College.[75] In the South, perhaps more than anywhere else, public speaking was to remain the domain of men.

Even at universities with the smallest enrollments literary and debating societies were popular during the Victorian Era. Queen's College, Galway and Aberystwyth both had active memberships. At Aberystwyth, "The aim of this Society is, by means of debates, to stimulate consideration of current questions and to afford practice in public speaking and discussion." Along with discussing literary or political subjects, the students also included "musical entertainments" in their meetings of the Debating and Literary Society, despite the existence of a separate Musical Society.[76]

Both women and men participated in the organization, with several women serving on the committee as well. Margaret Boyd took part in the Athenian Literary Society at Ohio University, the smallest U.S. campus in this study. They competed with their classmates who were members of the Philomatheans.[77] Boyd noted in her diary that on one night when "[t]he 'Philos' gained the contest" that she was unable to "give vent" to her "joy" for her friends, lest she betray her teammates.[78]

The popularity of literary and debating societies would largely become a Victorian phenomenon as newer extracurricular activities like intercollegiate athletics and Greek organizations started to gain in prominence in the twentieth century. Some colleges and universities noticed this trend in the 1890s and took steps to make sure that this valuable aspect of student life was not lost completely. Although it had previously been a student-driven activity, some administrators made moves to incorporate public speaking into the curriculum to make up for the loss of interest in the literary and debating societies. One institution that did this was the University of Missouri. Ostensibly their motivation was strictly the decline in the popularity of debating, yet once public speaking became something the students were graded on, it would have been less enjoyable as a pastime and that would have ended any likelihood that the literary societies could have rebounded in popularity.[79]

Athletics and Military Training

As discussed in previous chapters, many Victorians felt that a possible casualty of higher education was the physical health of women.[80] Secondary schools and universities alike made efforts to provide outdoor activities to head off the "melancholy effects produced by habits of indolence and inaction, especially in large towns."[81] At most universities some sort of provision was made for women's athletics as a means of countering the perceived negative effects of sedentary education. The increase in sporting facilities brought on by the concerns over the health of students led to the most visible segregation within higher education. As noted by the historian Barbara Miller Solomon, the "logic of separate spheres easily applied to athletics."[82] Men and women were certainly not able to share the same changing facilities, making a clear argument for either the duplication of space or the sharing of existing accommodations. Often at coeducational institutions money was not available to provide both male and female gymnasiums or swimming pools, making the evolution of women's access to athletics considerably slower than that of men.[83]

This support for physical activity as a necessary ingredient in education dominated the discourse between writers around the turn of the century. V. Sturge argued in "The Physical Education of Women" that much of the need for women to remain healthy was so that the human race might also remain strong.[84] Sturge supported education for women, as long as it was complemented by "systematic training" physically. This physical training needed to be organized for the women so that it would have the full and proper effects. Sturge argued that, without a formal structure, students would not seek sufficient exercise on their own. Another commentator on the subject was A. Lapthorn Smith who, in 1905, found that significant advancements in physical training for women had been made:

> One of the greatest objections to the higher education of women, namely, the interference with outdoor exercise, no longer can be raised, because the universities and boarding-schools have within the last ten years foreseen this danger and met it by special courses of instruction in athletics and the encouraging of girls to spend a good deal of time in outdoor sports.[85]

Despite Sturge's belief that students would not pursue exercise on their own, and Smith's decision to give credit to the institutions, rather than the students, much of the agitation for more organized athletic activity came from the men and women who would benefit from it.

As discussed in Chapter Three, the male students at all institutions that received Morrill Act funds were required to take part in military drill.[86] Many students at Alabama felt that the military drill they engaged in was not sufficient physical activity for them, and requests were made starting in 1874 for a gymnasium to be built on campus. Much of the appeal came from a desire to counteract the generally sedentary life of a student, and some can certainly be attributed to the desire for more social outlets that would be provided by sporting events.[87] Military training became so popular at the University of Missouri in the wake of the Civil War that the women asked if they could join the men. They were permitted to form their own company and wore modified uniforms.[88] The women's military training eventually evolved into a "girls' rifle team" that continued well into the twentieth century.[89]

The male students were not the only ones agitating for athletic facilities on campus. In 1882 the ladies' magazine at the University of Michigan, *The Amulet*, ran a story entitled "Our Share in Athletics" which outlined their desire for a gymnasium that they could use as well. Several sports were discussed that women might enjoy taking part in if they had the chance including archery, fencing, and lawn tennis. The further suggestion was made that the women unite in some sort of athletic association even before a gymnasium was built as a means of proving that they should have access to the new facility. As the article's unnamed author stated the "girls of Michigan University do not know of what they are capable." The need for exercise to promote physical health was clear to them, as they knew that this was one of the arguments against women's higher education. As students the only exercise they got at the time was "mounting half a dozen flights of stairs a day and going to a hop once a week."[90]

Ultimately the Waterman Gymnasium was built for the use of the male students in 1894, though it was reserved for women "during the forenoon hours of each day" until "a separate wing" could be built to accommodate their needs.[91] The Barbour Gymnasium, for the exclusive use of the women students, was built in 1897.[92] This structure was not merely an athletic facility but also had bathrooms, a dining room, parlors, and a lecture hall for the benefit of the women students. Some of the social functions, like dances, held there were for both the male and female students where they could be supervised in their time together.[93]

The campus gymnasium at the Pennsylvania State College was available for use by both men and women, though at different times of day. Conflicts arose over the times assigned for use in 1892, with women given the most sought after hour from four to five in the afternoon, just before "the supper hour."[94] Some of the male students were upset that this ideal time slot for exercising had been given to a small group of "about twenty ladies." This concern was a reversal of the sentiment expressed in *The Free Lance* the previous year when it was first announced that women would be allowed to use the gymnasium. At that point the male students supported the need for the women to have access to physical culture, something that the men could also receive through their required military training. Indeed, the men at that time argued that the women should have *more* hours set aside for them in the gymnasium.[95] The only drawback to the separate times was stated later by the male students: "We Regret... That the Co-Eds practice basket ball behind closed doors."[96]

At Wisconsin, especially after the Board of Visitors controversy in 1877 about the health of the female students, it was clear that physical activity was missing from the academic program. When a new gymnasium was built on campus in 1894, it was not accessible to the women students, who instead used a room in Ladies' Hall for exercise during specified times of the day. Makeshift facilities were also used for a time at Indiana University: the basement of Wiley Hall for "physical training" for women and an equivalent space in the basement of Owen Hall for men. It was thought that something more specifically designed for "educational gymnastics" would be beneficial, so in 1890 the Board of Trustees appropriated money to build a women's gymnasium on campus.[97] In 1896 a wooden building was completed that "served as an assembly hall and a gymnasium."[98] The university *Catalogue* explained "so that all danger from over-exertion is obviated" during their exercising, female students were supervised by Mrs. Harriet Colburn Saunderson, who would later become an instructor at Wisconsin. The classes she offered were optional, though "a large majority" of the women chose to take part in them.[99]

Physical training was usually offered for students once a gymnasium was built or other arrangements were made in existing facilities. A university gymnasium was provided at Manchester where instruction was given "according to the ages and the physical powers of the students."[100] In London the women of University College could take classes at the gymnasium at Alexandra House, which was a residence hall for women studying at the Royal College of Music in London.[101] Women at the Pennsylvania State College had their own instructor, whereas the men did not.[102] When physical training was made available for women, it made those universities offering it more appealing to potential students as increased competition between institutions meant that the athletic facilities could draw in students just as much, if not more, than academic programs.[103]

Showing that officials were concerned for the health of their students was a prime motivation when the gymnasium at Pittsburgh was completed as well. It was described as being "large and airy" with a changing room for the male students that was "fitted with modern conveniences, such as hot and cold water."[104] The students agreed that their gymnasium was "a very good one," but they still asked that changes be made to it. The simplest request was that a clock be installed, so that men who did not want to "carry" their watch (it being a pocket watch) while they exercised could keep track of the time.[105] A second, less serious request was that the gymnasium be opened "at all hours" so that students would not have to skip their lectures in order to have time to exercise. The tongue-in-cheek request showed a sincere concern over scheduling, but in this instance the students would have to solve their problem on their own.[106]

Some institutions benefited from their geographic location and had less need of building special facilities for athletic pursuits. The University of Durham's location on the River Wear provided an ample source of recreation for the students. This was seen as a key reason to attend the institution, as opposed to others that could have been selected by undergraduates. Indeed it was "the enjoyments of aquatic sport and river scenery which make Durham so attractive."[107] The students started a Swimming Club in 1879 for those who could swim or who wanted to learn. There were also "gardens and walks of some extent running down to the river" from Bishop Hatfield Hall.[108] Both the University College and Hatfield Hall had their own Boat Club as well, dating back to the 1840s.[109] The University of Edinburgh also had a Boat Club, as did the women at Aberystwyth, who are pictured at the start of this chapter with their male instructor.[110] The presence of water on or near a university campus was also beneficial in the winter months. Students at the Pennsylvania State College would ice skate on "the ice pond below Thompson's spring," and considerable effort was spent organizing sleighing trips to Bellefonte.[111] Similarly Lake Saint Mary, on the campus of the University of Missouri, was also used for recreation including ice skating in the winter and boating in the summer.[112]

The most popular individual sport internationally was golf. At St. Andrews students were able to play on the Royal & Ancient links, as they are today, and at Edinburgh students in their Golf Club played regularly at Musselburgh, another host of the Open Championship.[113] Golf links were built at the University of Missouri in 1900 as the sport gained in popularity.[114] The Golf Club at West Virginia University was coeducational and was the first and, for a time, only athletic organization that women would take part in there.[115] At Tennessee the director of physical training, Miss Anna M. Gilson, tried in 1901 "to build up an interest in the game of golf—perhaps of all forms of physical exercise, the most attractive and beneficial."[116] Other individual, though not solitary, sports that students found appealing included "boxing and fencing" and tennis, the latter being the only one pursued by women during the Victorian Era, though they also played tennis with the male students.[117]

Team sports for men in the United States included baseball, basketball, and (American) football.[118] In the United Kingdom cricket, football, and rugby were all played in university athletic unions or in individual groups based on each sport.[119] On each side of the Atlantic it was clear by the end of the century that football would be the most popular, though the American version of the sport and its place on university campuses would far surpass British interest in their football at an intercollegiate level. As early as the 1890s in the United States concerns were raised that football was taking over the campus and would, as a result, become a detriment to students' academic achievement.[120] The male sports were coeducational in the fact that they allowed men and women to interact as the women attended matches in support of their fellow students.[121]

Women's team sports were not viewed as being equally dangerous, though admittedly there were fewer for women to choose from. The women at South Carolina College and Tennessee both started a "Basket Ball Club" at the turn of the century, which was more like netball than modern basketball.[122] As Frances Melville pointed out in 1902, women usually played "hockey and golf instead of, say, football and cricket."[123] These noncontact sports were seen as more suitable for women, while at the same time instilling the values of teamwork and controlled competition. Of these women's sports, (field) hockey was by far the favorite in the United Kingdom with teams in most university cities.

Queen's College, Galway had a Ladies' Hockey Team, as did the women at Aberystwyth.[124] At King's College, London Lilian Mary Faithfull, the vice principal and secretary of the Ladies' Department, was avid about the sport of hockey for women and was the first president of the All England Women's Hockey Association that was formed in 1895. Beginning in 1896 this organization played an "annual international match" against "the Irish ladies" in either Dublin or London.[125] Since Miss Faithfull earned her M.A. in Dublin, the decision to compete there had much to do with her own personal connections in that city.[126] The immense popularity of hockey for women students extended into the twentieth century, and competitions between women at different universities became more common as transportation made travel easier and less expensive over greater distances. The only drawback to this "flourishing" was that "the men are getting greener and greener with jealousy at being outshone" in athletics by their female classmates.[127]

Student Government

Despite the successful mixing of male and female students in many areas of campus life, some areas still proved difficult for women to enter. Student government was one of the most awkward areas for integration on campus. This was largely due to the fact that the male students had organizations firmly in place prior to the admission of women. In addition, the limited amount of power given to students by the administration was not something to be shared freely between the sexes. These practical concerns of the male students were reinforced by Victorian arguments that giving women the right to self-governance would lead them into taking on the man's role in society.[128]

The primary governance of student affairs in Scottish universities was Students' Representative Councils. Such groups were regulated by Parliament in the 1889 Universities (Scotland) Act, and its subsequent ordinances. In June 1895 the University Commission issued Ordinance No. 60—General No. 22 on the topic of Students' Representative Councils. This ordinance does not specify that women should be admitted to the Students' Representative Council (SRC), only that "The Students' Representative Council in each University shall submit to the University Court for approval the regulations under which it has been formed or now exists."[129] At Glasgow the SRC, predated this ordinance having been established by the students in November 1886. In 1897, the SRC agreed to meet with representatives from Queen Margaret College (QMC) to consider the issue of women having representation on the university council. This move was somewhat late in coming, as the women had previously begun their own Students' Representative Council to parallel that of the men in 1893.[130]

The first president of the QMC SRC (which it was sometimes referred to as) was Marion Gilchrist, who was also one of the university's first women medical graduates. This group concerned itself with academic issues of the university. In January 1894 the QMC SRC sent the University Court a letter about the lack of training for women in botany.[131] This indicates the women students were trying to take an active role in their own education, and they were working toward it through their only official voice on the campus. The small amount of power given to the women through their SRC was not easily parted with, despite the university Students' Representative Council's offer of female representation on their board. The easiest way to skirt this problem was to continue the separate women's division, with the establishment of a "QM Section" of the SRC once the two representative councils merged.[132] In this way women were able to pursue the agenda issues they wanted to, and the male students retained control of the real power of student government.

No women were elected to top positions in Glasgow's SRC, though they maintained a high profile on various committees. Similarly at St. Andrews the Students' Representative Council was coeducational, with women winning some of the subsidiary elected positions not long after their admission to the university. In 1898 Miss Henderson-Roe was listed as one of the representatives of the United College, and Miss Hill was the representative of one of the campus societies.[133] The male students at Aberdeen had a relatively positive response to the introduction of women to their ranks, even if they expressed it in a less than serious manner. Mr. A. G. Anderson, speaking as a representative of the Students' Representative Council at the Quatercentenary Celebrations in 1906, noted that the "social life had been helped by the introduction of women students to the University," a statement which garnered both applause and laughter from those in attendance. He went on to say that the women "had done much to brighten their existence and helped it along in every possible way."[134] Undoubtedly the introduction of women to the university had changed the type and nature of social activities in campus, producing a greater variety of organizations and more pleasant distractions from study.

Women students at Aberystwyth were given all of the same rights as their male counterparts, including the right to take part in student government and university administration. As one commentator stated "it is worthy of note that in the new University of Wales every degree and office is open to women equally with men,—even that of chancellorship, there being no distinction whatever between the sexes."[135] Their Students' Representative Council was not established until 1900, but it was coeducational from the start. The membership was, in fact, almost evenly split between men and women.[136] Eventually there would be a Central Students' Representative Council for the University of Wales, with representatives from the three constituent colleges in Aberystwyth, Bangor, and Cardiff, the federated state of the administration thus being reflected in the organization of the students themselves.[137]

In the United States the most common form of government organization for female students was the Woman's League. There were groups with that name at Indiana, Michigan, Tennessee, and West Virginia, most started in the 1890s.[138] There were multiple reasons for forming these groups including bringing women together socially, as well as helping to allay fears about their supervision on campus. Florence V. Skeffington, the dean of the Woman's Department at the University of Tennessee, described the formation of their Woman's League in her Report in 1901: "The problem of discipline was solved by adopting the system of self-government, so generally favored in other colleges. To this effect the young women were organized into a woman's league to monitor their behavior within the university."[139] Because there was a general belief that the women would not wish to misbehave, the desire for self-government was often the result of a desire to show the outside world, including parents, that there was guidance for the women students in the area of self-discipline. Skeffington's mention of "other colleges" pointed toward another reason for establishing a version of the Woman's League on campus. If enough colleges or universities had such a group on campus, it was entirely possible that a prospective student's parents would prefer to send their daughter to an institution where she would be actively looked after.

On individual campuses the organizations focused on different subjects that were of particular interest to their student body and community. At Michigan the members were students "from all departments of the University" along with "wives of members of the faculties."[140] One of the key areas they decided to take responsibility for was helping new students find appropriate housing in Ann Arbor. At the start of each academic year they had members stationed in the Ladies' Reading Room in University Hall where they could answer questions for new students "in regard to rooms, board, and general University work."[141] As the number of women on campus increased they progressed to writing to incoming first-year students in the summer to arrange their rooms even before they arrived.[142] Just as officials wanted to put the minds of students and their parents at ease in Tennessee, the Woman's League at Michigan was doing the same thing in Ann Arbor.

At West Virginia the membership was also made up of female students, and associate membership in the League was extended to local Morgantown women. In this way there was increased contact between the students and women who could serve as role models to them. The group met every month, on the third Saturday, and arranged for speakers to come talk to them about topics relating to women's history and current events. For instance, the university president's wife spoke to them about "Jane Addams and Hull House" in 1898. Some of their events mixed study with entertainment as well. In February 1899 they held a "Colonial Tea" complete with costumes replicated from, or actually from, the 1770s.[143] In all cases the events or talks highlighted women's contributions to society in their traditional female sphere because that is what the students were being prepared for in their own lives.

The need to regulate women's presence in student government at the University of Wisconsin led to the emergence of a sex-segregated structure of student government, with separate male and female halves. The men's organization was the Wisconsin Student Organization, and the parallel women's group was called the Women's Self-Government Association (or WSGA). Like the Woman's League at other campuses, the WSGA aimed to

> regulate all matters pertaining exclusively to the undergraduate women of the University except those which fall under the jurisdiction of the faculty; to further in every way the spirit of unity among the women of the university; to increase their sense of responsibility towards each other; and to be a medium through which the social standards of the university can be made and kept high.[144]

Moreover, in one history of the women's organization, at the time of their fiftieth anniversary in 1948, it was stated that the WSGA was founded to "improve social relations between the men and the women on the campus." To this end they were specifically concerned with the regulation of visitors to Ladies' Hall and furthering the course of male-female social interaction on campus. At times this led to replication of duty, as the men's group was charged with representing all students on campus, not just the men. Their focus was also chiefly related to housing regulations, though they looked beyond Ladies' Hall to other housing in Madison.

Politics and Rectorial Elections

Despite their lack of political voice outside the university, there was a great deal of political advocacy present among the women students over time.[146] It is important to note that while participation in campus government and political groups can clearly be seen as providing practice for women's future use of the vote, the wider debates of suffrage did not permeate higher education as one might expect. Lynn D. Gordon concludes that the lives of women students in the United States between 1890 and 1920 "reflected off-campus political and social reform movements and closely linked collegiate activities to preparation for leadership in those areas."[147] At newly coeducational state universities, however, student government focused on intrinsically student issues, and larger societal issues only reached a campus forum when they related directly to students' lives. Though the increasing activism of women students in controlling their lives at university would be of practical benefit in their later lives, at the time they did not intentionally set out to gain this experience. Instead their energies were focused on their academic responsibilities and those issues which directly impacted on their lives while at university.

A key example of this can be found in British universities in the Rectorial elections held among the students. The Lord Rector was elected by all the matriculated students and acted as the official president of the University Court for a three-year term. Rectorial elections were taken extremely seriously by students, as it was their main opportunity to have a say in the workings of the university. In 1872, when the admission of women was being debated at the University of Edinburgh, Sir William Stirling-Maxwell was elected Lord Rector. The day he arrived in the city to accept his position, students met his train and took the opportunity to appeal to him to "not, as President of the University Court, favour the pretensions of the literary ladies."[148] Because the Lord Rector is a position elected by the students, they had every right to appeal to him as constituents.[149] Reports from the day indicate that Sir William was surprised that he had entered into such a controversy, and though he tried to calm the situation, his statements did not alleviate their concerns.

Students at the University of Aberdeen are arranged into nations for the purposes of organization and voting for the Lord Rector. These nations are determined by a student's hometown and are Buchan, Mar, Moray, and Angus (students who do not live in the areas covered by the first three nations are a part of Angus, including any international students).[150] The historian Lindy Moore reports that female students were active in the Rectorial elections in Aberdeen, just as they were elsewhere in Scotland. As early as 1896 a woman, Rachel Annand, served as the president of one of the candidate's supporter groups on campus. Other women worked actively on the campaigns in subsequent years; this shows that the women had a real interest both in the outcome of the elections and the process of campaigning.[151]

During the campaigns for Lord Rector in Scotland, students did considerable work on behalf of their chosen candidates, and the women at the University of Glasgow stepped into this part of their university life with remarkable ease. In November 1893 the QMC SRC held separate meetings to consider their choice for the Lord Rector of the university. The first such meeting was for women who supported the candidacy of Mr. Asquith, the Home Secretary. At the meeting, which was open to the male students also, "Miss Gilchrist made an efficient chairwoman. She recommended Mr. Asquith as a friend of teetotalers and labour." The meeting was attended by "about as many male as female electors."[152] The university SRC held their own meetings for support of various candidates, propelling the male supporters of a particular candidate to attend two meetings. The duplication of standard proceedings could also be found in the existence of two polling stations, one on Gilmorehill and one at Queen Margaret College. Faculty members were needed at both locations to regulate the voting; this resulted in further repetition of work. The results of the 1893 election were of more significance than some other years, since it was the first chance female students had to participate in the process. As reported in the *Glasgow Herald*, the 154 women students who voted in the election played an important part in deciding the outcome as the "ladies thus gave Sir John Gorst a clear majority."[153] There is no evidence that the male students resented the ability of the female students to determine the outcome of the election, but it was clear that future campaigning would need to take the female vote seriously.

In Glasgow, there were also student Liberal, Labour, and Unionist Clubs on campus that were led by men, though records indicate that women were allowed to attend meetings if they chose to do so. Ties to the national political parties were not emphasized by the students, and there is little evidence that they organized vote getting at the time of national elections. Significantly, the organization of these clubs centered on the Rectorial elections, with each selecting a candidate and holding dinners and debates for the furtherance of their candidate's campaign. This sporadic mobilization of the groups had a tendency to limit the participation to those most involved in the political world—men—with the women often left to hold separate luncheons for the candidates in question. As time progressed, women's functions for Rectorial candidates were often held on the main campus, though they remained of a more social, less political nature. The difference between male and female strategies for supporting their candidates was further reinforced by the tactics of the male students, which were often physically aggressive, something that the "ladies" would not participate in.[154]

Morality and Religion on Campus

The apprehensions about the secular nature of education at state-supported universities were common on both sides of the Atlantic. They were the most remarked on at the Queen's Colleges in Ireland because of the adamant opposition toward their founding by the Catholic hierarchy and the longstanding sectarian political divide within the nation. The accusation that these institutions were "Godless" was not unique though, as the term was specifically used to describe the education provided at the Universities of Alabama and Manchester as well, with more general concerns being raised from time to time at all institutions.[155] To head off such complaints, some universities required attendance at chapel in their early histories, though this led to a different sort of protest. At West Virginia students argued that this requirement was a violation of the "separation of church and state" established in the Bill of Rights. The students were successful in their complaint, and the Board of Regents lifted the order to attend chapel and replaced it with a daily "roll call" of students.[156] While the student enrollment remained small, it was possible for faculty members to guard the morality of their charges with direct supervision. At Ohio Margaret Boyd had an active social life, revolving mostly around prayer meetings and having dinner at the homes of faculty members in Athens.[157]

The desire to have an underpinning of Christianity in colleges and universities, and at the same time trying not to offend the tax payers who were funding the institution, resulted in the formation of Students' Christian Associations in Britain (often with separate Men's and Women's Branches), and the Young Men's and Young Women's Christian Associations in the United States. These groups were formed for the purpose of Bible study and to provide "an opportunity of engaging in Christian work."[158] By the 1890s there were chapters of the Y.M.C.A. on every U.S. campus in this study, making it "one of the country's foremost collegiate organizations."[159] The remit of these groups was one of combined service and social activities that saw to students' free time on campus and directed their interactions with the wider communities in which they lived. At Tennessee funds were raised for a Y.M.C.A. Building that "contained a bowling alley, ball cage, gymnasium, race track, lockers, and baths," as well as a reading room and greenhouse—everything that was needed to keep students occupied in their spare time, and it was not all imbued with religion as one might expect.[160] As historians of that university stated, "It would be injudicious to suppose that all students lived by a strict religious code, but a religious attitude was at least good form."[161]

At Michigan the Y.M.C.A. became simply the Students' Christian Association after women's admission, rather than founding a Y.W.C.A. as most other U.S. universities did.[162] Their decision followed the British model in this way, the men and women only meeting together once a month and handling much of their business separately. There was a woman vice president who took "entire charge of the work among the women members," all of which was done in conjunction with the men of the association.[163] A special section of the *Calendar of the University of Michigan for 1894–95* entitled "Aids to Moral and Religious Culture" was devoted to the opportunities students had to engage in Christian activities while in Ann Arbor.[164] The information provided included a description of the Students' Christian Association and their activities, and a list of the denominations that had local churches. There were also a number of guilds that welcomed students to their membership "for religious and moral culture and for social entertainment."[165] All of the information included was publicly available to students once they arrived on campus, but presumably it was added to the *Catalogue* as a means of reassuring parents who were considering sending their sons or daughters to the university.

Individual campuses also had their own, unique concerns about the morality of their students. It was common during the Victorian Era for students, both male and female, to be required to provide character references before they were admitted to study in the first place.[166] Maintaining a clean reputation was important for the students, their families, and the institutions themselves because any disrepute could be disastrous for all involved. Regulations were often put in place to protect "student morality," targeting specific activities that were favored by students in certain locations. For example, in Alabama "games of chance," and drinking or possessing alcoholic beverages were prohibited, and students had to sign a pledge that they would not carry weapons on campus (this was intended to prevent dueling as a means of settling disputes, a common concern at southern campuses).[167] Student disruptions at the Pennsylvania State College in the 1870s included such actions as "[h]issing and foot stomping during chapel, defacing walls, raiding orchards and gardens, putting the plumbing system out of commission, throwing water and even furniture from the upper floors of the main building."[168] And at West Virginia there were concerns about poker playing and other card games with "card sharks" taking advantage of their classmates.[169]

One of these issues, alcohol use, was of profound interest to the students in both the United States and the United Kingdom, due to the larger Temperance Movement that was gaining momentum in both countries. On some campuses national figures came to speak to students to stir up their belief in the dangers of drink. In 1898 Mrs. Clara C. Hoffman spoke to the students of the Pennsylvania State College as a representative of the Woman's Christian Temperance Union of which she was the national secretary. Her talk was titled "The Problem that Faces Us," and it was apparently delivered excellently.[170] In Britain Aberystwyth and St. Andrews had Total Abstinence Societies and there was an Edinburgh University Temperance Society, all of which were open to male and female students.[171] The students at Indiana University had perhaps the most interesting means of encouraging temperance work by awarding students for the two best essays written "on some subject connected with the temperance question." These "Temperance Prizes" were awarded in 1884 to William C. Mason (first prize) and Elmer E. Griffith (second prize).[172] In this way students were able to regulate their own behavior, and that of their classmates, both by setting a good example and by frowning on misbehavior. The skills developed in this area reflected the same evolution of thought in the wider community and would aid students in their transition into their lives after graduation.

Missionary and Settlement Activities

A final set of extracurricular groups to consider are the humanitarian associations that university students often took part in. Realizing that their education was largely designed to help them move into their communities as leaders and educators of future generations, the women on both sides of the Atlantic took this aspect of university life quite seriously. The British Settlement Movement, influenced greatly by similar efforts in the United States, was an attempt to increase interaction among societal classes which urbanization affected detrimentally. Martha Vicinus described the Settlement Movement as work that "emphasized the womanly virtues of public life." Lynn D. Gordon reaffirms this, noting the spirit of reform that allowed women to extend their private sphere to include domestic issues outside the family home.[173] One of the clearest examples of the benefits of coeducation for the students and their communities was the formation of settlements by men and women together, though women often established them on their own. For students with a more international perspective, missionary societies were established on campus that were intended to help prepare members for jobs in that field after graduation.

In Scotland the students were inspired by their moral philosophy lectures (which were the secular equivalent of theology courses) where they learned about the concepts of "social modelling" as a "continuation of organic evolution" which people could actively participate in. Education on the whole was thought to "evoke and educate the will," and any aid given to the less fortunate helped in the development of "moral character."[174] These teachings were further reinforced by the Rectorial Address made by Sir John Gorst in 1894 on " 'Settlements' in England and America."[175] Efforts to inspire students to apply their knowledge to their real lives in an effort to better their communities also came from their families. The historian Catherine Mary Kendall supports this in her thesis "The Queen Margaret Settlement 1897–1914: Glasgow Women Pioneers in Social Work." She found that the families of female students were active in a wide range of charities, and this family commitment to aiding society translated to a similar level of commitment in the Settlement Movement which was "a highly active and dedicated sector" of Glasgow society.[176] Traditionally women's domestic role entailed teaching children and caring for the sick, and their admission to universities provided them with opportunities to hone these skills.[177]

The majority of Victorian settlement work took place in urban areas where social distress was most apparent. In Glasgow activities of the Queen Margaret Settlement included giving milk to mothers and their children, opening an Invalid School, and setting up a Settlement House. In London most work was concentrated in the East End, although it was not necessarily done by men and women who were studying in London itself. Students from the Cambridge and Oxford colleges were actively involved in collecting donations, setting up settlement houses, and even establishing "children's country holiday funds" to enable young people living in poor homes to spend some time outside the city to get fresh air and exercise.[178] The University Settlement in Manchester, a coeducational group, organized both a Men's House and a Women's House for the purpose of "educational and recreative work" in metropolitan Manchester.[179] All of these actions were intended to ameliorate the conditions of the inhabitants of the city, but they had other benefits as well. The work of the students and graduates who participated in these settlements transmitted the name of their universities around the country, and in some cases around the world, while at the same time reinforcing the idea of the benefits of higher education to the students themselves because their academic accomplishments afforded them a level of respect and leadership in their communities that they could not attain any other way.

Another humanitarian-minded group was the Student Voluntary Missionary Union, which was active in both the United States and the United Kingdom. Unlike the settlement associations who were actively working to change domestic society while they were students, the missionary groups were about studying and preparing to change societies overseas. They upheld the membership declaration: "It is my purpose, if God permit, to become a Foreign Missionary."[180] Those former students who left "for the Foreign Field" were given honorary life memberships and were held in the highest esteem by upcoming students. The organization was coeducational, though on campuses like Edinburgh women were not elected to offices in the group.[181] At the University of Wisconsin the Y.M.C.A. and Y.W.C.A were affiliated with the Student Volunteer Mission Movement, and to this end they held weekly meetings "to consider questions related to their work" with "a number of members... preparing to engage in active missionary work in foreign fields."[182] Similarly in Glasgow the Queen Margaret chapter of the British College Christian Union held as a central career goal becoming a foreign missionary. This branch of the Christian Union sponsored talks by various people working as missionaries to encourage further interest in the profession, particularly as it was considered an acceptable field for women to pursue.

Conclusions

University educated women were in a unique position in society, able to transmit their knowledge to others in their community. The question of admitting women to universities clearly brought many implications with it. Some administrators felt that the new female presence would be distracting to the male students, while others argued that they might bring higher moral standards to the campus environment. As seen in the previous chapters on academics and campus facilities, administrators saw three main areas in which they could control students' lives on campus: academics, housing, and extracurricular activities. With time in the classroom limited and directed primarily at the subject of study to be considered, the extracurricular areas offered the best possible, yet most indirect, access to controlling students' behavior. As the historian Barbara Miller Solomon points out, the increased "involvement in myriad activities kept students constantly aware of their responsibilities as liberally educated women, giving them additional purpose as a result of their position in society."[183] In addition, this "flowering of extracurricular associations" provided a richer education for the students in university.

The gradual shift that began in the late nineteenth century of the university's role from traditional, intellectual bastions of society to that of a place for the cultural development of the nation's youth was profound. It was now thought requisite for universities to concern themselves with every aspect of a student's life, not just academic instruction. In his Baccalaureate Sermon on the occasion of Alabama's Centennial in 1931, Dr. George W. Truett summed up this transition in the purpose of a university education by saying, "The highest and best contribution that the students of these graduating classes can offer to the world is the gift of well-rounded and worthy lives."[184] The extracurricular activities of the men and women discussed in this chapter show that their education was "well-rounded" and that they intended to go on to make worthwhile contributions to society. This "production" of valuable young men and women by the universities was not a new argument. What was revolutionary was the belief that providing a strong academic curriculum would no longer sufficiently prepare students for life after they left academia.[185] Instead, a new generation of university students were arguing that higher education should be both social and academic. What can also be seen is an increase in agency among the student body itself. As the historian Helen Lefkowitz Horowitz notes: "Coming into a world created for them by others, students made the college their own."[186] Increasingly, the receivers of higher education were speaking out about what they wanted their university life to be.

CHAPTER SEVEN

Student Publications

No. 5—Vol. III. GLASGOW, JANUARY 14, 1891. Price One Penny.

Studies of Students in Black and White.

No. III.—TYPES FROM ST. MARGARET'S.

Image 7 Illustrations from the *Glasgow University Magazine* shortly before the admission of women to the institution in 1892.

Credit: With Permission of Glasgow University Archive Services.

As student organizations continued to flourish in the 1880s and 1890s, they needed ways to promote their activities and recruit new members. Equally important was a desire for students to comment on their lives and their perceptions of the university experience. This need for discourse among the student body led to the emergence of several types of student publications. Some recorded the events of student life and their activities, like athletic competitions or social gatherings. Others were vehicles for students to publish their writings, both fiction and nonfiction, to be shared with their classmates. Still others were intended as a means for talking with administrators and alumni before formal student government and alumni associations were formed. Because there were so many different types of student publications during the Victorian Era, some colleges or universities could have one or more publications. Each of these publications was intended for all students to read, though not all were written or edited by coeducational staffs.

Along with being vehicles for student communication that became campus traditions, most publications were intended as a way for students to pass essential information from one year to the next, with as little interference from the faculty as possible. Students at the University of Wisconsin commented on the emergence of new publications in their first yearbook, *Trochos*:

> Believing that all introductions are best made when a third party is present, the editors ask a moment's indulgence while they introduce to its readers this Annual.
>
> ...It introduces a new era in University life, and, we hope that each of the years to come may see the production of a creditable and flourishing Annual.[1]

This new era at universities, in both the United States and the United Kingdom, will be the focus of this chapter. The various types of "official" student publications and the type of information they sought to provide to their fellow students will be analyzed. While there were many informal or "underground" newspapers and magazines, these publications were not sanctioned by the universities and often did not reach university archives; this makes a full survey of them impossible. As a result of this limitation and as many of these underground publications were never intended for a wide audience, this study is being limited to the sources recognized by the institutions. The types of publications will be examined first, followed by a discussion of female contributors and student perceptions of coeducation.

University Calendars and Student Handbooks

One of the most essential items of reading for students in British universities was their university calendar, precursor of the modern prospectus. Although these were not written by the undergraduates, it is safe to say that they were written by faculty and graduates of the institutions who knew what information students would need as they embarked on their university careers. In Durham students were told "A Calendar should be procured on matriculating, if not before."[2] Not only did calendars include explanations of the university hierarchy, descriptions of student organizations, explanations of fees, and lists of courses and faculty, they included past examination papers that could aid students in their work.[3] As a student at Edinburgh, Norman Fraser, recalled, these "gave me a good idea of the nature of the questions which would be asked" and helped students focus their studies and learn which subjects their professors favored.[4] Looking from year to year, the examination subjects did not alter greatly, even if the wording of questions changed, so if students were able to acquire, keep, and pass on previous years' calendars, all students could benefit from the collected knowledge of their predecessors. Norman Fraser noted another way students could access calendars, saying it was common for students to do as he and take "as much advantage of the Reading Room as possible, looking over as many *Calendars* as I could."[5] In this way university officials encouraged students to value the information held in the calendars as well.

University calendars were also a valuable resource to society who wanted to know what happened at institutions in the country. Copies were available for purchase from the printers and book dealers; this made it possible for interested parties to compare universities and their programs of study. Victorian commentators remarked "such works as these constitute the permanent literature of the subject of university education . . . written from a rare and special standpoint" within the institutions themselves.[6] The importance given to various topics in a given university calendar showed students and the public what was valued by the administration and alumni and, to a greater extent, what the community at large put store in. This was especially important if one were moving from a culture they were familiar with to one they were less familiar with, either because no one in their family had previously attended university or because they chose to study in a foreign nation, something that women in particular might need to do to study the subjects they wanted to.

Taking a cue from the reliance students had on university calendars and wishing to introduce students to the informal side of higher education, students on many campuses began writing their own handbooks that were usually provided free of charge to incoming students. These publications were done under the auspices of the university and contained occasional contributions from the faculty and administration. Designed as a resource for students as they entered university, topics ranged from brief histories of the institution and information on using the library to instructions about how to dress for various social occasions. At universities that had student handbooks they quickly became one of the primary ways students were informed about campus life and social expectations. Another effect of these publications was that students were starting to establish their own standards of conduct, passed from class to class for the maintenance of order within the student body. These "regulations" were often found in student handbooks.

At the University of Glasgow, the first *Student Handbook* was published by the Students' Representative Council in 1893. A primary question raised by Principal Caird in the "Prefatory Note" were the changes brought to the campus by the recent legislation on women and universities. The handbook included "much valuable information by competent writers on the new regulations laid down by the Ordinances of the Universities Commissioners." Furthermore, because the "application of these regulations to particular cases may often prove a somewhat difficult question, and individual students may be in some perplexity," clear explanations would be given about the implementation of the new regulations at Glasgow. Since the handbook was published by the Students' Representative Council, a good deal of space was also set aside for the promotion of the council and their role on the university campus. Student organizations each received one to two pages they could utilize in any manner they saw fit, to explain their group and forthcoming activities. In 1893, this was generally done with a brief statement about the aims of the group, a note of the time and place of their regular meetings, a list of office bearers, and a syllabus of any lectures or debates they might have scheduled already for the upcoming year.[7]

At the University of Wisconsin, there was one handbook available to students. It was first "presented by the University Young Men's and Young Women's Christian Associations" in 1894. Like the *SRC Handbook* at Glasgow, this handbook was offered free of charge to students at the start of the academic year. It outlined the various student groups on campus and provided a calendar for the upcoming year, with events of particular interest (like university holidays) given special importance. And, as the SRC included a certain amount of self-promotion for their organization, so too did the Y.M.C.A. and Y.W.C.A. The writers of the handbook encouraged new students to ask the returning ones for advice "for they are always glad to assist."[8] Since the focus in the handbook was purely extracurricular, the students at Wisconsin left instruction on the academic aspects of campus life to university officials. The students did offer some general "pointers" on the academic side of things though. They provided information on the library opening times (both on campus and in the city), and they encouraged students to organize and begin their studies "as soon as possible."[9] Of course, equal weight was given to reporting the "Calling hours at Ladies' Hall" which might have been more useful to students on a daily basis, whether the faculty and administration wanted to admit that or not.

Student Newspapers and Newsletters

Many student publications began as outgrowths of the activities of the literary and debating societies on campus. While the students enjoyed expressing their ideas on important subjects verbally, they also wished to develop their persuasive writing skills. The format of the publications often had to do with the means available to students, and also to the audience they hoped to reach. It would be incorrect to assume that something referred to as a newspaper was printed daily or that a journal or magazine necessarily had longer features. All of the terms were very fluid at universities in the nineteenth century, just as there was variation in their meanings in wider society. Newspapers and newsletters focused on significant news from the university, along with notes on the social happenings among the student body or reports on meetings of social organizations. After coeducation men and women were both included in these updates to some extent, though men's activities—particularly sports toward the end of the century—were the most abundant. As noted in the records of the University of Wisconsin Archives: "The literary element was predominant; essays and poetry, texts of speeches and gossip were the main fare."[10]

One of the responsibilities taken on by the literary societies at Indiana University was the publication of the *Indiana Student*, a monthly paper. In the early 1870s the female Hesperian Society was "allowed to take part in the paper's operation," joining with the already established Athenian and Philomathean Societies.[11] At the Pennsylvania State College the Cresson and Washington literary societies worked together to publish *The Free Lance*, which was also published monthly from 1887 until 1904.[12] There was a potential for competition between the members of the societies for space in the papers, but no single group had enough membership or adequate funds to publish their own newspaper. Additionally it was better for sales if competition happened for space in the paper rather than among papers. Once women were admitted, there were more students vying for a voice in the publications, plus there were new subjects that needed to be covered. A further incentive would be added on many campuses around the turn of the century when salaries were established for editors, making the positions even more highly sought after. The money for the salaries came from sales of advertisements that were also increasingly common by 1900. In every way the students were trying to create publications that mirrored "real world" ones so that they could use their experience at university as practice for future careers.

In 1892 a student-produced, evening newspaper, *The Daily Cardinal*, was established at the University of Wisconsin. The editors (of which there were usually ten) were elected by the Cardinal Association, the student group responsible for publishing the paper. Reporters could be chosen from the entire student body, though competition was fierce as many students looked for work experience. In its first year *The Daily Cardinal*, heralded as the "first west college daily," was four-pages long, with four columns per page.[13] The cost of the paper was initially 3 cents, and it rose to 5 cents in 1894. Because it was printed and distributed daily, news issues were more short term in nature, those which were happening that evening or had happened the day before. The *Cardinal* was also limited to term-time publishing, with news stories that took place during the summer months unable to be covered promptly. In 1894 it became the "official paper of the university" and was given an office in University (later Bascom) Hall.[14] As at other campuses, the newspaper's target audience was not solely the student body. Faculty, staff, and members of the general public could also purchase copies.[15] In 1900, for example, the student body at Wisconsin numbered 2,422, while the circulation of the *Cardinal* was reported at approximately 3,000. This enlarged readership put more restraints on the newspaper editors, as they were conscious of not offending readers and of portraying themselves and their institution in a positive light. Much as the students wished to debate subjects in an entertaining way, the continued influence of the Board of Regents, who contributed $250 to the *Cardinal* annually, ensured the paper's "relatively moderate" stance on most issues.[16]

University students did not only express their opinions on campus life in the pages of their publications. They often took on major social and political issues of the day. For instance, in 1895 *The Free Lance* ran an article on "The Dangers of Socialism" which provided a very clear and succinct argument against government control of business and property and in support of freedom, an "essentially American characteristic which socialism seeks to destroy."[17] The points made in the article have a timeless quality to them, even though it was written long before socialism had gotten a foothold in any nation. That this topic should be of particular interest to students in "the nation's most industrialized state" is not surprising.[18] The added fact that Andrew Carnegie had just become one of their trustees would have also added to their wish to debate the capitalism versus socialism question, and the conclusion reached was simply that free market capitalism was better aligned with a free society than a government-run economy.

At Manchester the students in the Women's Department published a newsletter called *Iris* which was published semiannually from 1887 until 1894.[19] Like its U.S. counterparts the publication was produced by the Social Debating Society and was intended in the first instance to inform group members of past or upcoming activities. The wider appeal of the publication to other members of the university and the community meant that subjects would not be limited for long. The historian Mabel Tylecote noted that the publication showed "the corporate, pioneer spirit of the early academic women" at Manchester and recorded many of their accomplishments, while at the same time comparing those achievements with women at other colleges and universities in Britain.[20] The newsletter was dissolved after the male students invited the women to join them in publishing the *Owens College Union Magazine*. There were both benefits and drawbacks to this decision for the women. Although the magazine was published more often (monthly), it would not be devoted to women's issues on campus as the *Iris* had been.[21] The preference for inclusion in the life of the campus, along with the financial limitations on producing the newsletter, made the decision to discontinue the *Iris* relatively easy in the end.

University Journals and Magazines

It was more common in the United Kingdom for students to produce a monthly journal or magazine like the *Owens College Union Magazine*, than a newspaper or newsletter. Rather than being published by the literary and debating societies, they were often published by the Students' Representative Councils, as *College Echoes* was at St. Andrews; this changed their focus somewhat.[22] These publications also differed in some cases because they featured "contributions by past and present University members."[23] *The Durham University Journal* functioned in this manner, as did *The Students' Journal and Hospital Gazette* which was published in London, but included reports on all colleges and universities in Britain and Ireland. The latter publication promoted itself as being "the one organ of student opinion; its columns are open to any and all who desire to correct error, amend abuse, or call attention to administrative defects; while its chief object is to aid and assist the student in his legitimate endeavours to assist himself."[24] Despite the reference to *his* "endeavours," women were able to submit material to the publication if they wished to, and the emphasis on student empowerment and self-reliance would have made the publication appealing to those students studying in the various colleges of the University of London and at other institutions, like the Queen's Colleges in Belfast, Cork, and Galway, where there was not yet an in-house student publication.

The publication pictured at the start of this chapter, the *Glasgow University Magazine* (or *G.U.M.*), was first published by the Students' Representative Council in February 1889 at the cost of 1p. The first issue reported that there was a special meeting of the SRC "for the purpose of organising a University Magazine" as previous attempts to establish such a publication had failed. The editors of the new magazine argued that it would be successful because it was being approached differently than earlier versions: "In the first place it is the outcome of a general desire on the part of the Students themselves; it is a spontaneous movement, and not the ambition of any great mind or aspiring literary clique." To prevent any one "clique" from gaining control of the magazine, two editors were elected from each Faculty, so that all areas of concern would be addressed. The overall purpose of the magazine was to be "a medium of communication for Students," and it proceeded in the publication of articles, editorials and debates on a variety of topics.[25]

One additional element in the *G.U.M.* was the inclusion of illustrations like the one by "Madge Wildfire" at the start of this chapter.[26] Few students had or were willing to share such artistic ability, but when images were included, they added greatly to the commentary in the publication. The "Studies of Students in Black and White" were "freehand drawings, some caricatures, some exaggerations, some compositions, none photographs of varied aspects of student life."[27] Other "studies" in the series included "The Cap and Gown Question," which illustrated perceived problems with current students becoming fat, messy, and disheveled in their appearance, and "Two Sides of a Wall," which presented the pitfalls of male students cavorting and drinking into all hours of the night, when they should have been studying.[28] In each of these pairs of drawings extremes of behavior or appearance were presented to the readers. No additional narrative was included, with conclusions to be drawn from the pictures themselves. With this being the case, it is unclear if either extreme ever occurred at the university or if Wildfire was mocking both the hoped for best scenario and the feared worst.

In the United States students also had their share of journals or magazines, and depending on the year or location, they often had several of them. On some campuses publications were founded or folded fairly regularly because interests changed or because students or officials thought that launching a "new and improved" publication would give circulation a boost. The University of Michigan had a series of different publications that were each designed to appeal to certain interests of the students. In 1869 the *Chronicle* started work as "an eight-page fortnightly journal" which the historian Wilfred B. Shaw reported "flourished as a very creditable example of undergraduate journalism."[29] It would be replaced in 1890–1891 by two new publications, *The Michigan Daily* and *The Inlander.* The first was published as a daily newspaper, similar to *The Daily Cardinal* at Wisconsin, and the second was a monthly literary magazine.[30] There was also a humorous paper called *Wrinkle* that existed from 1893 until 1905 when it was replaced by the *Gargoyle.*[31] The students used each of these publications differently, with *The Michigan Daily* becoming the primary place for students to debate subjects of interest or to advocate for changes they felt were needed on campus. A good example of this was the desire for a student union building which was appealed for starting in 1892 and regularly afterward until after World War I when one was finally completed.[32]

Another function of student journals and magazines was the interchange of ideas they were able to have with similar publications on other campuses. The "exchanges," as they were typically called, became a regular feature of many publications during the nineteenth century. It is not always clear if students on one campus actually subscribed to the publications on another campus or if copies were simply exchanged as a common courtesy. The columns printed in one's home publications ranged from telling students what was happening at other campuses to evaluating the outside publications, as students were constantly looking to see how what they were doing measured up to what their contemporaries did. An early exchange was published by the women students of Vassar College in 1886 after they received a copy of Ohio University's *College Current.* The new midwestern publication was described as being "unusually well edited," and unlike some other student magazines it did not focus solely on campus events; this made it a cut above the rest in the minds of the eastern women.[33]

In Pittsburgh *The University Courant*, or *The Western University Courant* as it was eventually called, devoted a great number of column inches to their "Exchanges"; sometimes more than were devoted to their "Locals" or "Localettes" about happenings on their own campus or in their own city.[34] They placed a request for such communication with other institutions in their September 1888 issue: "We invite exchanges from all our companions in misery, and assure them that exchanges will not be viewed with fault-finding eye, but to find any hints which might be to our advantage, and to give any useful hints we can to others."[35] The desire for contact at Pittsburgh was considerable, as they had previously had a regular feature called "The Exchange Editor's Table" in their first student journal, the *Pennsylvania Western*.[36]

Similar to the aims and objectives of publications in other nations, at Aberystwyth the students produced the *U.C.W. Magazine* as a means of "interchange of thought" between the students, faculty, and "all the friends of the College" about what was happening in the various student societies. The publication received a start-up sum of £250 from the National Eisteddfod, a Welsh cultural festival that promotes literature and other arts, in 1878 and would rely on contributions and subscription rates in future years for subsistence.[37] In 1885 the cost of a subscription was "three shillings and sixpence" with free postage for people who wished to have it mailed to them or individual issues could be purchased for 7d.[38] The students were clearly not going out of their way to make a profit, preferring instead to reach as many people as they possibly could with news of their activities. Although the publication was successful, it was not as ambitious as some others in terms of the regularity of publication. According to the historian Iwan Morgan, the renamed *Dragon* "miraculously appears at least once per term" but suffered from a lack of "literary contributions for which the harassed Editor makes repeated appeals."[39] Morgan also criticized the publication of the early twentieth century because it lacked the "earnest seriousness" of its Victorian predecessor. The *Magazine* in this more serious state was popular in its day at Aberystwyth where all other publications that tried to find a footing on campus, including one written by the women students, were short lived because they could not compete with the more established one.[40]

Yearbooks

A uniquely American student publication during the nineteenth century was the yearbook. As implied by the term, each yearbook was intended as a record of one year's activities and personalities at a college or university, from the perspective of the students. Because these publications have always been intended as mementos to be saved by alumni for years to come, there tends to be a good deal of sentimentalism in them, rather than objective reporting on events which was more common in journals, magazines, or newspapers. Another aspect of yearbooks that can cause difficulty for modern historians is the large amount of "inside jokes" that editors included, some of which are well disguised as being facts, rather than fiction. Some level of humor became a common feature of all yearbooks across the country in time, and for some it was the primary goal as the motto of Wisconsin's first yearbook *Trochos*—"nothing dry in it"—testifies.[41] The final section in yearbooks that became standard at all institutions, once the technology made it possible, was the inclusion of photographic portraits of students. As Wilfred Shaw put it, the "annual yearbook, the *Michiganensian*... has a gallery of ancestors which at least establishes an ancient lineage."[42] It was up to each new generation of students to decide what their contribution to this lineage would be.

The naming of yearbooks was an important proposition, as the title needed to represent the student body and the institution over time and should be recognizable to people outside the local university community. This is one of the reasons *Trochos* was changed in 1888 to *The Badger*, the new mascot of the state and university. The name chosen for the yearbook at Indiana, *Arbutus*, came from the name of a flower that grew only on Arbutus Hill, east of Bloomington.[43] The students at the Pennsylvania State College chose the name *La Vie*, meaning "the life," for their annual in 1889.[44] In Mississippi a contest was held to name the yearbook in 1897, with the winning submission coming from Emma Coleman Meek who suggested "Ole Miss." The term soon became synonymous with the university itself and is still how people refer to the institution in the twenty-first century.[45] The fusion of yearbook names with mascots and symbols of the university can also be seen at Tennessee. In 1897 their yearbook, *The Volunteer*, was first published. In 1902 the term from the yearbook was used to refer to sports teams on campus, and in the spring of 1905 reporters at the *Knoxville Journal and Tribune* followed this trend and used the name for all the male sports teams, while the women are referred to as the "Lady Vols."[46]

A typical yearbook contained a survey of the academic year including events held by the campus organizations and athletic teams. Histories of the institution and various student groups were also common, with new traditions added to the old ones. Once photographs became affordable to include, group images of campus societies were also printed, or at least of the senior members who would be graduating. This was often the only way students would be able to have a photograph of themselves with their friends from university. As time progressed, they were increasingly used as showcases for students' artistic talents, whether drawings or writing. It was also normal to highlight student accomplishments like awards or honors granted by the university or statistics about the student body and listing the names of the students getting different degrees.

At the University of Alabama one of the standard items to include in the yearbook, *The Corolla*, was a description of an "Average Student at the University." They do not give their statistical methodology, so the accuracy of their findings may be suspect, but the information is interesting to a modern historian nonetheless. In 1896 they reported that the average Alabama student was "18 years, 3 months and 19 days" old.[47] With an entrance age minimum at the time of 16 for men and 18 for women, this is not a surprising average age when all students are taken into account.[48] Details were also given as to average height (just under 5 feet 8 inches) and weight (just over 139 pounds). Other averages were taken as well, like the average "retiring hour" which was "calculated to be at 10:10 p.m."[49] A more humorous take on "Statistics" was provided in South Carolina's yearbook, *The Garnet and Black.* They listed the categories of "Most Popular Co-ed, Brightest Co-ed, Prettiest Co-ed, Most Stylish Co-ed, Most Talkative Co-ed, Sweetest Co-ed, and Sweetest Flirt" as items that the students were surveyed on before publication.[50] As will be discussed later in this chapter, the term *co-ed* referred specifically to female students, but there is no indication if women were consulted about who should "win" each of these distinctions. The focus on women, as the newest and most appealing members of the university, was common in the yearbooks, just as it was in other types of publications on campus.

Humorous articles, poetry, and illustrations provide great insight into the perceptions students had of their role on campus and in life in general. A popular subject for discussion in coeducational yearbooks was male-female relationships; more specifically, dating. The types of men women "would like to marry" or the attractiveness of women on the campus were prominent "researched" features in *The Badger,* similar to those at Alabama and South Carolina. Whether these questions were really posed to the student body cannot always be ascertained, but comments were directly attributed to various students. One "Official Ballot" does survive from an election in January 1900. The "offices" on this ballot include: "Handsomest Man; . . . Biggest Dude . . . Biggest Flirt; . . . Champion Billiard Shark; . . . Most Ladylike Man; . . . Most Versatile Woman" and others.[51] "Elections" such as these became common in institutions throughout the United States, often making predictions about future status when asking who was "most likely to succeed" or "the couple most likely to marry." This theme was followed in yearbooks throughout the period, as students continued to ponder their role in society and to challenge the boundaries placed on them by the university. An underlying aspect of these categories was the reinforcement of traditional gender roles, even if the relationship between the students was increasingly questioned.

The Prospect of Admitting Women

Before the admission of women to their institution, the possibility of coeducation was discussed in several student publications. In some cases the topic came up as a point of interest because women would add to the social climate of the institution. *The Durham University Journal* contained one such anecdote in May 1883. A description was provided of a scene on the River Wear that took place on Ascension Day:

> a boat glided down the river sculled by a lady. In the boat sat a figure in cap and gown. Poor fellow! he wishes, we suppose, to seize the opportunity, and could spare no time to doff his gown; or did he regard the river as one of the 'public places'?[52]

Presumably the student in question would have preferred not to be seen by his classmates or by university officials, and the fact that he still had his academic robes on, which were required of students while at services in the cathedral and "in all public places" before one in the afternoon, made that impossible.[53] This scene may have also been noteworthy because the boat in question was being sculled by the woman, not the man, as would have been customary at the time.

Before the incorporation of Queen Margaret College with the University of Glasgow in 1893, there was a considerable amount of debate in the *Glasgow University Magazine* about the potential changes this would bring. Events at QMC, like the opening of the medical college, were covered at length in the *G.U.M.*, but the prospect of having women taking courses and earning degrees at the men's university would be even more newsworthy. One writer in 1890 expressed his own desire, and seemingly that of his fellow male students, that women be admitted to Scottish universities in the near future: "[W]e hope that the day on which Queen Margaret College will be affiliated with the University will soon dawn. It will be a noble day for Glasgow when she places women on the same platform as men, and gives them equal privileges and opportunities for study."[54] Support for women's admission to the university was not universal, and the voice of opposition was equally heard in the magazine. In March 1891 a complaint was registered by a "young lady" that the length of the Union Debate would keep them out past 11.00 "and mamma said it was not proper." The sarcastic reply followed: "Let it not be said that our University is wanting in courtesy to our fair visitors."[55] The lady's letter may, or may not have been written to the Union. It might simply have been a fabricated tale used to make a point by the writer, as was common in the magazine.

The male students at the Western University of Pennsylvania had numerous opportunities to debate the possibility of coeducation coming to their campus. Many of these debates were in the pages of the *Pennsylvania Western*.[56] One such debate was about the "tyranny" of single-sex education that needed to come to an end because it was an outdated mode of instruction. According to the editors, coeducation "is the order of the day." Furthermore, they argued that "it is a poor rule that will not work both ways and—Vassar must admit boys."[57] Their point was a reasonable one, and one that was not made often during the Victorian Era. The pressure for male universities to open their doors to women was in no way equaled by men demanding admission to women's colleges. Two issues later the subject of coeducation was raised again, this time in "The Exchange Editor's Table." The subject of the piece was the large number of coeducational institutions in the West as opposed to the small number in the East. The editor commented:

> It is generally supposed that the people of the Eastern and New England States are better educated than their Western brethren. Is this the reason co-education does not meet with favor in the East? If this be true, does high intellectual training necessitate the separate training of the sexes?[58]

No answers were provided for these questions as they were just throwing them out for the students reading them to think about. They also did not indicate whether they considered Pittsburgh to be in the East or West, which in the 1880s may have still been up for debate in such a dialogue. The impact women would have on their institution, and its reputation, was definitely on the minds of the students, or at least those who edited the journal.

Discussion of Coeducation and Women's Presence on Campus

Once women were admitted to a college or university, they instantly became the center of attention, with all manner of stories, illustrations, and verses dedicated to them. The Aberdeen students' magazine, *Alma Mater,* published a series of poems called "The Jack Daw of Rhymes" between 1888 and 1894. Some of these were then published as a collection known as *College Carols*, edited by John Malcolm Bulloch (M.A., Aberdeen, 1888).[59] One poem of particular interest was "Trim Little Maids at 'King's'," which was discussed briefly in Chapter Two. It was written shortly after the admission of women to the university. The poem begins:

> Comes a train of little ladies,
> Bajans by a new decree;
> Each a little bit afraid is,
> Wondering what the Quad, can be.[60]

Whether the author meant that the women were small in stature or just seemed to be because they were apparently meek or "afraid" is unclear. The specific reference to the "new decree," or Act of Parliament, that turned these women into Aberdeen students is an indication that this event was ample cause for comment by the male students. The poem continues with more discussion of the King's College quadrangle:

Is it only for our brothers,
Cigarette equipt?
Is it sanctioned by our mothers?
Were it better skipped?[61]

The question of sharing facilities was a standard one at all universities that decided to become coeducational, but the added references to cigarettes and the appropriateness of being in the quad along with the male students shows a deeper consideration than just where women would spend their time on campus.

After Queen Margaret College became a part of the University of Glasgow, there was a similar amount of comment on women's presence found in the pages of the *Glasgow University Magazine.* The front cover of the magazine, over a number of issues, featured an illustration by Madge Wildfire presenting the marriage of the women's college to the university. In this drawing a dour university official is shown presiding over the union in his academic robes, with the male figure representing the university itself graciously kissing the hand of his new "wife."[62] Marriage imagery was used frequently by the students to illustrate the union between the two institutions, with the male university subsuming the female college. Helen Nimmo, a student at the time, went so far as to refer to the endowment Queen Margaret College raised as "the dowry of our little College."[63] This allusion is quite astute indeed, as the endowment had to be raised before incorporation with the university could take place.[64]

This fixation with marriage imagery was carried through subsequent issues with Wildfire's illustrations like the one at the start of this chapter.[65] The two female students shown in the 1891 issue represent the expected effects of coeducation on women. Either a woman would study music and other feminine subjects like the one on the left and remain attractive, or she would study Plato, Aristotle, and Sophocles like the one on the right and become a truly unpleasant creature. The caricature on the right is further elaborated with items that were typical for students to have in their rooms while studying like a bottle of ink, a globe with Africa on it, and a cup labeled "coffee," which she has apparently been drinking in abundance. Like her more feminine counterpart, this student is presented wearing academic robes, though the lines highlighting them are much more severe, as are the features of the woman herself. Interestingly, the drawing also includes a sign on the wall that reads "WARNING MAN TRAPS." This sign illustrates a key point; many people thought that women in university were out to trick men into personal relationships, under the guise of attaining a university education.

A second set of drawings of "Queen Margaret types" followed in the next issue of the *Glasgow University Magazine,* indicating a high amount of fascination with this topic (by Wildfire or the student body or both).[66] Unlike the first set of drawings which had a number of different messages imbedded in the imagery, the second set focuses on one issue in particular. The student on the left has a timid and feminine posture and she appears to be standing under mistletoe, though that is not clear.[67] If it is mistletoe, this reinforces the idea that women in university were attempting to trap men into a personal relationship. The important element, however, is in the other half of the illustration. Here Wildfire has presented an exceptionally masculine female who is wearing an outfit that is half dress, half suit including a necktie and bowler hat. Most significantly, she is holding a paper entitled "DESCENT OF MAN." This use of Darwin's text gives an idea of the fear felt by some Victorians that caused them to oppose women's admission to university. The conclusion drawn from this illustration is that educated women, who undertake unfeminine pursuits, will eventually become men, or at least, they want to become men. And, as in the first set of drawings, the woman shown here is physically unattractive as a result of excessive education, probably making her unmarriageable.

The distinction between gender roles and the need to maintain them was another theme that was developed in many student publications. At Wisconsin, stories about students' travels or other interesting anecdotes were also included once *The Daily Cardinal* expanded in length. One such story, entitled "A True Hearted Woman," considered the entertaining strategies of two Wisconsin women, Miss Louise Lawson and Miss Jennie O'Neill Potter.[68] The women were presented as extremely worldly, as well as very popular in society. Their experience in cultural circles made them the ideal women of the 1890s, and the choice of the campus newspaper to highlight this type of woman indicates what the male students writing for the paper valued in female companions and, therefore, what the female students should aspire to become.

Individual students were often profiled in publications, either because they were the first to enter or graduate from the institution or simply because their lives made for an interesting story. The 1900 *Garnet and Black* included a profile of "The First Woman-Graduate of the South Carolina College," Mattie Jean Adams. She was described as having "an almost passionate fondness for books" and a "curious love for investigation," both of which served her well as a pioneer of coeducation.[69] Miss Adams' decision to attend South Carolina College rather than a women's college or a normal school was also considered. According to the author of the piece:

> Like all true students, Miss Adams felt that she needed a better education than the Southern female colleges afforded, and her thirst for knowledge led her to look with longing eyes to the South Carolina College, which, recognizing that the "fulness of time" had come, had generously invited the daughters of South Carolina to share the advantages given to her sons.... It was here that Miss Adams' nature

found the satisfaction of drinking more deeply at the fountain of knowledge.[70]

She earned a bachelor of arts in 1898 and went on to become an instructor of "Expression and English Literature" in Mississippi, showing herself to be entirely capable of handling the education provided at the formerly male institution.

Discussion of "Co-Eds" and Romance on Campus

It did not take long for the term "co-ed" to become "a fixture in the collegiate vocabulary" in the United States.[71] The term was initially somewhat derogatory, or more precisely as a slight on an institution, often made by eastern men's colleges against those in the middle of the country. For example, a New York book entitled *Wisconsin Wickedness* used the "coed" of Wisconsin to illustrate the distraction women caused if allowed to take part in university, creating great difficulty for men who wished to concentrate on their studies.[72] The historian Jennifer Stein, writing about "The History of the *Daily Cardinal* from 1892–1991," found numerous instances where the *Cardinal* writers "defended the co-ed against charges that she attended the university because it was a matrimonial bureau."[73] These arguments were accompanied in the pages of the *Cardinal* by headlines such as "Educated Women Make Best Wives," which indicate there was not a consensus about the marriage question. *The Inlander* at Michigan also articulated concerns about the use of the term *co-ed* to describe women students because no such slang title was used in other countries. Though not directly expressing a feeling of being insulted, the implication was clear that at least some of the women found the use of this term to show a lack of respect for them and their accomplishments.[74]

Some female collegians embraced the term *co-ed,* rather than holding it in disdain. The women at South Carolina College formed themselves into "The Co-Ed Club" whose motto was "Woman is the better man." One of the illustrations that accompanied their membership list in *The Garnet and Black* in 1899 showed a happy co-ed being walked home by a freshman. She may have been happy simply to be next to him, but it is far more likely that she is smiling because he is carrying both her book, and his own.[75] In the same yearbook there was a "College Alphabet" of verse, with illustrations by Miss Belle Davis. With the exception of a few names of students, the alphabet would translate to just about any U.S. college or university of the late 1800s. "E is Exams... G is Gymnasium" and so on. And, of course, "C is for Co-ed, that Creature divine(?), Oh, Lordy, I tell you they think they are fine!"[76] Perhaps as a response to this sentiment, the following year listed the name of the women's organization as the "Coquettish Co-Ed Club."[77] The decision to use the term *coquettish* to describe themselves would have been welcome to the male students, since the focus of their writings often focused on dating, or at least flirting with, their female classmates. As one issue of *The Western University Courant* put it in 1901: "Why is a co-ed like a copy of the Courant? Because every fellow should have one of his own and not borrow his neighbor's."[78]

The University of Missouri's yearbook, *The Savitar*, contained a great deal of material about the male-female relationships that were developing on campus.[79] In 1891 the members of the Clover Club recorded the toasts from their annual meeting including "How to Manage Two or More Sweethearts at Once" and "Marriage—A Failure."[80] Another piece in the same issue was a poem entitled "Love in a College Town."[81] The distraction provided by women students was included in a stanza from "To My College Girl" written in 1899:

> Oh a college girl—
> A college girl.
> There is just one glance
> Sets my head awhirl—
> Sets me dreaming while waking—
> Makes me dream while I'm waking.
> My college girl—
> My college girl.[82]

The author, J. D. Derelic, also had stanzas about the expenses of trying to woo his college girl and the time he had "lost" waiting to see her on campus or around town. For anyone who felt that the entirety of a student's attention should be spent on their studies, the sentiments included in this poem could be of concern. In addition, parents of female students may not want to think that male students might be pursuing their daughters so intently. The full-page poem does indicate that the romantic aspects of student life increased as the number of female students on campuses increased.

The *Glasgow University Magazine* showed yet another common expectation of the admission of women to the Scottish universities in 1892.[83] The illustration is titled "On Thin Ice" and shows a conversation between "Miss Marrywell (of Queen Margaret College)" and "Mr. Fargone Stonebroke (of Gilmorehill)." Many people assumed that the primary goal of women was to "Marrywell" and that the serious male students would be skating "on thin ice" as a result. Lindy Moore included a similar sketch, also entitled "On Thin Ice," in her study of Aberdeen University, showing this to be a common portrayal of women's admission to higher education in Scotland at the time.[84] Whether this type of illustration was used merely to entertain or as a more direct warning to male students to be on their guard is not easy to determine. The message that was sent to the student body was that the men's institution would never be the same again with the added element of romantic entanglements that would distract students from their studies.

Not surprisingly, Madge Wildfire initiated a debate about the possibility, or lack thereof, of Platonic friendships between the men and women of the University of Glasgow. Wildfire called for a "canon of etiquette" for students now that women had been admitted. The "mingling of the sexes" brought about by coeducation introduced the prospect of various types of friendships.[85] The idea of Platonic friendships was particularly important for those students who took courses together on the main Gilmorehill campus or participated in university organizations. The reply to the letter, by a Queen Margaret student, argued that Platonic friendships were fully acceptable in society at the time and questioned Madge Wildfire's assertion that the idea needed to be worked at before it would be widely accepted. A third letter on the topic warned about the development of friendships because "college and work of any kind [would become] a secondary consideration altogether."[86] Sometimes topics of debate in the pages of the magazine were debated verbally as well, such as one in 1893: "That mixed Classes are better than separate Colleges."[87] Since the topic was a long-running one in Glasgow, both male and female students wrote to the magazine on either side of the question in an attempt to persuade their classmates.

The question of women proposing was raised at the University of Alabama, though in a more lighthearted manner. In 1896, because it was a leap year, the students at Alabama passed a resolution, though undoubtedly a humorous rather than a serious one. They resolved, "[F]or the benefit of the opposite sex... for six months only, we will be open to all proposals for marriage, made by eligible girls." For added flourish they ended with "Come one; come all."[88] No criteria were given to indicate what would make a girl "eligible," but the standards were perhaps known to the students at the time. The worry that a woman might take on a man's role in society, because she received a university education in the same classrooms as men, underpinned incidents like this one. In poking fun at the concerns by some in society and supposedly not giving them credence, the views of society at large were being reinforced among the student body. As Jennifer Stein concludes, publications could often be found "mirroring sentiments prevalent in the larger society," despite their claims to challenge it.[89]

Gender roles were also inverted in the pages of *The Amulet*, a magazine published by the women at the University of Michigan in 1882. They took the standard debates about coeducation and phrased them from a female viewpoint, as if men were being admitted to *their* institution. "The higher education of men is no less important than that of women," they argued. Although women's "influence in the home" was the most important in society and required "a broad mental training," surely men would also need university training to provide financially for their families and to become politicians. "Is not this a sufficient reason why he should be admitted to all the privileges so generously provided for women in the University of Michigan? The co-ed is here; let him stay until some better argument is presented against him than has yet been adduced."[90] The humor is unmistakable, but the stinging reality is that these were the objections women had faced in their attempts to enter male universities. The final jab provided by the women in the piece regarded accusations about women's frivolous nature and their inability to handle university coeducation: "Now it is true enough that there are now and then silly young men even in college; but we maintain that their number is remarkably small, and that the same persons would be even more silly if they had remained at home or gone to a school where ladies are not admitted."[91]

Women as Contributors

Though there were some notable exceptions, the experience of working on campus publications was limited for women in coeducational universities, as men held most positions of authority and were the primary contributors. The historian Barbara Miller Solomon notes that many female writers got their start working on student newspapers, and found a way to assert their talents in a male-dominated environment. A prime example of this is Willa Cather, who cultivated her writing talents on the student newspaper at the University of Nebraska in the 1890s.[92] These students were exceptions, since most publications that were started by male students chose to relegate women to subsidiary and supporting roles or ones that emphasized traditionally "feminine" skills like art and poetry.

In 1899 the first issue of *The Garnet and Black*, South Carolina College's yearbook, had an editorial board of fifteen, two of whom were women. Assistant Editors Anne Fayssoux Davis and Beulah Gertrude Calvo were joined by three female artists, Belle Harper Davis, Laura Annie Bateman, and Jean Adger Flinn.[93] The institution's monthly journal, *The Carolinian*, had an all-male editorial board in the same year.[94] Previously it had printed contributions from Beulah Calvo in 1897, showing it was not an explicitly male publication; it was just that the men were the editors and made decisions about the content.[95] The student newspaper at the University of Alabama, *The Crimson-White*, also featured two women on its first Board of Editors in 1894—Bessie Parker and Anna Adams.[96] Miss Adams continued in her role as an associate editor in the 1894–1895 academic year and was pictured with the rest of the editorial board in a group photo in *The Corolla* that year.[97] The faculty controlled the publication and "selected all members of the editorial staff" so there is no way to know if the male students wanted the women to be editors.[98] A third southern institution, the University of Tennessee, had a newspaper, a magazine, and a yearbook—the *Orange and White, The Tennessee University Magazine*, and *The Volunteer*—all of which were coeducational publications.[99] And *The Monticola*, West Virginia University's yearbook, was also coeducational, though as at other institutions, the male contributors greatly outnumbered the female ones.[100]

A simple headcount of female members of editorial boards is not entirely helpful in assessing the contributions of women to their university's student publications. Since there were more men on campus, it followed that there would be more men who wished to contribute to them. Another difficulty is that contributions were often submitted anonymously or under pseudonyms, making the identification of the artist's or writer's identity virtually impossible to ascertain. The presence of women as editors was typically only discussed when there was dissatisfaction with it or when the women were trying to take on a larger role on a certain publication. At Missouri *The Savitar* Board of Editors was not coeducational for much of its early life, and the women students actually had to negotiate a position on it in 1909. The issue was considered to be of such significance that *The American Educational Review* ran a piece on it as evidence of the problems of coeducation on university campuses.[101] Interestingly the student newspaper at Missouri, *The University Argus*, had a coeducational editorial staff and had no need of the "negotiations" that the yearbook did.[102] As a side note, one of the regular contributors to the *Argus* was Miss Susan Alexander who wrote verses that were also published commercially while she was still a student in 1897.[103]

One of the ways that women were included in student publications was by having their own column or regular feature. Even before women were admitted to their institution, the male students at the Western University of Pennsylvania reached out to their nearest neighbors at the Pittsburgh Female College, or P.F.C. as they called it, and asked them to contribute to *The University Courant* since they knew the women subscribed to the journal and had "no college paper of their own."[104] After women were admitted, there were no women on the editorial board of *The Western University Courant* during the remainder of the century, although that did not preclude them from submitting articles. One story that was "promised" by one of the female students in 1896 was going to be titled "Cupid at the University." The piece never appeared despite the enthusiasm the male editors expressed at its potential contents, stating that there was "already a great demand for next month's issue."[105] The prospective author was not named, but in the same year Stella Stein had become the vice president of the Philomathean Literary Society which copublished the journal, so it is likely that she was the potential contributor.[106]

Since Queen Margaret College developed largely as a "sister" institution to the University of Glasgow, there were many connections between the two sets of students. "Queen Margaret" sent New Years' greetings to "Gilmorehill" through the *Glasgow University Magazine,* saying that they anticipated "much friendly and helpful intercourse" after the institution became coeducational. The *G.U.M.* addressed the needs of women students further when they established a Queen Margaret Column (sometimes referred to as "Notes from Queen Margaret") at the time of incorporation. Issues brought forward by the women were featured, almost always by a female contributor. The column generally covered the activities of student organizations at Queen Margaret, though sometimes they featured poetry or other creative writing. The magazine furthered their coverage of particularly female issues when they began to devote one issue per year to the contributions of women. This "Queen Margaret Number" usually appeared in the spring and coincided with a major occurrence at the women's college, such as the QMC Bazaar.[107] These issues still covered topics of relevance to the entire university, while providing more representation of women's views than usual.

The students who published *The Free Lance* at the Pennsylvania State College chose a similar approach to ensure their contributors were coeducational. Its editors described their decision as a way "to fairly represent the sentiment of the body of students at all times."[108] They enlisted various women to work as anonymous "Co-Ed Correspondents" whose primary purpose was to supply thinly veiled campus gossip and general reporting on campus events.[109] More serious contributions were also solicited from the women of the campus. In 1900 the editors pointed out, after looking at the publications of the women's colleges, that "girls can write stories—good ones, too." As a result they wondered why more female students did not contribute to *The Free Lance.*[110] The men were open to the possibility of having additional contributions from women, yet the women themselves were limiting their contributions to feminine gossip and social reports.

The final nation that has not been discussed at any length in this chapter is Ireland. Unfortunately students there would not find a regular voice on campus until after the Victorian period. Women at Queen's College, Galway were given a column known as "Ladies' Notes" that was included in the monthly magazine, *QCG: a record of college life in the city of the tribes,* which was first published in November 1902. It is unclear whether one writer was the sole contributor to this column or not, with signatures at the end ranging from "I am, NOBODY IN PARTICULAR" to "AS-YOU-LIKE-IT" to simply "SHE."[111] The topics included accounts of the Ladies' Hockey Team and events on campus, along with discussions of fashions and traditions of the college. There were so few students at any of the three Queen's Colleges during the nineteenth century that it is not surprising that they did not organize publications any earlier. They could also see how the controversies over religion and politics played out in their local and national newspapers and may have felt that they might overwhelm any student publication as well. Should they wish to engage in those debates, they certainly had the chance to send letters to the editors of those papers, and if they wanted to discuss specifically educational topics, they could always contribute to *The Students' Journal and Hospital Gazette* which ran regular features on happenings at Belfast, Cork, and Galway.[112]

Conclusions

Traditionally, the imposition of rules or conduct on university students, whether male or female, is thought to be the role of the institution itself. But, as can be seen in the realm of student publications, when the administrators failed or chose not to do so, an informal student culture of roles and expectations began to fill the void. The growth of importance in social life relates directly to the growth of importance of student commentary on this area of higher education. Increasingly there was also a great degree of irreverence exhibited by students about their academic life. As university officials relaxed their policies of control over the students, the students felt freer to criticize their faculty and administrators. As noted by Jennifer Stein in her work on Wisconsin's *Daily Cardinal,* there was a steady increase in "dedication to broad-based student inclusion and accountability to its student constituency as it gradually loosened itself from official university influence."[113]

The emergence of student publications on campus provided a new voice for grievances among the student body. Usually veiled with sarcasm or humor, student cartoons, poetry, and news articles reflected the opinions of students about their experiences. All the student publications examined in this chapter were readily accessible to university officials, and often they provided the only insight faculty or administrators would get into student actions, opinions, and concerns outside the classroom. This, coupled with the increasing power of student government and organizations discussed in the previous chapter, gave students a certain amount of agency and perceived control over their university experience. The description by the first editors of *Trochos* of "a new era in University life" is certainly no overstatement, as student publications became the vehicles for dialogue, not only between students, but also from students to campus officials, becoming one of their best means of lobbying for change. Unfortunately, as the historian Lynn D. Gordon comments, women still had "limited options" in the area of student publications. Their contributions showed evidence of "strong consciousness and accurate perceptions of the cultural and social barriers they faced on and off campus."[114] The perpetuation of women's traditional roles, as perceived by the students and exhibited through their writings on each campus, would carry students into their lives after graduation, prepared to take an appropriate place within their communities.

CHAPTER EIGHT

Life After Graduation

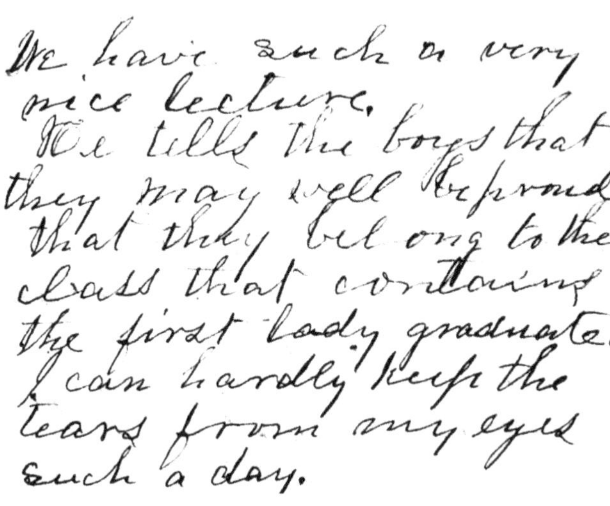
We have such a very nice lecture. He tells the boys that they may well be proud that they belong to the class that contains the first lady graduate. I can hardly keep the tears from my eyes such a day.

Image 8 Excerpt from Margaret Boyd's Diary, Sunday, June 22, 1873.

Credit: Boyd Family Collection, Mahn Center for Archives and Special Collections, Alden Library, Ohio University, Athens, Ohio 45701.

The Sunday before Margaret Boyd became the first female graduate of Ohio University, she reported in her diary that the preacher had told "the boys that they may well be proud that they belong to the class that contains the first lady graduate." This statement brought tears to her eyes, both because she was proud of her accomplishments and because others recognized her contribution to the history of the institution.[1] The support she found in the community in Athens did not extend equally to all coeducational colleges and universities, for despite women's accomplishments in the classroom, there was still apprehension about the roles the female graduates would take on in society. The varied outcomes of women's higher education, with graduates finding employment, while still pursuing traditional female roles, forced many of the remaining opponents to concede that university coeducation would not bring about the end of the human race.

Though much of the maintenance of women's traditional roles was due to their own views and instilled family values, the role played by the universities themselves was of the utmost importance. The historian Adele Simmons terms this the institutions' attempt "to turn out efficient housewives."[2] University efforts, whether conscious or unconscious, were efforts to guide women into traditional roles, despite those roles being broadened somewhat. In housing and extracurricular activities, university administrators on both sides of the Atlantic supported "appropriate" interactions between the male and female students of their institutions. Relationships between the male and female students were further encouraged by the tailoring of university social life, facilities, and academics to men and women, with the intention of guiding them into marriage. Though this directing came out of Victorian separate spheres ideology, its persistence in the twentieth century has as much to do with the students' wishes as with those of administrators or the community at large. This chapter will therefore consider both the perceptions and actual outcomes of students' lives after leaving university in their relation to three primary issues: marriage, careers, and society.

Students and Marriage

As noted above, the primary expectation of women in the nineteenth and early twentieth centuries was that they marry and start a family. The use of "race suicide" arguments by opponents of women's higher education, in the wake of Darwin, left their mark on academia for more than a century. Reinforced by personal expectations and peer pressure, women in coeducational universities in the United States and the United Kingdom generally felt that they had dual goals: gaining an education and gaining a husband. Statistical evidence to support this argument is difficult to obtain since marriages would have taken place in individual churches, probably in a student's hometown and not all institutions collected data from their graduates on their marital status during the nineteenth century. In 1904 *The Edinburgh Medical Journal* published an article by T. Claye Shaw of St. Bartholomew's Hospital in London on "The Collegiate Training of Women." He argued:

> If the statistics of marriage of female graduates could be obtained, they would almost certainly point to the conclusion that scholastic excellence is not a very valuable asset in the matrimonial market, not so much because men are afraid of women with diplomas, as because the very course of study and the trend of mind which it creates seem to displace in the woman the marriageable attributes, to give them

> a feeling of independence and the desire of relying upon their own efforts.[3]

By prefacing his statement with an admission that he was not basing his opinions on any sort of statistical evidence, Shaw's "medical" conclusions amounted to little more than speculation. He was in all likelihood reflecting a continuing perception in society that women who were trained in universities became unmarriageable, even if he stated it in a less salacious manner than earlier critics.

The fascination with the marital rates of female university graduates, and the desire to come to a definite conclusion about the effect of coeducation on this aspect of society, would lead to numerous statistical studies in the early twentieth century to determine the "marriageability" of university women. One such study, by B. L. Hutchins in 1912, found that "[w]omen from co-educational colleges, it may be noted, marry more frequently than those from women's colleges, no doubt owing to the greater opportunities of friendship and social intercourse with men."[4] A later study by Glen H. Elder on "marriage mobility" in the United States supports this conclusion as well, revealing that women in university in the 1920s and 1930s were more apt to marry "well" due to their higher levels of "educational attainment" often finding their marriage partners while they were students.[5] A more recent study of *Geographical Mobility and the Brain Drain* at the University of Aberdeen found that women who earned scientific or medical degrees were more likely to be mobile in their search for a job, unless marriage precluded their ability to choose. In this way many women were forced to decide between marriage and a career because of social pressures placed on them.[6]

Institutional data on marriage rates was not commonly tracked during the Victorian Era. One of the institutions that did keep track of their students' marriages was the University of Alabama, however. Many of the women who entered in the late 1890s were already married by the time *A Register of the Officers and Students of the University of Alabama* was published in 1901, though most did not marry fellow students in Tuscaloosa. Lucy Grace Archer, who entered in 1896, married William Nessler McKelvey in October 1898. Her address in the "Record of Students" was listed as "U.S. Navy" because he was a captain in that branch of the service.[7] Four students who entered the University in 1897 had also married. The first, Lulu Virginia Hosmer, married Professor James Angus Baxley in November 1900 and lived with him in Greensboro, listing no profession of her own.[8] One of her classmates, Mabel Eloise Bealle, married Dr. Nathan Herbert Carpenter in June 1899, and a third classmate, Lydia Peck Martin, married William Marshall of Ann Arbor, Michigan in December 1900.[9] The final student who entered in 1897 that married was Nela McCalla who became the wife of M. Neely Pride of Nashville, Tennessee.[10]

Alabama is also noteworthy because they had a number of married women students during the Victorian Era. In 1898 Mrs. Mary Strudwick

Nicholson matriculated, eight years after she had married Richard H. Nicholson in June 1890, making her the first married woman to attend the university.[11] She was joined the following year by Mrs. Cordelia Caroline Brock, who had only recently married George William Brock in July of that year.[12] The final married woman to attend Alabama in the 1800s was Mrs. Susie Fitts Martin Mayfield, the wife of James J. Mayfield, who entered in 1899, two years after her marriage.[13] Surely the fact that there were other married women in attendance made it easier for all of them. Aside from being referred to as Mrs. rather than Miss in student records and publications, it is unlikely that they would have been treated any differently than other women students at the time. The codes of chivalrous behavior that dominated at Alabama and were reinforced by military discipline for the male students would have guaranteed that there would be no lack of propriety for the married women.

Some women students decided to get married before completing their university careers. At the University of Missouri both Ellen McAfee and Susie Trimble were noted in *The Savitar* in 1891 as having married during the academic year.[14] The West Virginia University's yearbook, *The Monticola*, also made note of women that had gotten married while they were students. A member of the Class of 1899, Mrs. Edna D. Tyson, was listed as being in the Art Department and as having been married on September 7, 1898.[15] In all cases this change in status was seen by their fellow students as a good thing, though it is not clear from the yearbooks if the students completed their degrees or chose to leave university when they entered into marriage. Typically female students who did not wait until graduation to marry did not complete their educations. As B. L. Hutchins put it: "*[I]pso facto* her career as a student would usually be cut short."[16] Because colleges and universities of the nineteenth century did not have pressure on them to keep the retention rates of their students high, most administrators would have found a woman student's decision to marry as a job well done on the part of the institution. They had, after all, done all they could to prepare her for a successful life after graduation.

One of the main benefits of being at a coeducational institution was the access men and women had to each other to form permanent relationships. Marriage between male and female students even happened at universities where there were very few women on campus. At the Western University of Pennsylvania one of the first women students, Margaret Stein, married one of her classmates, John Colvin Fetterman, who became a dean of the college.[17] Succeeding generations of women students would see such a relationship as an example that they should follow if they could, and the administration at some universities would even promote past marriages between classmates as the ideal student relationships. At the University of Wisconsin many community leaders and their wives were educated together at the university, spending four years as classmates and friends, before deciding they would be compatible life partners. Senator Robert LaFollette and his wife Belle Case LaFollette in 1881 are the most cited example of this practice, along with Helen and John Olin who married in 1880.[18] Each of these couples met and married as a result of the coeducational policies at the University of Wisconsin and the opportunity it gave them to meet members of the opposite sex who had similar interests and intellects. Additionally, married university graduates were thought "to be personally more contented with the conditions of their marriage" since they had made an educated choice in agreeing to it.[19]

For other women, their marriage and family became their greatest accomplishment and contribution to the world. Flora Hamilton, a graduate of Queen's College, Belfast, went on to marry Albert Lewis in 1894.[20] Her education was a sense of pride for her and her sons. Warren Hamilton Lewis, described his mother as a "brilliant mathematician" and further stated that "had her youth been lived in the period of female emancipation, [she] would almost certainly have taken a good degree."[21] This comment has more to do with the lack of respect some held for the Queen's Colleges, particularly in comparison with Oxford where his younger brother Clive was a Fellow of Magdalen College.[22] As the author of *The Chronicles of Narnia*, C. S. Lewis remembered his early training from his mother with fondness and commented that education was one of the legacies she had given him before she died in 1908 when he was just ten years old.[23] Although Flora Lewis' academic accomplishments may have been of secondary importance in her own life, the fact that they gave her the confidence to make a suitable match and raise intelligent children made her a great example of the goals of university coeducation for women during the Victorian Era.

It was common for officials and supporters of the admission of women to universities to comment on the success of female students in finding husbands. In this way they were able to demonstrate that coeducation had not adversely affected the structure of society and that women were as ready as ever to take on traditional roles and responsibilities. In 1909 James Coutts wrote an authorized *History of the University of Glasgow*. In it he reflected the persistent concern about the life to be led by women after graduation:

> Of the women who have graduated in Medicine not a few have obtained public appointments, and the number of women graduates who have entered into marriage has been remarkable.[24]

His surprise regarding the success of university-educated women in finding husbands after graduation indicates that he was not an avid believer in university coeducation. Also, by first discussing the success of Glasgow's medical women graduates, he placed that ahead of marriage in importance, or at least saw the two issues as parts of what was believed to be a successful life for women at the time. Similarly in Aberystwyth, A. Wallis Myers remarked on the fact that "the University of Wales has turned out some admirable specimens of bright and intelligent British girls—and girls who, let it be said, turn out excellent wives and devoted mothers."[25] First and foremost the women were intelligent because of their university education, and as a happy coincidence, they also happen to be good wives and mothers.

Administrators also made more subtle attempts to guide their students in making plans for life after graduation. Michigan developed a special "Cap and Gown Collection" of books in the Reading Room of their library. The 300-item collection included "books on the art of living, advice to young people,... helps to the choice of a career, manuals of hygienic living, college lore, reminiscences of life at various universities, collections of college verse and prose, fiction and songs."[26] That the selection of works ranged to the "art of living" was an indication that they were trying to prepare students for more than just their academic studies or their professional careers; they were trying to make them ready for daily life, and for some, running a "hygienic" home. The extent to which students thought about the reading materials they were being presented with is unknown, but being surrounded by pressure to meet society's expectations was apparent to the women at the time. In her semi-autobiographical novel *An American Girl and Her Four Years in a Boys College*, Michigan alumna Olive San Louis Anderson described the difficulty her main character, Wilhemine "Will" Elliot, had choosing between a career in medicine and falling in love: "[N]ow here I am drawn off the track like any other girl... so I shall do the natural and proper thing of forsaking a professional career for the one I love, and give up dreams of fame as master of the healing art."[27] Sadly in this case when she told her male admirer of her desire to study medicine, he was disgusted that she would want to dissect anything and their romance came to an abrupt end.

Other students also attempted to make up their own minds about their place in society, both through serious discussions and humorous ones. In Glasgow the list of debate topics chosen by the Queen Margaret College Literary and Debating Society gives a good indication that the question of male-female relationships was predominant. Marriage issues such as "That clever men prefer unintelligent women" or "That cooking is the most important thing for a girl to learn" were considered from opposing angles by the students, requiring them to see both sides of the issues and leaving them to choose what they believed.[28] The women at West Virginia University were encouraged to become housewives with enough regularity that they spoofed the notion in *The Monticola* in 1901 by listing female students who were working toward the "Degree of B. K. M." which stood for "Bachelor of Kitchen Mechanics."[29] Just as the women at Glasgow were taking an active part in deciding their future roles in society and not just accepting what society expected of them, the women in West Virginia were contemplating the same issues in a more amusing manner. What was consistent for all women at coeducational universities was that they knew how society wanted them to conduct themselves, and the choice remained if they would comply with the wishes of the community at large or act in a fashion that was contrary to expectation.

Students and Careers

In 1883 Mercy Grogan wrote a book entitled *How Women May Earn a Living*. In it she listed all the colleges and universities that offered degrees to women at the time. She described the purpose of her work as a response to "the wants of the immense number of ladies who have to depend upon their own exertions for their support."[30] The shift in demographics during the nineteenth century, which led to there being more single women who needed to find employment, was also one of the arguments used by those in favor of women's higher education. These changes brought with them shifting views in society about the place of women in university and life afterward and also coincided with shifts in the nature of university study as new subjects, such as the social sciences, made their way into higher education. This process was part of a larger evolution in universities that began in the mid-nineteenth century. As the enrollment of institutions grew significantly, both in size and makeup, universities lost their exclusive status with more students from lower-income families gaining access to higher education. This alteration, along with the need for many teachers to attain degrees before getting jobs, also allowed women the chance to follow traditional female roles in a traditionally male university.

To what extent these academic changes enabled women's access to colleges and universities and to what extent women's access propelled the acceptance of newer subjects in the curriculum remains difficult to ascertain. As *The Missouri Alumni Quarterly* put it:

> Whether the growth of the University has determined the number of Co-Eds, or whether the Co-Eds have been the cause of the growth of the University would be an interesting subject for debate; but this much is certain, that these two facts, co-education and college development have come to be synonymous terms.[31]

Unquestionably the first generations of women students primarily studied "feminine" subjects—arts, literature, music, and teaching. Once women proved themselves to be successful in these fields, it became much easier to gain access to the male fields of science, economics, law, engineering, agriculture, and so on. Whether women chose to pursue these new paths open to them was not always completely up to them. The efforts of faculty and administrators remained dedicated to guiding men and women into fields traditionally suited to their gender. The addition of stereotypically female areas, such as health care or education, to the public sphere of society brought with it the inclusion of women to the public sphere as well. In particular, the introduction of new career opportunities for women, most notably social work, allowed society to educate women without greatly altering the pervading gender roles in the workplace.[32]

This traditional gender divide is seen in both anecdotal and statistical evidence from coeducational universities during the nineteenth and early twentieth centuries. Looking at the University of Wisconsin in the first fifteen years following women's entrance, female graduates had a clear preference to pursue careers in teaching at various levels, with 72 percent of those employed working in an education-related field. This early emphasis on education stems in part from women's admission to the University of Wisconsin originating in the Normal School, but the primary factors have more to do with the student's own knowledge of what jobs would be available to them and acceptable to undertake after graduation. The expansion of the country and increase in population meant that teaching was becoming a prominent profession for women to enter. The list of female graduates who became teachers, at the primary, secondary, or tertiary level, is considerable in that with few exceptions, it is the only profession women chose to undertake. In Alabama Mamie Augusta Bullock, Leila Harris, Katie Louise Powers, and Alyce Wildman all became teachers in Tuscaloosa.[33] Fannie Ingersoll and Julia Trent Royall became teachers in the Birmingham Public Schools.[34] Lucye Marion Wilson returned to her hometown of Quanah, Texas, to teach.[35] And finally, Annie Greer Turk became a teacher in the City Schools of Ennis, Texas. She married "Prof. Graham" in August 1901 and was still listed as a teacher in the 1901 alumni directory.[36]

Another teaching option was to work at normal schools or women's colleges. Parker worked first for the Georgia Normal and Industrial College in 1896–1897; moved on to the State Normal School at Jacksonville for the 1897–1898 academic year; then went to Converse College in Spartansburg, South Carolina from 1898 to 1901; and finally moved on to work at Ward's Seminary in Nashville, Tennessee in 1901. She then moved on to work at Ward's Seminary in Nashville, Tennessee, in 1901.[37] DeBardeleben worked at the Normal College in Livingston, and Searcy taught at both A. C. F. College and T. F. College.[38] Only twelve women had graduated from the University of Mississippi by 1890 when Edward Mayes wrote his *History of Education in Mississippi*, a fact that was not that dissimilar from other universities at the time.[39] Two, Sallie Vick Hill (1885) and Mattie James Smythe (1888), were valedictorians of their respective classes and five had gone on to teach at colleges elsewhere in the South.[40] Harriet "Hattie" Lyon, the first female graduate of West Virginia University, became a teacher after she completed her degree in 1891.[41] She worked at Parkersburg High School and Broaddus Female College, but would spend the majority of her life in Fredonia, New York, where her husband, Franklin Jewett, became the head of the Science Department at Fredonia State Normal School.[42]

Some of the early female university graduates in the United States found work in an itinerate fashion, moving from one location to the next until permanent employment was available. Ohio University had one such employee in 1888, Miss Kate A. Findley, who was originally from Andover, Massachusetts. According to an advertisement she placed looking for work, she was an "Instructor in Elocution, Reading and Rhetoric" in Athens, having worked previously in both her home state and Pennsylvania.[43] Ohio's first female graduate, Margaret Boyd, also had to move to find a teaching job, even though she did not want to leave her family and friends. Her first teaching job was in the town of Monroeville, Ohio, in the north central part of the state. She had to be encouraged to apply for the job and was reticent to actually take up the position when offered it.[44] When she left for Monroeville in August of 1873, she said, "I feel awful to say good bye this morning. I only say good bye to Mother and the children. Poor Mother, I do not know as I do right to leave her."[45] In the end she did not want to let down those who had supported her in earning her degree, and she did not want to be a financial burden on her family. Two other women who were perhaps itinerant teachers were Ruth Alma Bishop, who became a teacher in Akron, and Abby Hogan Hazard, who became a teacher in Orlando, both of whom graduated from the University of Alabama.[46]

The story in the United Kingdom was somewhat different. Although teaching was still a main career choice for women, there was a considerable variety of careers among graduates, with medical-related fields being very common. The historian Sarah V. Barnes found that approximately 69 percent of women who graduated from Owens College in Manchester during the Victorian Era went on to become teachers.[47] They had notable medical graduates as well, including Catherine Chisholm who earned her B.A. in 1898 and then went on to get her medical degree, also in Manchester, in 1904.[48] The primary categories of employment for Queen Margaret women graduating between 1894 and 1908 show that 40 percent of employed graduates listed medical occupations including practitioners, assistants, and inspectors, and another 40 percent of women graduates pursued teaching, either in the United Kingdom or abroad. Though lists of graduates do not always indicate it, a significant number of these women would have, at one point or another, practiced medicine or worked in another health-related capacity outside the United Kingdom.

The heightened interest in medical studies in Britain, as opposed to the emphasis on educational careers in the United States, is attributable to an added factor in the United Kingdom—the British Empire. One of the arguments used by women in the establishment of the Queen Margaret College Medical School in 1890 was the need for female doctors in India because it was culturally improper for women there to be "medically examined by men."[49] The historian Wendy Alexander notes in her study of "Early Glasgow Women Medical Graduates" that the humanitarian consideration of women's medical rights in India came "at a crucial juncture" in women's quest for medical education in Britain in the 1870s and 1880s.[50] Another way women wound up working overseas was in government jobs in British colonies. Both men and women could study for the British Civil Service at King's College, London. This included training for telegraph operation, one of the rapidly expanding fields of employment in the Victorian Era.[51]

In some cases working overseas was necessary for women, because people were not willing to hire them to do the same jobs in their home nations. In the medical fields, women were not permitted to practice in many hospitals in the United Kingdom itself. In Edinburgh opponents of women's presence in the medical field "succeeded in shutting them out of the Infirmary," though some professors let them get in some clinic hours when the male students were there.[52] In Glasgow as well, the Western Infirmary near campus was closed to women, though they were able to do clinical practice at the Royal Infirmary in the city center.[53] Just as legislation had been needed to admit them to universities, more legislation would be needed to open local hospitals and infirmaries. Even after such legislation was passed, it was still common for women to work in hospitals that specialized in the care of women and children. One of the University of Edinburgh's most well-known graduates was Dr. Elsie Inglis. When she completed her undergraduate work, she moved on to do her medical training at Glasgow. After working for a time in London and Dublin, she returned to Edinburgh to found a maternity hospital with her former classmate, Jessie MacGregor. She was also active in the Scottish Women's Hospitals that were established in France, Russia, and Serbia during the First World War and were staffed by numerous female university graduates.[54]

One of these women was Agnes Forbes Blackadder (Savill), the first female graduate from St. Andrews.[55] Like Elsie Inglis she chose to study at the Queen Margaret College Medical School in Glasgow. Her Master's thesis was entitled "Six cases of Acute Dilatation of the Heart occurring in Children," a topic that would have been entirely acceptable for a woman to study.[56] In her professional life, along with her time treating wounded soldiers in France, she worked alongside her husband Dr. T. D. Savill in the field of dermatology and was based in London at the Children's Hospital.[57] She also did pioneering work in the use of X-rays, advising "fewer exposures and shorter doses" to prevent illness caused by radiation.[58] Another female medical pioneer was Dora Allman, who earned her medical degree from the Royal University in 1898 after studying at Queen's College, Cork. The following year Dr. Allman became superintendent of the Armagh Mental Hospital and was the first woman in the United Kingdom to become "chief officer" of such a facility. An article printed at the time of her resignation from this post in 1936 noted that when she was a student, women were not able to do their clinical training in any hospital in Cork, so Miss Allman and her friends and supporters began to agitate for a change to the regulations.[59]

At Aberdeen women were most praised for their work as educators, "pursuing the avocations for which they are fitted, the equals of the men in acquirement and skill, their superiors in language, delicacy and grace and in all gentle accomplishments."[60] Their university education therefore enhanced their femininity, while at the same time preparing them to earn a living. In total, twenty-three women earned degrees at Aberdeen by 1900, with two, Miss Jane Ellen Harrison and Miss Anna Swanwick, earning their L.L.D.s in 1895 and 1899, respectively. Miss Harrison became a "classical archaeologist" who taught at Newnham College, Cambridge, and Miss Swanwick an "educationist."[61] Female medical students also made a place for themselves at Aberdeen by the end of the Victorian Era, with the first, Myra Mackenzie, earning her M.B. and Ch.B. in 1900. "Miss Mackenzie" went on to become the "House Physician and Surgeon" at the Royal Aberdeen Hospital for Sick Children, working as well at a children's hospital in Sheffield and spending time on the medical staffs of two other hospitals in England.[62] Like their contemporaries, the women graduates of Aberdeen worked in the community in areas that were not a threat to their femininity.

Teaching and medicine were not the only fields pursued by early women graduates of coeducational universities. Along with becoming the first female member of the English faculty at her alma mater, Indiana University, Sarah P. Morrison also worked as an author, and for a time as a "minister" within the Woman's Christian Temperance Union.[63] Her most well-known books were a two volume set entitled *Among Ourselves: To A Mother's Memory* that were published by the Society of Friends, otherwise known as the Quakers.[64] Her speaking skills were on display at her commencement ceremony. Reports of the occasion described her as "perfectly collected and dignified" and spectators were surprised that she did not need to use a document, but chose instead to be "unincumbered by any manuscript whatever."[65] This impressive command of the material would have made her well suited to speak in front of a religious gathering, as would her appropriate demeanor.

Music was another field that women could enter without a great deal of resistance. Marian Millar, a Manchester graduate, was one of the first women to earn a music degree in England in 1894 and the first to do so at a coeducational university.[66] She was soon followed by Marian Ursula Arkwright at Durham, who received her degree in 1895.[67] Careers for these women could include teaching, but musical composition and performance were also common. In *The Englishwoman's Year Book* for 1899 advertisements were placed for "Ladies' Amateur Bands" and prize competitions for "Organ Composition" and submissions were welcomed for a number of different "Musical Magazines."[68] Marian Millar was noted for writing the libretto of the newly composed "Historic opera, Harold" in 1897.[69] Marian Arkwright worked as a "composer of orchestral and chamber music."[70]

Bookkeeping and stenography were also fields chosen by women in both the United States and the United Kingdom, along with "Merchant" and "Bank" and various types of inspecting work in factories.[71] Although some might see these occupations as being masculine, rather than feminine, an alternative explanation would suggest that they actually do reinforce proper Victorian gender roles. This view is most easily explained through the example of Queen Margaret College secretary Janet Galloway's life, as she like many women of her day, worked for a time keeping the books for her father's business.[72] As a result of family obligation, then, it was possible for women to engage in nontraditional forms of employment. Each of these forms of employment would have been within the bounds of female propriety and would have allowed women to extend themselves to help their families and communities without taking on any dramatically masculine traits.

Community-minded employment was also an extension of the settlement work that many women undertook as students. When they graduated, they could continue working in the same settlement houses or in others, or they could pursue missionary work overseas. At the University of Glasgow only one female graduate reported that her career was settlement work during the period 1894–1908, but twenty women had become missionaries. The fact that this field supported the female gender roles of caring and social consciousness made it desirable to many women. The availability of positions within the British Empire made it easier to find suitable work, though the Y.W.C.A. and various religious denominations in the United States also regularly engaged in international work. University of Michigan president Angell noted in 1882 "Several women who graduated from the Medical Department are in heathen lands discharging the varied and responsible duties of medical missionaries."[73] This type of placement was facilitated by the Student Volunteer Movement for Foreign Missions in the United States that tried "to bring together carefully selected delegations of students and professors from all important institutions of higher learning in North America."[74] In 1898 there were delegates from Indiana, Michigan, Missouri, Ohio, South Carolina, West Virginia, and Wisconsin.[75] All of these associations were used to gain a "better understanding of social problems" that all university graduates would be asked to help correct as the leaders of their generation.[76] Because these groups were open to men and women, having previous experience working in concert with each other while they were students enabled the members of these groups to function more effectively in society.

A final common occurrence for early female graduates was to work on behalf of the further university education of women. At the University of Tennessee female students decided to form the Southern Association of College Women, which was affiliated with the national Association of Collegiate Alumnae. At the Western University of Pennsylvania the five Victorian female graduates formed the Collegiate Alumnae Association in 1898. The group, which would later change its name to the University Alumnae Association, was described in official histories as "for the most part a social affair" with the women usually meeting for luncheon.[77]

Through these organizations and institutional alumni associations, women who graduated from colleges and universities could take on roles as leaders and advocates of higher education for women, along with raising funds for scholarships that would assist students in the future. The social networking possible for female graduates was also a benefit of these groups. Since university-educated women were still a small group within society, they found that the friendships and support provided by others who had shared experiences with them were essential elements of their lives. As the motto of the Alumnae at the Pennsylvania State College stated: "Always Together and Always for the Women of Penn State."[78] This sentiment could be applied to every college and university, as all graduates felt a loyalty not only to their institution but to future generations of students who joined their ranks as alumni. In this way they were also able to bridge the gap between their lives as students and their lives as citizens by working within a familiar university framework to help society at large.

Students and Society

Other developments in universities during the period of this study focus on the decentralization of power and an increase in student agency. In both countries, through either internal (administrative) or external (parliamentary or federal) legislation, the administrative functions of government-supported universities shifted toward the faculty and students. This newly "democratic" aspect of higher education had a tendency to make access to courses better, as many faculty members were intellectually liberal in their support of social equality. A second outcome of this relinquishing of administrative control was the increase it provided in student agency. As the heavy hand of discipline weakened, students began to take an increasing amount of control over their own lives. The expansion of extracurricular activities gave students the opportunity to make leadership decisions with little or no interference from faculty and administrators. The growth of student societies and unions gave an organized voice to student views that had not been possible earlier. The confidence students gained through this process was transferred to their academic careers, with a steady increase in students questioning the education they were receiving. This affected both male and female students, but as women had previously had relatively little access to decision-making processes in society, the impact on their lives was greater than that of their male counterparts.

The final issue to address is the influence of women in universities on greater rights for women. While direct links are hard to defend, it can be argued that the admission of women to higher education helped promote women's enfranchisement and general emancipation from reliance on men.[79] The evolving view of women as being equal to men did make the idea of women's suffrage more palatable to some in the male electorate. This did not mean that there were not still concerns about whether women would be traditional in nature. In a poem from the Aberdeen students' magazine *Alma Mater* called "The Whirling Wheel," a sentimental approach was taken to the women who were no longer spending their time making clothing and instead spent their time outside "her sex's sphere" and otherwise challenging social norms.[80] The author lamented these changes to some extent, while at the same time accepting the progress that was taking place:

> The girl of to-day is of mental mould,
> And traverses man's career;
> She scorns, 'tis said, to be now cajoled
> By the epithets, "darling," "dear;"
> At times she stands as a pamphleteer;
> And learns from Bain what it is to feel:
> But, as for me, I must still revere
> The witching wile of the whirling wheel.[81]

The emphasis placed on the mental development of women, and their intrusion into some men's fields was only part of the problem at hand. What was of greater concern was the scorn some educated women had started to show for traditional endearments that men still wished to call them. This shift in gender relations, if not in gender roles, was another aspect of concern faced by university-educated women.

The initial gains made by women in the nineteenth century through their admission to male universities were not followed quickly by advances in other areas of life. In 1901 the issue of whether women should be allowed to enter the legal profession was raised in Scotland. A primary argument for women's capabilities in this area was the success of her "sisters" in universities, particularly in the field of medicine. "Cockie Law! Is a Girl a Person?" is a poem written at this time, discussing the problem faced by women who felt they should be able to study law. In two stanzas it makes the comparison with women already admitted to universities:

> Like the girls who practise medicine,
> Teach and write, and clerk and draw,
> She aspired to make her living
> From the pickings of the law.
> So she mastered Bell and Rankine,
> Climbing up the hills of brass,
> Till she thought she was a "person"
> Duly qualified to pass.
>
> She had seen her little sisters
> Capped M.A.'s, with applause,
> And she wished to climb life's ladder

As a Bachelor of Laws.
So she asked to be examined,
And to pass, if pass she could,
Forth into the black profession
Of the pleading brotherhood.[82]

The rest of the poem centers on the question of whether a woman was a person equal to a man in the eyes of the law. In this case the woman lost her appeal, and the fight to enter law school would continue for a few more years. This case does shed some light on the progress of the women who *had* gained admission to their chosen education and profession, though. Many areas were still unattainable for women thus the equality they had found was still seriously limited in wider society.[83] Regardless of this, those women who chose to attend coeducational universities made the most of their opportunities.

Despite the small number of women who chose to attend the Queen's Colleges in Ireland, their experience of coeducation was used in the early twentieth century as a key argument in favor of the admission of women to University College, Dublin (UCD), the final holdout in the higher education of women.[84] In 1905 the Irish Association of Women Graduates and Candidate Graduates wrote a memorial to UCD officials asking that lectures be opened to women, citing the success of women at the Queen's Colleges in Belfast, Cork, and Galway.[85] The organization was made up of women who had graduated from the Royal University of Ireland, whether they had attended a Queen's College or not, so there was no inherent bias toward those institutions. The leaders of the association at the time were Isabella Mulvany and Mary Hayden, both of whom were affiliated with Alexandra College, Dublin. There were Queen's College alumna who served as representatives of their "provincial districts," including Agnes Perry for Connaught in 1905.[86]

Just as the arguments made by supporters of women's admission to University College, Dublin used the success of women graduates of the Queen's Colleges as an exemplar, those pushing for Oxford and Cambridge to admit women to degrees used the success of women at the colleges of the University of London in their appeals. In *The Admission of Women to Universities* W. Le Conte Stevens discussed "the success of the experiment at University College, London," saying that it was proof that coeducation was possible in any setting, whether it be a city or a small town.[87] The University of Durham's first female graduate, Ella Bryant, was also held up as an example of why more support should be given to women's education. In 1900 the Iron and Steel Institute was offered a sum of money by Andrew Carnegie to establish a medal and scholarship that would be awarded for student research in any university in connection with the Institute. One of the stipulations to the award was that it be given "irrespective of nationality and sex." To prove that this was a reasonable demand, the "work done by Madame Curie, by Mrs. Ayrton, and by Miss Ella Bryant" was cited as evidence that women could indeed compete for the new honor.[88] The Members of Council of the Institute agreed to Mr. Carnegie's terms and conditions and thanked him for his "munificent gift."

At the University of Michigan the Board of Regents also reported on the success of women's admission to their institution and the social impact it was having, predicting that coeducation would soon become standard practice. "As we have now for thirteen years, without the least embarrassment, admitted women to all the privileges of instruction in the University, we cannot but observe with gratification how rapidly public opinion, both in Europe and America, is coming to approve the granting of substantially the same opportunities for education to women as to men."[89] They believed that their decision to pursue coeducation was not only correct, it was making the community a better place because it raised both men and women to the highest level of educational attainment in a responsible and economically efficient way, unlike single-sex colleges that provided education in the classroom, but not in the ways of the world. That the students had not caused "the least embarrassment" indicated the worries held by some before women were admitted to the institution. The Regents were satisfied, in the end, that society was a better place as a result of university coeducation.

In South Carolina the opening of the state college to women started a trend of university coeducation in the area. Furman University and the Medical College of Charleston soon followed the precedent set, extending "the way of educational light" to more of South Carolina's women. The accomplishments of educated women, like Miss Sallie Allen who graduated first in her class at South Carolina College, similar to those of women in Britain, were used as arguments for further opportunities for women. In South Carolina specifically women were barred from employment as state officials, with the exception of the state librarian, Mrs. Caroline Leconte, and "a few clerical offices."[90] Women's ability and desire to serve as health inspectors or on school boards was noted in 1895 by the president of the Equal Rights Association of South Carolina, Virginia D. Young, who felt that a well-trained woman could do the job as well, if not better, than any man.

Educated women were in a position to transmit their knowledge and understanding to the wider community in which they lived. In Glasgow the Queen Margaret students provided evidence of just such an inclination as they interacted with the community of Glasgow through their Settlement Association. Their ever-widening sphere of influence reached across community and class boundaries and received praise throughout Britain. Mrs. Campbell wrote about the achievements of female graduates in her article, "The Rise of the Higher Education of Women Movement in Glasgow" in *The Book of the Jubilee: In Commemoration of the Ninth Jubilee of the University of Glasgow, 1451–1901*. She said that their accomplishments show that "the Queen Margaret students have earned their position, and that University education does not deteriorate womanhood, but rather lends to it a special grace."[91] The beneficial results of women's higher education did continue to propel the further expansion of women's traditional sphere. As noted by Wendy Alexander, the "separate spheres ideal had itself undergone considerable modification by the early years of the [twentieth] century."[92] Though societal gender roles were still limiting for women, this "modification" did give them an increasing number of choices in their lives. And the confidence gained through studying at a coeducational university, to enter a male-dominated world, made it all the more likely that women would choose the life that they wanted to lead.

Conclusions

Initially women were admitted to various forms of higher education for two main reasons, to train as teachers or to attain a level of cultural refinement. The range of institutions, including women's colleges in America, England, and Ireland and normal colleges throughout Britain and the United States, provided many options for those women who sought and could afford higher education. What these institutions did not provide, however, was the status of a university education and degree which women were prevented from attaining before the mid-nineteenth century. Sarah V. Barnes refers to this as "the presence of the 'dual market' for higher education in the late nineteenth century, combined with the increasing importance attached to university degrees."[93] If women wanted to move into professions and other areas of public life, they would need the education, titles, and respect that only male universities could provide. Society's belief that a woman's place was in the home, raising children, made it difficult for those supporters of women's higher education to prove that equal access was necessary. Detractors of women's higher education remained convinced that any young lady who chose to pursue a university degree would necessarily not marry and have children, and thus negate her primary purpose in society. The students of the Queen Margaret College Literary and Debating Society considered this issue twice between 1899 and 1905: "That it is undesirable for married women to engage in professions." This ever-present consciousness of the educated woman's potential place in society, as opposed to her accepted place, was a key point of concern for female students themselves.[94]

The statistical and anecdotal evidence provided in this chapter highlights several key areas of development in women's quest for equal higher education. The first students to attend and graduate from coeducational universities found work in a wide array of fields. While this may be attributed to the opportunities available in their communities at the time, it is far more likely that the career goals of women who achieved university admission in this period were as ambitious as the idea of attending university at all. In later years, the type of woman who attended university began to change, as did society's expectations of her. Though many women would still pursue careers in male-dominated fields like medicine, the increasing need for teachers and the specific desire to have women hold those positions helped to guide women into education. As the historian Amy McCandless points out in her study of women in southern U.S. universities: "College was to be an enhancing, not a transforming experience, one which would widen a young woman's horizons but not remove her from her separate sphere, one which would harmonize with her nature, not corrupt it."[95] This desire to broaden but not change a woman's place in society was fundamental to the experience women had at coeducational universities in the Victorian Era. In Scotland supporters of women's higher education used similar language, arguing that university training was a tool to make women "more fitted for the duties of life, not only at home, but broadened out to their fellow-creatures."[96] The development of care-giving fields like social work in the early twentieth century, would provide a new outlet for women graduates of universities, who soon realized that they could have a career which was both personally fulfilling and socially acceptable.

CHAPTER NINE

Drawing Conclusions

Image 9 Life Drawing Class at University College, London, 1881, from the *Illustrated London News*. *Credit*: Mary Evans Picture Library, London, England.

The historic differences between institutions in the United States and the United Kingdom caused universities to approach change in a variety of ways. Particularly in Britain's ancient universities, tradition and precedent were major obstacles to the new, more egalitarian views that emerged in the nineteenth century. In the United States, social democracy mixed with economic necessity propelled increased access to universities for various sections of society, though this did not come with the same conflict to tradition. What was common was the desire for all universities to be respected by society and their competitors, and the inclusion of women was thought to cause doubt about an institution's reputation and the potential for "the 'feminization' of the university."[1] Furthermore, the reputation of the female students needed to be protected at all costs. There were "numerous evils" and "anticipations of disaster" predicted by officials, commentators, and students alike, with most fears centering on having men and women in the same classrooms.[2] Some colleges and universities felt it was prudent to keep the sexes apart and control the materials women learned. Even in London, the first of the English universities to admit women to degrees, significant differences existed between the experience of women at King's College and University

College. As pictured on the previous page, "the daring experiment of mixed classes of men and women" included life drawing of a scantily clad subject.[3] The "Life Class" at King's College, on the other hand, had students draw and paint "models in costume" and was for women only.[4]

A common conclusion drawn by historians and commentators when considering university coeducation during the nineteenth century is that older universities were more resistant to change because they were fighting many years of tradition. In England this determination was based on a comparison between Cambridge and Oxford who were "essentially aristocratic, slow of growth and conservative" and the University of London that was "a young institution, unhampered by the prejudices and restrictions of its elders."[5] Durham and Manchester, as institutions with nineteenth-century origins, were similarly unencumbered by history, presumably making their transition to coeducation easier than that at older universities.[6] The fact that women were admitted to the ancient universities in Scotland before Durham and more thoroughly than they were at King's College, London means that such a basic distinction is not supported by the evidence. By broadening the comparison beyond Britain it becomes apparent that decisions to admit women to male universities were driven largely by local needs and demands, rather than national or international arguments about women's rights, with government influence helping to enable change in some instances. The admission of women to colleges and universities was also not done as a means of altering gender roles or characteristics, and every effort was made to ensure that this did not happen. Throughout the period of this study, the wider influence women did gain in society remained in traditionally "feminine" areas. The bending and reshaping of the female world through education was always done within the established boundaries of acceptability.

This book illustrates the contrasts between universities in the United Kingdom and the United States while at the same time highlighting the many similarities present at the institutions included in this study. The comparative aspect of this research sets it apart from many of its predecessors in the field. It attempts to bridge the gaps between the main areas of higher educational historiography by examining more fully the events and experiences at a dozen institutions on each side of the Atlantic. The evolution of society's perceptions of women, their abilities, and their options in terms of education and employment was remarkably consistent regardless of location. In each nation medical, moral, and religious opposition to educating women needed to be overcome if any access were to be granted to universities. The desire for an equivalent education to that of men, while maintaining the reputations of the institution and the female students, perpetuated an amount of academic and social separation within the university. Despite this, women proved themselves to be worthy of meeting the same academic standards of their male counterparts, and increasingly they were accepted as valuable members of the institution.

An answer to the primary theoretical question posed in this book—whether women's education worked with or against prevailing societal beliefs—seems clear. The evidence from both sides of the Atlantic illustrates the differences in university life for male and female students. Additionally, the sources gathered from the student perspective indicate that they themselves wished to have a certain amount of gender separation in their academic and social lives. My proposal that there was a polarization of the curricular and extracurricular college worlds on the basis of gender and that this separation did not evolve in conflict with the prevailing society is crucial to the understanding of this period of higher educational history. Though previous historians of the subject, most notably Carol Dyhouse, have contributed to my assessment of this issue, the depth of material provided in particular case studies is essential to get at the root of the students' own understanding of their place in society. The question posed by Dyhouse in the title of her book—*No Distinction of Sex?*—is answered negatively through her study of British higher education, and this assessment, I believe, is reinforced unequivocally through my research as well.[7] The underlying presence of separate male and female spheres in Victorian society and in the early twentieth century helped to direct the activities of both men and women in universities. Despite the new opportunities given to women, the structure of the institutions and the expectations of family and community continued to guide them into traditionally female roles, particularly those as wives and mothers.

By the end of the Victorian Era, women at many universities were thought to be capable of handling higher education, while others still struggled with full inclusion. Women may have been admitted to the courses and degrees of a university, but that did not always mean that they were welcomed to the institution or that they were seen as equals by their classmates or professors. Some of the challenge came from simple issues like word use. As Edward Sanford noted in a speech to the Alumni Association of the University of Tennessee, the "new and double academic language" caused difficulties for those who were accustomed to referring only to male students. Sanford continued, "[I]f, in what I have yet to say, I shall fail to make mention of the young women every time I speak of the young man, I can only" apologize.[8] This was not an isolated concern as the "gendered rhetoric" used by officials and commentators was common in all nations. Judith Harford notes the "terms 'fair sex' and 'gentler sex' were used throughout the period" as distinction was made between men and women in universities, and deference continued to be paid to women in society.[9]

In the four Scottish universities in Aberdeen, Edinburgh, Glasgow, and St. Andrews issues of mixing with male students and the desire of women to have a continued separation from them would persist into the twentieth century. One area where sex segregation would become more distinct was in the student unions built at the universities around the turn of the century. At Aberdeen a union was built at Marischal College that included "a debating hall, a concert room, a luncheon room, a billiard and smoking room," and an art gallery, while a "ladies' room and a pavilion" were added at King's College to accommodate women.[10] Similarly, the Glasgow Union contained "a Common Room, a Library, a Debating Hall, Committee Rooms, and a Refectory," but was not opened to the women of the Queen Margaret after their acceptance to the university. They would finally get their own building in 1908.[11] Women at St. Andrews would not have their own union until 1904 when Mrs. Andrew Carnegie donated money for one, and in Edinburgh women would get their own building in 1905.[12] As a result of this physical separation between men and women at the institutions, women's full inclusion would not happen for several generations, but this was due to the women's own choice as much as anything. Ultimately the most important issue for the women was to have a choice in their university experience as they worked toward the same degrees as men, which meant that they had accomplished their initial goal.

The decision to maintain separate spheres for male and female students highlights that another key aspect for the process of inclusion to be complete for women who were admitted to male universities was for the women themselves to feel at home there. Resentment directed at female students by male students would limit the likelihood that coeducation would be deemed a success during the nineteenth century. Because the resentment was often handled in a subtle or subversive manner, it is not always possible to determine whether the perception that male students did not want female students to attend an institution was based in fact. Male students who were vehemently opposed to the idea of coeducation may simply have attended one of the many all-male colleges or universities that still existed at the time. At Michigan President Angell did not feel that men avoided coming to the university because it was coeducational, although he had no evidence to support his assertion. He also declared that the "presence of women has not lessened 'college feeling' or *esprit de corps*, if we use those terms in their worthy sense."[13] Angell's statements were backed up in 1889 when *The Cosmopolitan* magazine ran a profile of "Student Life in the University of Michigan" which reported on a smooth inclusion of women students since their admission in 1870: "No one ever looks upon the girls now with curiosity, as they did in the early days; no one ever discusses their rights in the University, for long ago they were firmly established."[14] Importantly, in the long run, the attachment women felt to their alma maters was no less strong than that felt by men. Margaret Boyd wrote in her diary with a sentimental sadness that her time as a student would soon be over when she graduated from Ohio University, referring to "Many a pleasant hour have I spent within its walls."[15]

At the other universities in the midwestern states of Indiana, Missouri, and Wisconsin, and the eastern state of Pennsylvania, a similar sense of inclusion was reported by the end of the Victorian Era. The inclusion of women in universities did not bring the catastrophic problems that had been predicted at some of these institutions, and in general it was "demonstrated that bringing the sexes together for educational purposes stimulates both to closer study and to more careful deportment."[16] The benefit to developing states in having more university-educated people could also not be ignored, even by those who did not support coeducation. The fact that all teachers, both male and female, were able to receive university training increased public "confidence in their fitness to give instruction in secondary schools."[17] This made the states more attractive to migrants and could bring other types of development to them as well. For those advocates who were fighting for greater legal and political rights for women, the ability of women to handle university education, in the same classrooms with men, was an effective argument for increased inclusion of women in the public sphere of society. May Wright Sewall, writing about the "Position of Women in Industry and Education in the State of Indiana," concluded that university coeducation was the culmination of a "long pending struggle and the welcome victory which has lifted girls and women a little nearer to that equality of opportunity toward which the womanhood of this century so painfully strives."[18] Even though many of the early supporters of women's admission to male universities saw that goal as an important end in itself, the bigger picture for others turned it into an even more profound accomplishment for womankind.

Administrators expressed their pride in the women who graduated from their institutions and "secured conspicuous positions" at other colleges and universities. These high profile graduates and their successes would reflect well on the institution and "the wisdom of the Regents, who opened to them the opportunities for a thorough collegiate training." The women, like their male classmates, were "doing their full part in winning a reputation for the University."[19] University officials were not the only ones to see the success of female graduates as being a positive addition to society. In 1876, Ella Boyd (Margaret's niece) was one of the speakers that year at commencement, along with Ohio Governor Rutherford B. Hayes who had recently been nominated by the Republican Party as their candidate for the presidency. Hayes was so struck by Miss Boyd's speech that he commented on her during his, saying she represented "the thought qualities which lend success to every single life"; a statement which reportedly received much applause from the crowd.[20] Women need not be held to a different standard than men, as they could all become leaders of society.

In Alabama, Mississippi, South Carolina, Tennessee, and to some extent West Virginia, changes came more "slowly in the crystalized society of the South than in the newer western states."[21] The education of women at the University of Alabama was featured as one of the sketches performed by students during the Centennial Celebrations in 1931. The moment in time focused on was the opening of the Julia S. Tutwiler Annex, which provided housing for women on campus for the first time. The story included a discussion of "certain regulations for young women" that "the faculty found necessary to pass for your guidance."[22] These regulations focused mostly on conduct while on campus, but did include suggestions that women never walk alone, or with young men, when in the city of Tuscaloosa. The fact that the university's officials did not want their male and female students "to appear in public" together was an attempt by the university to maintain their own reputation just as much as the reputations of the students. Another interesting aspect to the story was that the female students were told that their presence at Alabama was still an experiment. They were warned that their behavior would be scrutinized, and even the "slightest imprudence committed by one of the young lady pupils—even with the most innocent intentions—would put an end to the movement, certainly for many years, perhaps for ever."[23] This seems overly dramatic, but it did reflect the persistent concerns of over university coeducation.

The Irish and Welsh experiences were more limited, because the enrollments of the four institutions in those nations were small for both men and women. Judith Harford concluded that the admission of women to the Queen's Colleges in Belfast, Cork, and Galway "was largely symbolic" because so few chose to attend, preferring women's colleges or religiously affiliated ones.[24] This statement has merit, but also unduly minimizes the experience of the women who did study at the institutions during the Victorian Era. Assessments of Aberystwyth's "experiment of mixed education" are more enthusiastic and highlight the fact that "men and women, staff and students alike, meet in class-rooms, share in debates, entertainments, concerts, with excellent results."[25] One can argue that the small size of these institutions kept serious opposition to women's presence at bay. Certainly the smaller numbers of female students would have minimized the threat posed by them to the college community.

The expansion of higher education during the nineteenth century and the increasing inclusion of women was a process that would continue well into the twentieth century at many of the institutions discussed in this book and at other male universities that began admitting women after 1900. Angie Warren Perkins, the acting dean of the Woman's Department at Tennessee at the turn of the century, concluded in "the transitional period of the last quarter of a century old educational methods have fallen into disuse; some new ones have been established; and others are in progress of abandonment or adoption."[26] The ongoing process at the first dozen male institutions in the United States to admit women would be used as a model elsewhere as pressure to become coeducational intensified. The full inclusion of female students at some coeducational institutions was also yet to be accomplished. As noted by the historian Cynthia Eagle Russett, it would not be until after the First World War that

> higher education for women and coeducation of the sexes were no longer controversial issues; they had become part of the birthright of the middle class. Educators were no longer prone to propose tailoring women's education to their intellectual mediocrity.[27]

The Victorians who had opposed women's higher education in the United States and the United Kingdom were proven wrong by the eventual successes of those women who entered university. As George Romanes concluded in 1887, the decision to admit women to higher education relied on the support of society to ensure its success. The evolution of thought and implementation of the coeducational systems at the universities in this study, though uneven at times, helped to lead other institutions in their progress toward equal education of the sexes. As can be seen through the responses of these universities, coeducation proved to be a successful mode of education for both men and women. Indeed, each academic community benefited in their choice to "give her the apple, and see what comes of it."[28]

NOTES

One: The Process of Inclusion

1. Many colleges and universities that are comprehensive in the twenty-first century provided a more limited curriculum in the nineteenth century. While the education provided at institutions that focused on agriculture and mechanical subjects or at normal schools that provided teacher training is extremely important to the overall history of higher education, it is not the purpose of this book. For more on these institutions see Christine A. Ogren, *The American State Normal School: "An Instrument of Great Good"* (New York: Palgrave Macmillan, 2005).
2. Patricia Albjerg Graham, "Expansion and Exclusion: A History of Women in American Higher Education," *Signs* 3 (Summer 1978), 759. The classic history of women's educational history in the United States is Thomas K. Woody, *A History of Women's Education in the United States*, 2 vols. (New York: Octagon Books, 1929). Though out-of-date in many respects, it is still valuable to look at.
3. Historians have contested the concept of separate spheres in recent years. This debate will be addressed fully in Chapter Two.
4. GUA, Volume of Presscuttings of Queen Margaret College, 1891–1894, "Queen Margaret College—Opening of Session," *North British Daily Mail* (November 3, 1891).
5. Graham, "Expansion and Exclusion," 760, and Mary Caroline Crawford, *The College Girl of America and the Institutions Which Make Her What She Is* (Boston: L. C. Page & Company, 1905), 258. See also Christine A. Ogren, "Where Coeds Were Coeducated: Normal Schools in Wisconsin, 1870–1920," *History of Education Quarterly* 25, 1 (Spring 1995), 21; and Konrad H. Jarausch, *The Transformation of Higher Learning, 1860–1930: Expansion, Diversification, Social Opening, and Professionalization in England, Germany, Russia, and the United States* (Chicago: University of Chicago Press, 1983).
6. Isaac N. Demmon, ed., *History of the University of Michigan, by the Late Burke A. Hinsdale, with Biographical Sketches of the Regents and Members of the University Senate from 1837 to 1906* (Ann Arbor, MI: Published by the University, 1906), 59.
7. *Report of the Commissioner of Education for the year 1894–95 Volume 1* (Washington, DC: Government Printing Office, 1896), 907.
8. A. Wallis Myers, "Women Students in Wales," *The Ludgate Illustrated Magazine Vol. VIII* (London: F. V. White & Co., 1899), 138.
9. "Girl Graduates," *The Students' Journal and Hospital Gazette* (August 30, 1879), 207.
10. Trinity College was granted a charter by Elizabeth I. Carol J. Summerfield and Mary Elizabeth Devine, eds., *International Dictionary of University Histories* (London: Taylor & Francis, 1988), 480.
11. For more see Rachel Holmes, *The Secret Life of Dr. James Barry: Victorian England's Most Eminent Surgeon* (Stroud, Gloucestershire: Tempus Publishing, 2007).
12. Emily Davies, "Women in the Universities," in *The Educators: Female Education*, ed. Marie Mulvey Roberts and Tamae Mizuta (London: Routledge/Thoemmes, 1995), 191, 193. See also Gillian Sutherland, "The Movement for the Higher Education of Women: Its Social and Intellectual Context in England, c. 1840–80," in *Politics and Social Change in Modern Britain*, ed. P. J. Waller (Brighton, Sussex: Prentice Hall/Harvester Wheatsheaf, 1987), 44.
13. Six institutions in particular are referred to as "red-brick" universities. These are the Universities of Birmingham, Bristol, Leeds, Liverpool, Manchester, and Sheffield. For more

see Edgar Allison Peers, *Redbrick University Revisited* (Liverpool: Liverpool University Press, 1996).

14. Sarah V. Barnes, "Crossing the Invisible Line: Establishing Co-education at the University of Manchester and Northwestern University," *History of Education* 23, 1 (1994), 38. Other institutions, like University College, Bristol, were coeducational from their foundation; thus the question of adding women to an existing male structure was a non-issue.
15. There is some evidence that women were admitted to lectures at the University of London as early as the 1840s, though more research needs to be done on this front. For more see Sarah J. Smith, "Retaking the Register: Women's Higher Education in Glasgow and Beyond, c. 1796–1845," *Gender & History* 12, 2 (2000), 310–335.
16. John M. Hall, *England: An Account of Past and Contemporary Conditions and Progress* (Detroit: Bay View Reading Club, 1906), 125. See also Barnes, "Crossing the Invisible Line," 41.
17. *British Universities: Notes and Summaries Contributed to the Welsh University Discussion by Members of the Senate of the University College of North Wales* (Manchester: J. E. Cornish, 1892), 50.
18. Katharine Lake, ed., *Memorials of William Charles Lake, Dean of Durham 1869–1894* (London: Edward Arnold, 1901), 129.
19. Elizabeth Cady Stanton, Susan B. Anthony, and Matilda Joslyn Gage, eds., *History of Woman Suffrage, Vol. III 1876–1885* (Rochester, NY: Charles Mann, 1887), 982.
20. J. T. Fowler, *Durham University: Earlier Foundations and Present Colleges* (London: F. E. Robinson, 1904), 119. George William Kitchin had previously been the dean of Winchester from 1883 to 1894 and was made an honorary student of Christ Church College, Oxford in 1896. For more see *Who's Who, 1901: An Annual Biographical Dictionary* (London: Adam & Charles Black, 1901), 655.
21. Lake, *Memorials of William Charles Lake,* 129.
22. Emily Janes, *The Englishwoman's Year Book and Directory 1900* (London: Adam and Charles Black, 1900), 5.
23. Harold Silver and John S. Teague, *The History of British Universities 1800–1969, excluding Oxford and Cambridge: A Bibliography* (London: Society for Research into Higher Education, 1970), 136, 153, 166, 172, 178. For more on all of London's constituent colleges see Negley B. Harte, *The University of London, 1836–1986: An Illustrated History* (London: Athlone, 1986), and F. M. L. Thompson, *The University of London and the World of Learning, 1836–1986* (London: Hambledon Press, 1990).
24. *The Calendar of King's College, London for 1896–97* (London: Published by the College, 1896), 274, 402.
25. Maria G. Grey, "The Women's Educational Movement," in *The Woman Question in Europe: A Series of Original Essays*, ed. Theodore Stanton (New York, London, and Paris: G. P. Putnam's Sons, 1884), 60. See also W. Le Conte Stevens, *The Admission of Women to Universities* (New York: Press of S. W. Green's Son, 1883), 22–25.
26. Alice Zimmern, *The Renaissance of Girls' Education in England: A Record of Fifty Years' Progress* (London: A. D. Innes & Company, 1898), 134.
27. Grey, "The Women's Educational Movement," 60.
28. Joseph Thompson, *The Owens College: Its Foundation and Growth; and its Connection with the Victoria University, Manchester* (Manchester: J. E. Cornish, 1886), 31–32, 44–45, 138. The establishment date cited by the twenty-first-century University of Manchester is 1824, using as its opening date the founding of the Mechanics' Institute in the city, which merged with the Victoria University of Manchester in 2004. For more see H. B. Charlton, *Portrait of a University 1851–1951* (Manchester: Manchester University Press, 1951), 13–15, and Jamil Salmi, *The Challenge of Establishing World Class Universities* (Washington, DC: World Bank, 2009), 43–44.
29. Ibid., 120, 122.
30. Ibid., 145–146, 211.
31. Ibid., 387–388, 405, 416–417, and William Jack, "The New English University," in *Macmillan's Magazine Vol. XLIII November 1880, to April 1881* (London: Macmillan and Co., 1881), 112–113.
32. Thompson, *The Owens College*, 492.
33. Ibid., 495–498, and Barnes, "Crossing the Invisible Line," 39–40.
34. Thompson, *The Owens College*, 511, 517, 533, 536–550, and J. E. G. De Montmorency, *The Progress of Education in England: A Sketch of the Development of English Educational Organization from Early Times to the Year 1904* (London: Knight & Co., 1904), 266. A supplemental charter

was granted in 1883 to give the university the power to confer degrees in medicine. For more on these institutions see A. N. Shimmin, *The University of Leeds: The First Half-Century* (London: Cambridge University Press, 1955), 11–29, and David R. Jones, *The Origins of Civic Universities: Manchester, Leeds and Liverpool* (London: Routledge, 1988). The institutions would go their separate ways after another Act of Parliament in 1903 that made them each individual universities.

35. Mabel Tylecote, *The Education of Women at Manchester University 1883 to 1933* (Manchester: Manchester University Press, 1941), 17–52. See also Barnes, "Crossing the Invisible Line," 43, 45–46.
36. It was at this point that the Republic of Ireland and Northern Ireland became separate political entities. For the purposes of this study, it is not important to explain the split, but those interested can read Tim Pat Coogan, *Ireland in the Twentieth Century* (New York: Palgrave Macmillan, 2006).
37. Thomas A. Boylan and Timothy P. Foley, *Political Economy and Colonial Ireland: The Propagation and Ideological Function of Economic Discourse in the Nineteenth Century* (London: Routledge, 2002), 49, and Tadhg Foley, *From Queen's College to National University: Essays on the Academic History of QCG/UCG/NUI, Galway* (Dublin: Four Courts Press, 1999), 16–21.
38. Robert Esler, *Guide to Belfast, The Giant's Causeway, and the North of Ireland* (Belfast: Wm. Strain & Sons, 1884), 23, and F. S. Dumaresq de Carteret-Bisson, *Our Schools and Colleges Vol. II: For Girls* (London: Simpkin, Marshall & Co., 1884), 203–204.
39. *The Public General Statutes Passed in the Forty-Fourth and Forty-Fifth Years of the Reign of Her Majesty Queen Victoria, 1881: With a Copious Index, Tables, &c.* (London: George Edward Eyre and William Spottiswoode, 1881), 211–212.
40. Esler, *Guide to Belfast*, 23.
41. Susan M. Parkes and Judith Harford, "Women and Higher Education in Ireland," in *Female Education in Ireland 1700–1900: Minerva or Madonna*, ed. Deirdre Raftery and Susan M. Parkes (Dublin and Portland, OR: Irish Academic Press, 2007), 123.
42. Justin McCarthy, *A Short History of Our Own Times from the Accession of Queen Victoria to the General Election of 1880 in Two Volumes, Vol. I* (New York: Frederick A. Stokes & Brother, 1888), 94, and Judith Harford, *The Opening of University Education to Women in Ireland* (Dublin and Portland, OR: Irish Academic Press, 2008), 77.
43. "Irish Education," *Anglo-Celt*, June 8, 1867, 2.
44. Harford, *The Opening of University Education to Women in Ireland*, 77.
45. "County Items," *The Nation*, November 4, 1876, 2.
46. QUB, Earliest surviving photograph of Queen's College, Belfast staff and students, c. 1886, and "Irish Education," *Anglo-Celt*, June 8, 1867, 2. See also Eibhlin Breathnach, "Women and Higher Education in Ireland (1879–1914)," in *The Irish Women's History Reader*, ed. Alan Hayes and Diane Urquhart (London: Routledge, 2001), 46.
47. "Irish Education," *Anglo-Celt*, June 8, 1867, 2. See also Parkes and Harford, "Women and Higher Education in Ireland," 105, and Harford, *The Opening of University Education to Women in Ireland*, 77.
48. *Report of the Board of Curators of the State University of the State of Missouri to the XXXIst General Assembly* (Jefferson City, MO: Tribune Printing Company, 1881), 14.
49. Logan Esarey, *A History of Indiana* (New York: Harcourt, Brace and Company, 1922), 108–110, and James Albert Woodburn, *Higher Education in Indiana* (Washington, DC: Government Printing Office, 1891), 80–81.
50. Ivy Leone Chamness, "Indiana University," *The Lyre* 25, 3 (April 1922), 260. See also Theophilus A. Wylie, *Indiana University, Its History from 1820, When Founded, to 1890, with Biographical Sketches of Its Presidents, Professors and Graduates, and a List of Its Students from 1820 to 1887* (Indianapolis, IN: Wm. B. Burford, 1890), 111–112.
51. Samuel Bannister Harding, *Indiana University, 1820–1904: Historical Sketch, Development of the Course of Instruction, Bibliography* (Bloomington: Indiana University, 1904), 17.
52. May Wright Sewall, "The Education of Woman in the Western States," in *Woman's Work in America*, ed. Annie Nathan Meyer (New York: Henry Holt and Company, 1891), 72. For more on Isaac Jenkinson see Ivy Chamness and Burton D. Myers, eds., *Trustees and Officers of Indiana University 1820 to 1950* (Bloomington: Indiana University, 1951), 269–272.
53. "The First 'Coeds,'" in *Indiana University Alumni Quarterly Vol. IX—1922* (Indianapolis, IN: C. E. Pauley and Co., 1922), 216, and Wylie, *Indiana University, Its History from 1820*, 75.

54. Harding, *Indiana University, 1820–1904*, 176.
55. Grace Smith, "Indiana University," *Kappa Alpha Theta* 17, 2 (January 1903), 105.
56. Andrew Cunningham McLaughlin, *History of Higher Education in Michigan* (Washington, DC: Government Printing Office, 1891), 67.
57. W. L. Smith, *Historical Sketches of Education in Michigan* (Lansing, MI: W. S. George & Co., 1881), 73.
58. Demmon, *History of the University of Michigan*, 132, and Wilfred B. Shaw, *A Short History of the University of Michigan* (Ann Arbor, MI: George Wahr, 1937), 49.
59. McLaughlin, *History of Higher Education in Michigan*, 67.
60. *The President's Report to the Board of Regents for the Fiscal Year Ending June 30, 1870* (Ann Arbor, MI: Published by the University, 1870), 22. Haven resigned to become president of Northwestern in 1869. For more see Ruth Bordin, *The University of Michigan: A Pictorial History* (Ann Arbor: The University of Michigan Press, 1967), 13.
61. H. B. Hutchins, "The University and Co-Education," *Michigan Alumnus* 17, 160 (January 1911), 182.
62. Bordin, *The University of Michigan: A Pictorial History*, 12–13.
63. James and Vera Olson, *The University of Missouri: An Illustrated History* (Columbia: University of Missouri Press, 1988), vii, 3, 5.
64. Marshall S. Snow, *Higher Education in Missouri* (Washington, DC: Government Printing Office, 1898), 40, and John E. Swanger, comp., *Official Manual of the State of Missouri for the Years 1907–1908* (Jefferson City, MO: Hugh Stephens Printing Company, 1907), 231.
65. Charlotte Wronker, "Co-Education in the 'Varsity," *The Missouri Alumni Quarterly*, (December 1905), 28. The prohibition on attending chapel was lifted in 1870, at which point the women were relegated to the gallery and the men sat on the main floor. For more see Olson, *The University of Missouri*, 13–14.
66. Ibid., 28.
67. "Woman at Wisconsin: A Chronology," *The Wisconsin Magazine*, (March 1916), frontispiece. See also Ogren, "Where Coeds Were Coeducated," 2, 5, and Jean Rasmusen Droste, "Women at Wisconsin," (M.A. thesis, University of Wisconsin, 1967), 21.
68. Helen R. Olin, *The Women of a State University: An Illustration of the Working of Coeducation in the Middle West* (New York and London: G. P. Putnam, 1909), 4–5.
69. Ibid.
70. UWA, *Minutes of the Board of Regents* (1866 through 1876), 103, 118, 163, 180, 194–198, 201. Chadbourne was followed in the position by John H. Twombly. Then, in 1874, Twombly's tenure was abruptly terminated due to "irreconcilable differences of opinion," and he was replaced by John Bascom. Pages 196–197 of the *Regents' Minutes* detail the firing of President Twombly due originally to "incompetency…he possessing neither the learning to teach, the capacity to govern, or the wisdom to direct." They subsequently amended and softened this statement greatly.
71. Henrietta Wood Kessenich, " 'Twas Long, Long Ago," *Wisconsin Alumnus*, (1938), 306–309.
72. Thomas N. Hoover, *The History of Ohio University* (Athens: Ohio University Press, 1954), 1, 19, and James J. Burns, *Educational History of Ohio: A History of its Progress Since the Formation of the State Together with the Portraits and Biographies of Past and Present State Officials* (Columbus, OH: Historical Publishing Co., 1905), 358. See also Charles M. Harvey, "A Hundred Years of Ohio," *The World's Work: A History of Our Time*, 5, (November 1902 to April 1903), 3237.
73. George W. Knight and John R. Commons, *The History of Higher Education in Ohio* (Washington, DC: Government Printing Office, 1891), 13.
74. Hoover, *The History of Ohio University*, 37, 142–143.
75. "Educational Intelligence," *The Ohio Educational Monthly; Organ of the Ohio Teachers' Association and The National Teacher* 5, 7 (July 1880), 234. This increase in income was followed in 1881 when the state legislature appropriated "$20,000 for repairing the buildings of Ohio University." For more see Hoover, *The History of Ohio University*, 147.
76. *Annual Catalogue of the Ohio University 1885* (Athens, OH: Published by the University, 1885), 75.
77. Hoover, *The History of Ohio University*, 141–142. President Scott was also an alumnus of Ohio University.
78. William D. Fulton, *Ohio General Statistics for the Fiscal Year Commencing July 1, 1917, and Ending June 30, 1918 Volume IV* (Springfield, OH: Springfield Publishing Company, 1919), 528.

79. OUMC, Margaret Boyd Diary (1873), 4, in Ohio Memory: An Online Scrapbook of Ohio History, www.ohiomemory.org/index.html (accessed June 21, 2009). The diary was given to her by her sister Kate with the provision that she write in it every day so that in "after years" she could look back on her time as a university student.
80. Ibid., 7.
81. Oberlin opened in 1833.
82. *The Pennsylvania State College Alumni Directory, 1861–1935* (State College, PA.: Penn State Alumni Association, 1935), vii, and Michael Bezilla, *Penn State: An Illustrated History* (University Park and London: The Pennsylvania State University Press, 1985), 21.
83. Bezilla, *Penn State,* 4.
84. Ibid., 11–12.
85. Ibid., 35, 54–55.
86. Harriet A. McElwain, "Ladies' Department," in *Report of the Pennsylvania State College, for the year 1888* (Harrisburg, PA: Edwin K. Meyers, 1889), 45, and Bezilla, *Penn State,* 38.
87. Harriet A. McElwain, "Ladies' Department," in *Annual Report of the Pennsylvania State College, for the year 1894* ([Harrisburg, PA]: Clarence M. Busch, 1895), 148.
88. George M. P. Baird, "Fragments of University of Pittsburgh Alumni History," *Western Pennsylvania Historical Magazine* 1, 1 (January 1918), 133, and Agnes Lynn Starrett, *Through One Hundred and Fifty Years: The University of Pittsburgh* (Pittsburgh: University of Pittsburgh Press, 1937), 60, 213, 219–223.
89. *Catalogue of the Western University of Pennsylvania for the year ending 1896 with detailed statements of the courses of instruction* ([Pittsburgh]: Western University of Pennsylvania, 1896), 5, 18.
90. Starrett, *Through One Hundred and Fifty Years,* 189–190.
91. William Jacob Holland, *History of the University of Pittsburgh* (Pittsburgh: University of Pittsburgh, Digital Research Library, 2006), 19.
92. "Editorials," *The Western University Courant* 11, 1 (September 1895), 1.
93. *Catalogue of the Western University of Pennsylvania for the year ending 1896,* 18, and Holland, *History of the University of Pittsburgh,* 18–19. See also Robert C. Alberts, *Pitt: The Story of the University of Pittsburgh, 1787–1987* (Pittsburgh: University of Pittsburgh Press, 1986), 41–43, and Starrett, *Through One Hundred and Fifty Years,* 203.
94. W. H. Venable, *The Beginnings of Literary Culture in the Ohio Valley: Historical and Biographical Sketches* (Cincinnati: Robert Clarke & Co., 1891).
95. James Morton Callahan, *Semi-Centennial History of West Virginia* ([Charleston]: Semi-Centennial Commission of West Virginia, 1913), 229–232.
96. *West Virginia University, Morgantown, Catalogue 1889–90: Announcements for 1890–91* (Charleston, WV: Moses W. Donnally, 1890), 8. For more see *Catalogue of the Officers and Students of West Virginia University for the Year 1872–73* (Morgantown, WV: Morgan & Hoffman, 1873), 13–22, and William T. Doherty, Jr. and Festus P. Summers, *West Virginia University: Symbol of Unity In a Sectionalized State* (Morgantown: West Virginia University Press, 1982), 4–13.
97. Doherty and Summers, *West Virginia University,* 13. After the Morgantown Female Seminary burned down in 1889, the need for women to attend the university was heightened. For more see "Women's Rules and Regulations at WVU–Pre–World War II," *WVU Women: The First Century* (Morgantown: WVU Women's Centenary Project, West Virginia University, 1989).
98. W. P. Willey, "West Virginia's Wrong to Womankind," *The West Virginia School Journal* 7, 6 (June 1888): 6, and Geo. W. Atkinson and Alvaro F. Gibbens, *Prominent Men of West Virginia: Biographical Sketches of Representative Men in Every Honorable Vocation, including Politics, the Law, Theology, Medicine, Education, Finance, Journalism, Trade, Commerce and Agriculture* (Wheeling, WV: W. L. Callin, 1890), 550.
99. Willey, "West Virginia's Wrong to Womankind," 7.
100. The state Senate voted down a resolution on coeducation in 1889, deciding that the Board of Regents had the responsibility under the terms of the university's charter. It is interesting to note that separating from the Commonwealth of Virginia was one of the best things that could have happened for the progress of university coeducation in West Virginia. Women would not be admitted to the College of William and Mary in Williamsburg until 1918, and the University of Virginia would not admit women in all fields of undergraduate study until 1970. For more see Doherty and Summers, *West Virginia University,* 45; Amy Thompson McCandless, "Maintaining the Spirit and Tone of Robust Manliness: The Battle against Coeducation at Southern Colleges and Universities, 1890–1940," *NWSA Journal* 2,

2 (Spring 1990), 199–216; and Annabel Wharton, "Gender, Architecture, and Institutional Self-Preservation: The Case of Duke University," *South Atlantic Quarterly* 90, 1 (Winter 1991), 182, 194–197.

101. *The Universities (Scotland) Act, 1889 together with Ordinances of the Commissioners under the said Act, with relative Regulations & Declarations and University Court Ordinances made and approved subsequent to the expiry of the Powers of the Commissioners. With an Appendix containing the Universities (Scotland) Act, 1858* (Glasgow, 1915), xviii.
102. *The Aberdeen University Calendar Part I* (Aberdeen: A. King & Co., 1898), 1, and John Malcolm Bulloch, *University Centenary Ceremonies* (Aberdeen, 1893), v. See also William Robbie, *Aberdeen Its Traditions and History* (Aberdeen: D. Wylie & Son, 1893), 165–166, 436–437, and *Aurora Borealis Academia: Aberdeen University Appreciations 1860–1889* (Aberdeen: University Printers, 1899), ix, 1, 6. See also Robert Sangster Rait, *The Universities of Aberdeen: A History* (Aberdeen: James Gordon Bisset, 1895), 338–353.
103. John Malcolm Bulloch, *A History of the University of Aberdeen 1495–1895* (London: Hodder and Stoughton, 1895), 209. See also William Watt, *A History of Aberdeen and Banff* (Edinburgh and London: William Blackwood and Sons, 1900), 386.
104. Ja. F. Kellan Johnstone, "Tuesday, 25th September," in *Record of the Celebration of the Quatercentenary of the University of Aberdeen*, ed. P. J. Anderson (Aberdeen: Aberdeen University Press, 1907), 70. The same observer noted that the women walked "quietly and proudly... with brighter eyes and more pleasant and happier expression than their brother students, as if conscious that their welcome presence marks the early stage of a new and hopeful epoch in the life of the University."
105. Mary A. Marshall, "Medicine as a Profession for Women," in *The Woman's World: Volume I*, ed. Oscar Wilde (London, Paris, and Melbourne: Cassell & Company, 1888 and London: Source Book Press, 1970), 106. See also Alice Horlock Bennett, *English Medical Women: Glimpses of Their Work in War and Peace* (London, Bath, New York, and Melbourne: Sir Isaac Pitman & Sons, 1915), 10–19.
106. *The Edinburgh University Calendar 1871–72* (Edinburgh: Edward Ravenscroft, 1871), 147–148.
107. Marshall, "Medicine as a Profession for Women," 106.
108. Mrs. Campbell, "The Rise of the Higher Education of Women Movement," in *The Book of the Jubilee: In Commemoration of the Ninth Jubilee of the University of Glasgow, 1451–1901*, ed. Students' Jubilee Celebrations Committee (Glasgow: James Maclehose and Co., 1901), 131.
109. Ibid., 134.
110. Catherline Mary Kendall, "The Queen Margaret Settlement 1897–1914: Glasgow Women Pioneers in Social Work" (M.A. thesis, University of Glasgow, 1993), 40.
111. James Coutts, *A History of the University of Glasgow: From its Foundation in 1451 to 1909* (Glasgow: J. Maclehose and Sons, 1909), 458. See also Sheila Hamilton, "Women and the Scottish Universities circa 1869–1939: A Social History" (Ph.D. thesis, University of Edinburgh, 1987), 114. Mrs. Elder made one major stipulation along with the gift; that the college could not have the deed until they had raised a £20,000 endowment. One bit of irony is attached to the previous owner of North Park House. Mrs. Elder purchased the property from John Bell of Bell's Pottery, who was a reputed misogynist. He lived with his brother, and neither allowed women enter the premises. For more see Campbell, "The Rise of the Higher Education of Women Movement," 135, and Johanna Geyer-Kordesch and Rona Ferguson, *Blue Stockings, Black Gowns, White Coats: A Brief History of Women Entering the Medical Profession in Scotland in Celebration of One Hundred Years of Women Graduates at the University of Glasgow* (Glasgow: University of Glasgow, Wellcome Unit for the History of Medicine, 1994), 44.
112. Campbell, "The Rise of the Higher Education of Women Movement," 136. Geyer-Kordesch and Ferguson, *Blue Stockings, Black Gowns, White Coats*, 44. See also GUA, Frances Melville, "Presentation Address," On the occasion of the first award of the Frances Melville Medal in Philosophy on the final closure of the College (November 1935), 2, and GUA, Olive Checkland, "Women in Glasgow University: Queen Margaret's College, Hall, Settlement and Union," (Typescript. July 1979), 3. The Medical School was housed in the basement kitchen of North Park House, and Mrs. Elder agreed to cover the expenses for two years. A medical college building followed in 1895 for lectures on the campus. Mrs. Elder also endowed a chair of naval architecture for £12,500 in 1883 and gave £5,000 toward the

endowed chair of engineering the same year. See Coutts, *A History of the University of Glasgow*, 449, and GUA, Booklet of views of interior, exterior and grounds of Queen Margaret College (6 copies—some negatives), n.d.

113. A. L. Brown and Michael Moss, *The University of Glasgow: 1451–1996* (Edinburgh: Edinburgh University Press, 1996), 35.
114. James Maitland Anderson, ed., *The Matriculation Roll of the University of St. Andrews 1747–1897* (Edinburgh and London: William Blackwood and Sons, 1905), xi–xiii, xvi, xxvi, 296–302. Although women were not prevented from entering St. Mary's College, they did not do so initially. See also Isabel Maddison, ed., *Handbook of Courses Open to Women in British, Continental and Canadian Universities* (New York: The Macmillan Company, 1896), 108.
115. Ibid., 135. Miss Garrett was listed as studying "Anatomy, etc." See also Jo Manton, *Elizabeth Garrett Anderson* (London: Methuen, 1987).
116. Ibid., lxvii.
117. Miss Mary Ann Baxter and John Boyd Baxter, *Deed of Endowment & Trust of the University College, Dundee* (Dundee: John Leng & Co., 1882), 8. The Deed of Endowment stipulated that the college would be "for promoting the education of persons of both sexes." See also "Recent Removals," in *The United Presbyterian Magazine Vol. II* (Edinburgh: Andrew Elliot, 1885), 93, Michael Shafe, *University Education in Dundee 1881–1981: A Pictorial History* (Dundee: University of Dundee, 1982), and Donald Southgate, *University Education in Dundee: A Centenary History* (Edinburgh: Edinburgh University Press, 1982), 9, 21–35, 123.
118. Graham Balfour, *The Educational Systems of Great Britain and Ireland* (Oxford: The Clarendon Press, 1898), 303, and Grey, "The Women's Educational Movement," 57.
119. "University of St. Andrews. Higher Education for Women, with Title of L.L.A., Equivalent to M.A. for Men," *The Educational Times, and Journal of the College of Preceptors* 36, 271 (November 1, 1883), 300. See also Southgate, *University Education in Dundee*, 16.
120. Grey, "The Women's Educational Movement," 57.
121. *The St Andrews University Calendar for the year 1899–1900* (Edinburgh: William Blackwood and Sons, 1899), 115, 118.
122. James B. Sellers, *History of the University of Alabama* (Tuscaloosa: University of Alabama Press, 1953), 474.
123. Doherty and Summers, *West Virginia University*, 36–37.
124. McCandless, "Maintaining the Spirit and Tone of Robust Manliness," 200, 216.
125. Linda K. Kerber, *Toward an Intellectual History of Women* (Chapel Hill: The University of North Carolina Press, 1997), 230–232. Despite being the state's land-grant institution, the University of Georgia did not accept women for graduation until 1918. For more see F. N. Boney, *A Pictorial History of the University of Georgia* (Athens: University of Georgia Press, 2000), vii, 29, 94, 97.
126. Thomas Chalmers McCorvey, "V. Henry Tutwiler, and the Influence of the University of Virginia on Education in Alabama," *Transactions of the Alabama Historical Society* 5 (1904), 83–84, 96. Tuscaloosa used to be the capital of Alabama and was "the political as well as the educational center of the State."
127. Sellers, *History of the University of Alabama*, 473.
128. Ibid., 474.
129. McCorvey, "V. Henry Tutwiler," 85–86, 88. Henry Tutwiler, professor of ancient languages, was one of the first members of the faculty to be appointed at the University of Alabama.
130. Sellers, *History of the University of Alabama*, 477.
131. Ibid., 478. All three men were also graduates of the university. See Thomas Waverly Palmer, comp., *A Register of the Officers and Students of the University of Alabama 1831–1901* (Tuscaloosa: The University of Alabama, 1901), 25, 96, 115.
132. Palmer, *A Register of the Officers and Students of the University of Alabama*, 380, 395, 417, 423. Thomas Chalmers McCorvey was also married to Netta Lucia Tutwiler who was probably in favor of university coeducation as well.
133. *The Corolla of Ninety-Four* (Tuscaloosa: Published by the Students of the University of Alabama, 1894), 162, and *The Corolla '96* (Tuscaloosa, AL: W. H. Ferguson, 1896), 62–63.
134. Sellers, *History of the University of Alabama*, 473.
135. James Allen Cabaniss, *A History of the University of Mississippi* (University: University of Mississippi, 1949), 6, and *Historical Catalogue of the University of Mississippi 1849–1909* (Nashville, TN: Marshall & Bruce Company, 1910), 5–8.

136. Dunbar Rowland, *The Official and Statistical Register of the State of Mississippi 1912* (Nashville, TN: Brandon Printing Company, 1912), 219.
137. David G. Sansing, *The University of Mississippi: A Sesquicentennial History* (Jackson: University Press of Mississippi, 1999), 136–137, and Wharton, "Gender, Architecture, and Institutional Self-Preservation," 198.
138. Cabaniss, *A History of the University of Mississippi*, 101–102.
139. Sansing, *The University of Mississippi: A Sesquicentennial History*, 137, and Cabaniss, *A History of the University of Mississippi*, 129. The term *Ole Miss* was chosen as the name of the student yearbook in 1897 and had been submitted as part of a contest by Emma Coleman Meek. It soon became a term synonymous with the university itself.
140. *Historical Catalogue of the University of Mississippi*, 86. Barnard was chancellor from 1859 to 1861 and had previously been the president of the University of Mississippi from 1856 to 1859. He is also the namesake of Barnard College in New York. For more on his life see John Fulton, *Memoirs of Frederick A. P. Barnard, Tenth President of Columbia College in the City of New York* (New York: Macmillan and Co., 1896).
141. Frederick A. P. Barnard, *Should American Colleges Be Open to Women as Well as to Men? A Paper Presented to the Twentieth Annual Convocation of the University of the State of New York, at Albany, July 12, 1882* (Albany, NY: Weed, Parsons and Company, 1882), 15.
142. E. J. Watson, *Handbook of South Carolina: Resources, Institutions and Industries of the State* (Columbia, SC: The State Company, 1908), 177–179, and Daniel Walker Hollis, *University of South Carolina Volume II. College to University* (Columbia: University of South Carolina Press, 1956), 3–4.
143. Yates Snowden, ed., *History of South Carolina Volume II* (Chicago and New York: The Lewis Publishing Company, 1920), 1161. For more on the goals of Winthrop Normal see *The Revised Statutes of South Carolina Vol. 1 Containing the Civil Statutes, Approved by the General Assembly of 1893* (Columbia, SC: Charles A. Calvo, Jr., 1894), 398–399.
144. *Catalogue of the South Carolina College 1904–1905* (Columbia, SC: The R. L. Bryan Company, 1905), 41.
145. McCandless, "Maintaining the Spirit and Tone of Robust Manliness," 201.
146. Edwin L. Green, ed., *A History of the University of South Carolina* (Columbia, SC: The State Company, 1916), 122–123, and McCandless, "Maintaining the Spirit and Tone of Robust Manliness," 203.
147. Edward T. Sanford, *Blount College and the University of Tennessee: An Historical Address Delivered Before the Alumni Association and Members of the University of Tennessee* (Knoxville, TN: Published by the University, 1894), 3, 23, 38, 63, 74, and T. C. Karns, "The University of Tennessee," in *Higher Education in Tennessee*, ed. Lucius Salisbury Merriam (Washington, DC: Government Printing Office, 1894), 63. See also James Riley Montgomery, Stanley J. Folmsbee, and Lee Seifert Greene, *To Foster Knowledge: A History of The University of Tennessee 1794–1970* (Knoxville: The University of Tennessee Press, 1984), 9–13.
148. Miss Johnson, "Higher Education of Women in the South," in *Proceedings of the Eleventh Conference for Education in the South* (Nashville, TN: Published by the Executive Committee of the Conference, 1908), 130.
149. Montgomery, Folmsbee, and Greene, *To Foster Knowledge*, 11–13. See also Stanley J. Folmsbee, "The Early History of the University of Tennessee: An Address in Commemoration of its 175th Anniversary," *The East Tennessee Historical Society's Publications* 42 (1970), 15–16. The importance of these women to the history of coeducation at the University of Tennessee is undeniable, since Barbara Blount and Polly McClung, as well as Jennie Armstrong and Kittie and Mattie Kain, have had women's residence halls named after them at various points in the institution's history. See also Stanley J. Folmsbee, *History of Tennessee, Volume 1* (New York: Lewis Historical Publishing Co., 1960), 234, 237, and "Strong Women Remember," *Torchbearer: The Alumni Information Source of the University of Tennessee* 47, 2 (Summer 2008), www.utk.edu/torchbearer/4702/strong/index.shtml.
150. *The Volunteer Published by The Students of the University of Tennessee Vol. IV* (Knoxville: Bean, Warters & Gaut, 1900), 97.
151. Thomas Lloyd, Julian Orbach, and Robert Scourfield, *The Buildings of Carmarthenshire and Ceredigion* (New Haven, CT: Yale University Press, 2006), 98.
152. John Vyrnwy Morgan, *A Study in Nationality* (London: Chapman & Hall, 1911), 394.

153. Lloyd, Orbach, and Scourfield, *The Buildings of Carmarthenshire and Ceredigion*, 98. See also *The Calendar of the University College of Wales, Aberystwyth, Fourteenth Session, 1885–6* (Manchester: J. E. Cornish, 1885), 28, 32–33, 35; "Opening of the Central Block of the University College of Wales, Aberystwyth," in *Journal of Education: A Monthly Record and Review, Volume XX* (London: William Rice, 1898), 695; Kenneth O. Morgan, *Rebirth of a Nation: A History of Modern Wales* (Oxford: Oxford University Press, 1989), 107; and Iwan Morgan, *The College by the Sea (A Record and a Review): "Nid Byd Byd Heb Wybodaeth"* (Aberystwyth: Published by the Students' Representative Council in Collaboration with the College Council, 1928), 52.
154. Day Otis Kellogg, ed., *New American Supplement to the latest edition of the Encyclopædia Britannica, Volume I* (New York and Chicago: The Werner Company, 1898), 27.
155. Morgan, *The College by the Sea*, 7, 22, and T. Mortimer Green, "University College of South Wales, Aberystwyth," *Journal of Education* 2, 22 (October 1, 1900), 544. See also Gwyn A. Williams, *The Welsh in Their History* (London: Croom Helm, 1982), 158, 165, 169.
156. Balfour, *The Educational Systems of Great Britain and Ireland*, 270.
157. Zimmern, *The Renaissance of Girls' Education in England*, 140.
158. James Laughlin Hughes and Louis Richard Klemm, *Progress of Education in the Century* (Toronto and Philadelphia: Linscott Publishing Company, 1907), 79.
159. Sutherland, "The Movement for the Higher Education of Women," 91.
160. Helen Lefkowitz Horowitz, "Does Gender Bend the History of Higher Education?" *American Literary History* 7, 2 (1995), 344.
161. See in the first instance Joan Wallach Scott, *Feminism and History* (Oxford and New York: Columbia University Press, 1996).
162. Rita McWilliams-Tullberg, *Women at Cambridge* (Cambridge: Cambridge University Press, 1998), ix, xiv–xvi.
163. Carol Dyhouse, *No Distinction of Sex? Women in British Universities 1870–1939* (London: Routledge, 1995).
164. For more see Joan Perkin, *Victorian Women* (New York: New York University Press, 1995), and Jane Roland Martin, *Reclaiming a Conversation: The Ideal of the Educated Woman* (New Haven, CT: Yale University Press, 1985).
165. Barbara Miller Solomon, *In the Company of Educated Women: A History of Women and Higher Education in America* (New Haven, CT, and London: Yale University Press, 1985), xix. See also Alison Mackinnon, "Male Heads on Female Shoulders? New Questions for the History of Women's Higher Education," *History of Education Review* (Australia) 19, 2 (1990), 39, 41–42.
166. Ibid.
167. Lynn D. Gordon, *Gender and Higher Education in the Progressive Era* (New Haven, CT, and London: Yale University Press, 1990), 2. See also Mackinnon, "Male Heads on Female Shoulders?" 43.
168. Ibid. The University of Chicago began as a coeducational university in 1892, and the University of California admitted women within two years of its founding in 1868. For more see Leslie Miller-Bernal and Susan L. Poulson, eds., *Going Coed: Women's Experiences in Formerly Men's Colleges and Universities, 1950–2000* (Nashville, TN: Vanderbilt University Press, 2004), 5, 22.
169. Andrea G. Radke-Moss, *Bright Epoch: Women and Coeducation in the American West* (Lincoln: University of Nebraska Press, 2008), 1–4.

Two: Victorian Views of Coeducation

1. "On the Poetry of the Present Age," in *The London University College Magazine Vol. I* (London: H. K. Lewis, 1849), 145. For more see Michael Sanderson, *The Universities in the Nineteenth Century* (London: Routledge and Kegan Paul, 1975), 171–172, and Margaret Birney Vickery, *Buildings for Bluestockings: The Architecture and Social History of Women's Colleges in Late Victorian England* (Newark: University of Delaware Press, 1999), xi.

2. "Girl Graduates," *The Students' Journal and Hospital Gazette* (August 30, 1879), 207. See also George Gissing, *The Odd Women*, ed. Arlene Young (Peterborough, ONT: Broadview Press, 1998), 344, and Morton Luce, *A Handbook to the Works of Alfred Lord Tennyson* (London: George Bell and Sons, 1906), 6, 222–258.
3. GSA, W. S. Gilbert, *Songs of a Savoyard* (London: George Routledge and Sons, 1894), 45.
4. Judith Harford, *The Opening of University Education to Women in Ireland* (Dublin and Portland, OR: Irish Academic Press, 2008), 90.
5. John Malcolm Bulloch, ed., *College Carols* (Aberdeen: D. Wylie and Son, 1894), 25, Lines 22, 25–26, "The Mikado" was produced following the relative failure of "Princess Ida" and quickly became one of Gilbert and Sullivan's most popular works. For more see *The Complete Annotated Gilbert & Sullivan*, ed. Ian Bradley (Oxford: Oxford University Press, 2001), 553.
6. Bulloch, *College Carols*, 26, Lines 29, 31.
7. Alison Mackinnon, "Male Heads on Female Shoulders? New Questions for the History of Women's Higher Education," *History of Education Review* (Australia) 19, 2 (1990), 36–37.
8. "Co-Education of the Sexes in Colleges," *Indiana School Journal* 25, 8 (August 1880), 421, 422.
9. Carroll Smith-Rosenberg and Charles Rosenberg, "The Female Animal: Medical and Biological Views of Woman and Her Role in Nineteenth-Century America," *Journal of American History* 60 (1973), 334. See also Martha Vicinus, ed., *A Widening Sphere* (Bloomington: Indiana University Press, 1977).
10. Alison Mackinnon has found in her research that this "British" influence extended equally to Australia, Canada, and New Zealand as well. See Mackinnon, "Male Heads on Female Shoulders?" 38–40.
11. Lindy Moore, *Bajanellas and Semilinas: Aberdeen University and the Education of Women* (Aberdeen: Aberdeen University Press, 1991), 23. See also Maris Vinovskis and Richard Bernard, "Beyond Catharine Beecher: Female Education in the Antebellum Period," *Signs* 3 (1978), 856–869, and Martha Vicinus, ed., *Suffer and Be Still: Women in the Victorian Age* (Bloomington: Indiana University Press, 1972).
12. Michael Sanderson, *Education, Economic Change and Society in England 1780–1870* (Cambridge: Macmillan Press, 1995), 55, and Frank K. Prochaska, *Women and Philanthropy in Nineteenth-Century England* (Oxford: Clarendon Press, 1980), 2. Prochaska cited the evidence provided by mid-century census results. See also Robert Woods, *The Population of Britain in the Nineteenth Century* (Cambridge: Cambridge University Press, 1995), 28, 32–34, and Smith-Rosenberg and Rosenberg, "The Female Animal," 345.
13. J. A. and Olive Banks, *Feminism and Family Planning among the Victorian Middle Classes* (Liverpool: Liverpool University Press, 1964), 128–129. See also Mackinnon, "Male Heads on Female Shoulders?" 43.
14. Sanderson, *Education, Economic Change and Society in England*, 55. See also Tom Begg, *The Excellent Women: The Origins and History of Queen Margaret College* (Edinburgh: John Donald Publishers, 1994).
15. W. H. Fraser and R. J. Morris, eds., *People and Society in Scotland: Volume II, 1830–1914* (Edinburgh: John Donald Publishers, 1990), 302.
16. Joseph A. McCullough, "Alumni Address. South Carolina College and the State," in *Proceedings of the Centennial Celebration of South Carolina College, 1805–1905* (Columbia, SC: The State Co., 1905), 187.
17. S. B. Elkins, *Address Delivered Before the Literary Societies of the West Virginia University, June 11th, 1888* (New York: Styles & Cash, 1888), 21. For more on Elkins see Jerry A. Mathews, "Stephen B. Elkins," in *Twenty Years in The Press Gallery: A Concise History of Important Legislation from the 48th to the 58th Congress*, ed. O. O. Stealey (New York: Publishers Printing Company, 1906), 270–274, and William T. Doherty, Jr. and Festus P. Summers, *West Virginia University: Symbol of Unity in a Sectionalized State* (Morgantown: West Virginia University Press, 1982), 76.
18. "Editorials," *The University Courant* 4, 4 (April 1890), 39.
19. H. R. L., "Co-Education," *The Free Lance* 2, 8 (February 1889), 127–128.
20. "Co-Education of the Sexes in Colleges," 422.
21. Linda Kerber, "The Republican Mother: Women and the Enlightenment—An American Perspective," *American Quarterly* 28 (1976), 205.
22. Eileen Breathnach, "Women and Higher Education in Ireland (1879–1914)," *Crane Bag* 4, 1 (1980), 47. Breathnach refers to "a rigorous moral code that 'simultaneously idealised and repressed women.'"

23. Mrs. Campbell, "The Rise of the Higher Education of Women Movement in Glasgow," in *The Book of the Jubilee: In Commemoration of the Ninth Jubilee of the University of Glasgow, 1451–1901*, ed. the Students' Jubilee Celebrations Committee (Glasgow: J. Maclehose and Sons, 1901), 127. See also Smith-Rosenberg and Rosenberg, "The Female Animal," 335.
24. *Report of the Scottish Institution for the Education of Young Ladies with an Appendix containing separate reports, by the different teachers, of the course of instruction, and the system pursued, in their respective classes* (Edinburgh: Oliver & Boyd, 1835), 10. See also Smith-Rosenberg and Rosenberg, "The Female Animal," 337, 352.
25. Breathnach, "Women and Higher Education in Ireland (1879–1914)," 47.
26. "Should University Degrees be given to Women?" in *The Westminster Review Vol. CXV January–April, 1881, American Edition* (New York: The Leonard Scott Publishing Company, 1881), 239.
27. W. Gareth Evans, *Education and Female Emancipation: The Welsh Experience, 1847–1914* (Cardiff: University of Wales Press, 1990), 15.
28. Ibid.
29. George J. Romanes, "Mental Differences between Men and Women," *Nineteenth Century* 21 (1887), 658–659. Romanes also noted that women were naturally given many of the good qualities encouraged in the Christian religion: "[T]he meritorious qualities wherein the female mind stands pre-eminent are, affection, sympathy, devotion, self-denial, modesty; long-suffering, or patience under pain, disappointment, and adversity; reverence, veneration, religious feeling, and general morality." See also Joan N. Burstyn, "Religious Arguments Against the Higher Education for Women in England 1840–1890," *Women's Studies* 1, 1 (1972), 111–131.
30. Adele Simmons, "Education and Ideology in Nineteenth Century America: The Response of Educational Institutions to the Changing Role of Women," in *Liberating Women's History: Theoretical & Critical Essays*, ed. Bernice A. Carroll (Urbana: University of Illinois Press, 1976), 123.
31. Maria G. Grey, "On the Special Requirements for Improving the Education of Girls," in *The Education Papers: Women's Quest for Equality in Britain, 1850–1912*, ed. Dale Spender (London: Routledge & Kegan Paul, 1987), 171–185. See also Carol Lasser, ed., *Educating Men and Women Together: Coeducation in a Changing World* (Urbana: University of Illinois Press, 1987), and Felicity Hunt, ed., *Lessons for Life: The Schooling of Girls and Women, 1850–1950* (Oxford: Blackwell, 1987).
32. Josephine Butler, "The Education and Employment of Women," in *The Education Papers*, 79. See also M. G. Fawcett, "The use of higher education to women," *Contemporary Review*, November 1886, 719–728, and Sir Alexander Grant, *Happiness and Utility as Promoted by the Higher Education of Women: An Address* (Edinburgh: Edmonston and Douglas, 1872).
33. Johanna Geyer-Kordesch and Rona Ferguson, *Blue Stockings, Black Gowns, White Coats: A Brief History of Women Entering the Medical Profession in Scotland in Celebration of One Hundred Years of Women Graduates at the University of Glasgow* (Glasgow: University of Glasgow, Wellcome Unit for the History of Medicine, 1994), 5, 8, 12. More specifically to women, arguments that prostitution was literally the enslavement of women tied the two subjugated groups more closely together. For more see Jean Fagan Yellin, *Women & Sisters: The Antislavery Feminists in American Culture* (New Haven, CT, and London: Yale University Press, 1989), and Eileen Janes Yeo, *Radical Femininity: Women's Self-representation in the Public Sphere* (Manchester and New York: Manchester University Press, 1998).
34. Helen Lefkowitz Horowitz, *Alma Mater: Design and Experience in the Women's Colleges from their Nineteenth-Century Beginnings to the 1930s* (Amherst: University of Massachusetts Press, 1993), 10.
35. GUA, "WOMEN. Position American," extract from an American Supplement of *Encyclopedia Britannica* (c. 1889), 910.
36. Romanes, "Mental Differences between Men and Women," 656. See also Smith-Rosenberg and Rosenberg, "The Female Animal," 337.
37. J. F. A. Pyre, *Wisconsin* (New York: Oxford University Press, 1920), 227.
38. Lynn D. Gordon, *Gender and Higher Education in the Progressive Era* (New Haven, CT, and London: Yale University Press, 1990), 190.
39. W. Le Conte Stevens, *The Admission of Women to Universities* (New York: Press of S. W. Green's Son, 1883), 24.
40. "Report from The Owens College, Manchester," in *Education Department Reports from University Colleges 1899* (London: Wyman and Sons, 1899), 245.

41. F. S. Dumaresq de Carteret-Bisson, *Our Schools and Colleges, Vol. II: For Girls* (London: Simpkin, Marshall & Co., 1884), 180, and James Heywood, "The Owens College, Manchester, and a Northern University," *Journal of the Statistical Society* 41, 3 (September 1878), 544. All students at the University of Mississippi also had to provide a "certificate of good moral character" to be admitted. For more see *Catalogue of the University of Mississippi at University P. O., Near Oxford, Miss.: Prepared this Year with Special Reference to the Schools of English and Belles Lettres. Thirty-Ninth Session 1890–'91* (Oxford, MS: Published by the University, 1890), 26.
42. R. D. Anderson, *Education and Opportunity in Victorian Scotland: Schools and Universities* (Oxford: Clarendon Press, 1983), 1–2, 10–11, 24–26, 159–161.
43. R. D. Anderson, *Education and the Scottish People 1750–1918* (Oxford: Clarendon Press, 1995), 17–24, 50–53, 100. See also Helen Corr, "Dominies and Domination: Schoolteachers, Masculinity and Women in 19th Century Scotland," *History Workshop Journal* 40 (Autumn 1995), 154–155.
44. Patrick O'Sullivan, *The Irish in the New Communities* (Leicester: Leicester University Press, 1992), 165.
45. Lilian Daly, "Women and the University Question," *The New Ireland Review* 17 (March 1902–August 1902), 74.
46. "The R. U. I. Examinations: Brilliant Success of a Macroom Young Lady Student," *Southern Star*, August 7, 1897, 3; L. M. Little, "Women's Education: Forty Years Ago and Now," *Irish Independent*, June 1, 1906, 4; Susan M. Parkes and Judith Harford, "Women and Higher Education in Ireland," in *Female Education in Ireland 1700–1900: Minerva or Madonna*, ed. Deirdre Raftery and Susan M. Parkes (Dublin and Portland, OR: Irish Academic Press, 2007), 110, 113, 115; and Harford, *The Opening of University Education to Women in Ireland*, 80–81. Magee College was open to men and women and was affiliated with the Presbyterian Church. See also Alison Jordan, *Margaret Byers: Pioneer of Women's Education and Founder of Victoria College, Belfast* (Belfast: The Institute of Irish Studies, The Queen's University of Belfast, 1991), 9–16.
47. Parkes and Harford, "Women and Higher Education in Ireland," 105.
48. L. U. Reavis, *Saint Louis: The Future Great City of the World with biographical sketches of the representative men and women of St. Louis and Missouri* (St. Louis: C. R. Barns, 1876), 263, and *Catalogue of the Western University of Pennsylvania for the year ending 1895 with detailed statements of the courses of instruction* ([Pittsburgh]: Western University of Pennsylvania, 1895), 20. Students at Pittsburgh were required to attend "a brief religious service" every morning at 9:00 a.m.
49. Thomas Chalmers McCorvey, "V. Henry Tutwiler, and the Influence of the University of Virginia on Education in Alabama," *Transactions of the Alabama Historical Society* 5 (1904), 96; *A Memorial of the Seventy-Fifth Anniversary of the Founding of the University of Michigan: Held in Commencement Week June 23 to June 27, 1912* (Ann Arbor, MI: Published by the University, 1915), 47; and *Report of the Commissioner of Education for the year 1897–98 Volume 2. Containing Parts II and III* (Washington, DC: Government Printing Office, 1899), 1527. In 1897 a survey was done of U.S. colleges and universities to see if Bible study was undertaken formally or informally on campus. The results were given to the commissioner of education and were included in his office's annual report.
50. *The Corolla of Ninety-Five, Volume III* (Tuscaloosa: Published by the Students of the University of Alabama, 1895), 177.
51. *The Corolla '96* (Tuscaloosa, AL: W. H. Ferguson, 1896), 26.
52. McCorvey, "V. Henry Tutwiler," 105–106. Admittedly, the male students at Alabama were subject to military discipline and were less likely to act up as a result. For more on the military discipline at Alabama see James B. Sellers, *History of the University of Alabama* (Tuscaloosa: University of Alabama Press, 1953), 486–513.
53. *The Corolla '96*, 115–116.
54. OUMC, Margaret Boyd Diary (1873), 57, in Ohio Memory: An Online Scrapbook of Ohio History, www.ohiomemory.org/index.html (accessed June 21, 2009).
55. Anderson, *Education and the Scottish People*, 17–24, 50–53, 100. The 1872 Education Act in Scotland, which was similar to the 1870 Act in England, stated that all children from the ages of five to thirteen must receive elementary education. The central skills that needed to be learned were reading, writing, and arithmetic, and children could be exempted from some schooling (before age 7) if they already had these abilities.
56. R. A. Houston, "Scottish Education and Literacy, 1600–1800: An International Perspective," in *Improvement and Enlightenment: Proceedings of the Scottish Historical Studies Seminar, University of Strathclyde, 1987–88*, ed. T. M. Devine (Edinburgh: J. Donald, 1989), 43–61.

57. J. T. Fowler, *Durham University: Earlier Foundations and Present Colleges* (London: F. E. Robinson, 1904), 119.
58. Sellers, *History of the University of Alabama*, 552.
59. Wendy Alexander, *First Ladies of Medicine: The Origins, Education and Destination of Early Women Medical Graduates of Glasgow University* (Glasgow: Wellcome Unit for the History of Medicine, University of Glasgow, 1987), 1–2. See also Smith-Rosenberg and Rosenberg, "The Female Animal," 332–337.
60. Geyer-Kordesch and Ferguson, *Blue Stockings, Black Gowns, White Coats*, 14–17, 20.
61. Cynthia Eagle Russett, *Sexual Science: The Victorian Construction of Womanhood* (Cambridge, MA: Harvard University Press, 1989), 122–123. See also Smith-Rosenberg and Rosenberg, "The Female Animal," 350–351; Alexander C. J. Skene, *Education and Culture as Related to the Health and Diseases of Women* (Detroit: G. S. Davis, 1889); and John Thorburn, *Female Education from a Physiological Point of View* (Manchester: Cornish, 1884).
62. Woods, *The Population of Britain in the Nineteenth Century*, 32–34. See also Carol Dyhouse, "Social Darwinistic Ideas and the Development of Women's Education in England, 1880–1920," *History of Education* 5, 1 (1976), 41–58, and Vickery, *Buildings for Bluestockings*, 149–153.
63. Joan N. Burstyn, "Education and Sex: The Medical Case Against Higher Education for Women in England, 1870–1900," *Proceedings of the American Philosophical Society* 117, 2 (April 1973), 79. See also Lasser, *Educating Men and Women Together*, 56–62.
64. Romanes, "Mental Differences between Men and Women," 654–672. See also Emma Wallington, "The Physical and Intellectual Capacities of Woman Equal to Those of Man," *Anthropologia* 1 (1874), 552–565.
65. Romanes coined the term *comparative psychology* among other things. The men met and became friends while at Cambridge. See also Robert J. Richards, *Darwin and the Emergence of Evolutionary Theories of Mind and Behavior* (Chicago: University of Chicago Press, 1989), 334–381.
66. Romanes, "Mental Differences between Men and Women," 654–655. For a complete discussion of the Social Darwinist arguments regarding women's higher education see Russett, *Sexual Science*, 41–44, 88–89, 100–103, 122–125.
67. Burstyn, "Education and Sex," 79.
68. Romanes, "Mental Differences between Men and Women," 656.
69. Andrew Cunningham McLaughlin, *History of Higher Education in Michigan* (Washington, DC: Government Printing Office, 1891), 68. See also J. McGrigor Allan, "On the Real Differences in the Minds of Men and Women," *Anthropological Review* 7 (1869), 195–215; "Are Men Naturally Cleverer than Women?" *Englishwoman's Journal* 2 (1858), 336; T. S. Clouston, *Female Education from a Medical Point of View* (Edinburgh: Macniver & Wallace, 1882); and W. L. Distant, "The Mental Differences between the Sexes," *Journal of the Anthropological Institute* 4 (1875), 78–87.
70. GUA, *Encyclopedia Britannica*, 908–913.
71. Ibid., 908.
72. A. Lapthorn Smith, "Higher Education of Women and Race Suicide," *Popular Science Monthly* 66 (March 1905), 467, 470–471.
73. Ibid. See also Ely Van de Warker, *Woman's Unfitness for Higher Coeducation* (New York: Grafton Press, 1903).
74. Geyer-Kordesch and Ferguson, *Blue Stockings, Black Gowns, White Coats*, 18.
75. Russett, *Sexual Science*, 120. See also Eliza B. Duffey, *No Sex in Education; or an Equal Chance for both Girls and Boys: Being a Review of Dr. E. H. Clarke's "Sex in Education"* (Philadelphia: J. M. Stoddart & Co., 1874), and Julia Ward Howe, ed., *Sex and Education: A Reply to Dr. E. H. Clarke's "Sex in Education"* (Boston: Roberts Bros., 1874).
76. Olive San Louis Anderson, *An American Girl and Her Four Years in a Boys' College*, ed. Elisabeth Israels Perry and Jennifer Ann Price (Ann Arbor: University of Michigan Press, 2006), 119.
77. Burstyn, "Education and Sex," 79–89. See also Smith-Rosenberg and Rosenberg, "The Female Animal," 340, and Simmons, "Education and Ideology in Nineteenth Century America," 118.
78. Henry Maudsley, *Sex in Mind and in Education* (Syracuse, NY: C. W. Bardeen, 1884), 24–25. Maudsley's initial article, "Sex in Mind and in Education," *Fortnightly Review* from June 1874, and several of the replies to it, can be found in Katharina Rowold, ed., *Gender & Science: Late Nineteenth-Century Debates on the Female Mind and Body* (Bristol: Thoemmes Press, 1996). See also Anne E. Walker, *The Menstrual Cycle* (London: Routledge, 1997), 39, and E. G. Anderson, "Sex in Mind and Education: A Reply," *Fortnightly Review* 15 (1874), 582–594.

79. Edward H. Clarke, *Sex in Education; or, A Fair Chance for Girls* (Boston: James R. Osgood and Co., 1873), 121–122, 145, 154. Clarke conceded that "Two or three generations, at least, of the female college graduates of this sort of co-education must come and go before any sufficient idea can be formed of the harvest it will yield." For more see Elizabeth Seymour Eschbach, *The Higher Education of Women in England and America, 1865–1920* (New York: Garland, 1993), 83–86; Dorothy Gies McGuigan, *A Dangerous Experiment: 100 Years of Women at the University of Michigan* (Ann Arbor, MI: Center for Continuing Education of Women, 1970), 53–58; Horowitz, *Alma Mater*, 4, 16; and Simmons, "Education and Ideology in Nineteenth Century America," 118–120.
80. Breathnach, "Women and Higher Education in Ireland (1879–1914)," 49.
81. George Van Derveer Morris, *A Man for a' That* (Cincinnati: Jennings & Pye, 1902), 15, 19.
82. Ibid., 19.
83. GUA, Volume of Presscuttings of Queen Margaret College, 1884–1890, *The Scotsman*, April 26, 1887.
84. Ibid.
85. Ibid.
86. Nathan Sheppard, *Before an Audience; Or, The Use of the Will in Public Speaking. Talks to the Students of the University of St. Andrews and the University of Aberdeen* (New York and London: Funk & Wagnalls Company, 1886), 19.
87. The speech apparently happened in the early 1870s, though Sheppard does not provide a precise date in his later work, compiling speeches he gave at Aberdeen and St. Andrews.
88. Sheppard, *Before an Audience*, 119.
89. David Staars, *The English Woman: Studies in Her Psychic Evolution*, trans. and ed. J. M. E. Brownlow (London: Smith, Elder, & Co., 1909), 308.
90. David G. Sansing, *The University of Mississippi: A Sesquicentennial History* (Jackson: University Press of Mississippi, 1999), 137, and James Allen Cabaniss, *A History of the University of Mississippi* (University: University of Mississippi, 1949), 129.
91. Dunbar Rowland, *The Official and Statistical Register of the State of Mississippi 1912* (Nashville, TN : Brandon Printing Company, 1912), 106, and Irene Harwarth, Mindi Maline, and Elizabeth DeBra, *Women's College in the United States: History, Issues, and Challenges* (Darby, PA: Diane Publishing Company, 1997), 11.
92. Romanes, "Mental Differences between Men and Women," 672. The entire sentence of this quote states: "In now again reaching forth her hand to eat of the tree of knowledge woman is preparing for the human race a second fall."
93. Ibid., 666–667, 672. Romanes also noted "women's colleges are springing up like mushrooms in all quarters of the kingdom."

Three: Administration and Legislation

1. PSU, "The College Government," *La Vie '93 published by the Junior Class* (State College, PA: Published by the University, 1893), 27.
2. Amy Thompson McCandless, "Maintaining the Spirit and Tone of Robust Manliness: The Battle against Coeducation at Southern Colleges and Universities, 1890–1940," *NWSA Journal* 2, 2 (Spring 1990), 213.
3. D. I. Mackay, *Geographical Mobility and the Brain Drain: A Case Study of Aberdeen University Graduates, 1860–1960* (London: George Allen and Unwin, 1969), 35.
4. *Catalogue of the Western University of Pennsylvania for the year ending 1895 with detailed statements of the courses of instruction* ([Pittsburgh]: Western University of Pennsylvania, 1895), 20.
5. Jane Rendall, "The Citizenship of Women and the Reform Act of 1867," in *Defining the Victorian Nation: Class, Race, Gender and the Reform Act of 1867*, ed. Catherine Hall, Keith McClelland, and Jane Rendall (Cambridge: Cambridge University Press, 2000), 122.
6. Sarah J. Smith, "Retaking the Register: Women's Higher Education in Glasgow and Beyond, c. 1796–1845," *Gender & History* 12, 2 (2000), 310–335. See also M. J. Tuke, *A History of Bedford College for Women, 1849–1937* (London: Oxford University Press, 1939), and Elaine Kaye, *A History of Queen's College, London 1848–1972* (London: Chatto and Windus, 1972).
7. Sheila Hamilton, "Women and the Scottish Universities circa 1869–1939: A Social History" (Ph.D. thesis, University of Edinburgh, 1987), 15.

8. Sarah V. Barnes, "Crossing the Invisible Line: Establishing Co-education at the University of Manchester and Northwestern University," *History of Education* 23, 1 (1994), 39, 41; Susan M. Parkes and Judith Harford, "Women and Higher Education in Ireland," in *Female Education in Ireland 1700–1900: Minerva or Madonna*, ed. Deirdre Raftery and Susan M. Parkes (Dublin and Portland, OR: Irish Academic Press, 2007), 111–112, and Judith Harford, *The Opening of University Education to Women in Ireland* (Dublin and Portland, OR: Irish Academic Press, 2008), 79. See also Rhama D. Pope and Maurice G. Verbeke, "Ladies' Educational Organizations in England, 1865–1885," *Paedagogica Historica* 3 (1976), 336–361.
9. *First Annual Report of the American Woman's Educational Association. May, 1853* (New York: Kneeland, 1853).
10. There was a general committee of thirty-seven ladies and an acting committee of ten. James Coutts, *A History of the University of Glasgow: From its Foundation in 1451 to 1909* (Glasgow: J. Maclehose and Sons, 1909), 458. See also Catherine Mary Kendall, *The Queen Margaret Settlement 1897–1914: Glasgow Women Pioneers in Social Work* (Master's thesis, University of Glasgow, 1993), 40; GUA, Frances Melville, "Presentation Address," on the occasion of the first award of the Frances Melville Medal in Philosophy on the final closure of the College, November 1935, 1; and GUA, Glasgow Association for the Higher Education of Women, General Committee meeting minutes with presscuttings, re: inaugural meeting (4 April 1877).
11. Princess Louise's title may have been honorary, but it did lead to her visiting Queen Margaret College in 1888 along with HM Queen Victoria. The princess' mother-in-law, the Duchess of Argyll, headed the executive committee of the Edinburgh Association, also an honorary position. GUA, Presscuttings book on visits of Queen Victoria and HRH Princess Louise, 1888–1890. For more see Jehanne Wake, *Princess Louise: Queen Victoria's Unconventional Daughter* (London: Collins, 1988), and Carol Dyhouse, *No Distinction of Sex? Women in British Universities 1870–1939* (London: Routledge, 1995), 16.
12. Mrs. Campbell, "The Rise of the Higher Education of Women Movement in Glasgow," in *The Book of the Jubilee: In Commemoration of the Ninth Jubilee of the University of Glasgow, 1451–1901*, ed. the Students' Jubilee Celebrations Committee (Glasgow: J. Maclehose and Sons, 1901), 130. For more on Scottish Ladies' Educational Associations in St. Andrews and Aberdeen see Hamilton, "Women and the Scottish Universities," 119–125.
13. Campbell, "The Rise of the Higher Education of Women Movement," 129. Professor Nichol continued to give several lectures to women at Glasgow between 1868 and 1877. See also Johanna Geyer-Kordesch and Rona Ferguson, *Blue Stockings, Black Gowns, White Coats: A Brief History of Women Entering the Medical Profession in Scotland in Celebration of One Hundred Years of Women Graduates at the University of Glasgow* (Glasgow: University of Glasgow, Wellcome Unit for the History of Medicine, 1994), 11, 37, and GUA, Frances Melville, "Presentation Address," 1. Mrs. Campbell's support of the college continued for thirty years, though her "failing health" caused her to take on a smaller role after 1890. She took it on herself to organize the raising of the endowment fund for the new college and approached Mrs. Elder for her first contribution. Mrs. Campbell also invited the students of Queen Margaret College to Tullichewan Castle, on Loch Lomond, for "at Home" recreation. Unlike some of her widowed counterparts, Mr. Campbell was also a part of the Glasgow Association for the Higher Education of Women formation (though he was not significantly active). James Campbell commented that he felt both men and women alike would benefit from women's admission to higher education. See GUA, Janet Galloway, "Historical sketch of the movement for the Higher Education of Women in Glasgow and Queen Margaret College," On the occasion of the golden wedding anniversary of Mrs. Jean Campbell of Tullichewan, May 1896.
14. GUA, Glasgow Association for the Higher Education of Women, inaugural meeting, *Glasgow News*, April 4, 1877. See also Geyer-Kordesch and Ferguson, *Blue Stockings, Black Gowns, White Coats*, 39. Over 1,500 of the 2,000 copies of the first prospectus were sent to families in Glasgow and its suburbs. The remaining 500 were kept by the publisher to be given to people who might ask for them. Advertisements in the *Glasgow Herald*, the *Mail*, and the *Glasgow News* appeared once each week from late September onward. The close connection with the university was maintained as Principal Caird provided the introduction to the first lecture.
15. "The Medical Society," *The Durham University Journal* 5, 7 (February 17, 1883), 77.
16. Ibid., 77–78. To provide a contemporary paradigm for this view, in Elizabeth Gaskell's novel *Wives and Daughters* (which was in print throughout the latter part of the nineteenth century) her heroine, Molly Gibson, is considered to be an atypical woman of the Victorian period

because she shares her future husband's interest in natural science. For more see Elizabeth Gaskell, *Wives and Daughters: A Novel* (New York: Harper & Brothers, 1866), 34–35, 64.

17. H. R. L., "Co-Education," *The Free Lance* 2, 8 (February 1889), 128.
18. Lewis Campbell, *On the Nationalisation of the Old English Universities* (London: Chapman and Hall, 1901), 224, and Emily Janes, *The Englishwoman's Year Book and Directory 1900* (London: Adam and Charles Black, 1900), 5.
19. Mary A. Marshall, "Medicine as a Profession for Women," in *The Woman's World: Volume I,* ed. Oscar Wilde (London, Paris, and Melbourne: Cassell & Company, 1888 and London: Source Book Press, 1970), 106.
20. Ibid., 106–107.
21. Sophia Jex-Blake, "Appendix: A Brief Summary of the Action of Declarator Brought By Ten Matriculated Lady Students Against the Senatus of Edinburgh University, 1872–1873," in *Medical Women: A Thesis and a History* (Edinburgh: Oliphant, Anderson, & Ferrier, 1886), 5.
22. Sir William Stirling-Maxwell, *Miscellaneous Essays and Addresses* (London: John C. Nimmo, 1891), 433–434.
23. Lindy Moore, *Bajanellas and Semilinas: Aberdeen University and the Education of Women* (Aberdeen: Aberdeen University Press, 1991), 4.
24. *Hansard's Parliamentary Debates, 38° Victoriæ, 1875, Volume CCXXII (222). Comprising the Period from the Fifth Day of February 1875, to the Seventeenth Day of March 1875, First Volume of the Session* (February 8, 1875), 142.
25. Margaret Todd, *The Life of Sophia Jex-Blake* (London: Macmillan and Co., 1918), 413.
26. "Punch's Essence of Parliament," *Punch* 68 (March 13, 1875), 110.
27. *Hansard's Parliamentary Debates, 38° Victoriæ, 1875,* 1149–1150.
28. Coutts, *A History of the University of Glasgow,* 427–428, 434–435, 441.
29. Suzanne Le-May Sheffield, *Women and Science: Social Impact and Interaction* (New Brunswick, NJ: Rutgers University Press, 2006), 119.
30. Apparently Mr. Cowper-Temple and Mr. Russell Gurney were "the kind of friends with whom one would go tiger-hunting." For more see Todd, *The Life of Sophia Jex-Blake,* 429.
31. Geyer-Kordesch and Ferguson, *Blue Stockings, Black Gowns, White Coats,* 5; Wendy Alexander, *First Ladies of Medicine: The Origins, Education and Destination of Early Women Medical Graduates of Glasgow University* (Glasgow: Wellcome Unit for the History of Medicine, University of Glasgow, 1987), 5; and Hamilton, "Women and the Scottish Universities," 96–97. See also "The Education of Girls: Their Admissibility to Universities," *Westminster Review* 109 (January 1878), 56–90.
32. Although the "equal protection" clause in the Fourteenth Amendment has been used to deal with educational issues in the past, the word education never appears in the U.S. Constitution or its amendments. As such, the Tenth Amendment that stipulates "The powers not delegated to the United States by the Constitution, nor prohibited by it to the States, are reserved to the States respectively, or to the people" includes the subject of education. For more see Kern Alexander and M. David Alexander, *American Public School Law* (Belmont, CA: Thomson/West, 2005), 67–70.
33. Michael Bezilla, *Penn State: An Illustrated History* (University Park and London: The Pennsylvania State University Press, 1985), 1–2.
34. Merle Curti, *The Social Ideals of American Educators* (Paterson, NJ: Pageant Books, 1959), 3. Curti was quoting comments made by Thomas Jefferson in 1779.
35. Ibid., 57.
36. Bezilla, *Penn State,* 1–2.
37. George N. Rainsford, *Congress and Higher Education in the Nineteenth Century* (Knoxville: The University of Tennessee Press, 1972), 96.
38. Winifred Bryan Horner, "Nineteenth-Century Higher Education: The Scottish-American Connection," in *Scottish Universities: Distinctiveness and Diversity,* ed. Jennifer J. Carter and Donald J. Witherington (Edinburgh: Edinburgh University Press, 1992), 38.
39. For more on the history behind the passing of the Act see Rainsford, *Congress and Higher Education in the Nineteenth Century,* Chapter 7.
40. Bezilla, *Penn State,* 2.
41. Other legislation passed in 1862 included the Pacific Railroad Act and the Homestead Act, both of which would help to move northern citizens to the west, thus boxing in the Confederacy. For more see David B. Danbom, *Born in the Country: A History of Rural America* (Baltimore, MD: Johns Hopkins University Press, 2006), 112.

42. James Riley Montgomery, Stanley J. Folmsbee, and Lee Seifert Greene, *To Foster Knowledge: A History of The University of Tennessee 1794–1970* (Knoxville: The University of Tennessee Press, 1984), 76.
43. Bezilla, *Penn State,* 46, and James and Vera Olson, *The University of Missouri: An Illustrated History* (Columbia: University of Missouri Press, 1988), 19.
44. Federal Writers' Project, *Indiana: A Guide to the Hoosier State* (New York: Oxford University Press, 1947), 81.
45. Montgomery, Folmsbee, and Greene, *To Foster Knowledge*, 88.
46. William T. Doherty, Jr. and Festus P. Summers, *West Virginia University: Symbol of Unity in a Sectionalized State* (Morgantown: West Virginia University Press, 1982), 36, 52.
47. Jean Rasmusen Droste, "Women at Wisconsin" (Master's thesis, University of Wisconsin, 1967), 12 and 21. See also Adele Simmons, "Education and Ideology in Nineteenth Century America: The Response of Educational Institutions to the Changing Role of Women," in *Liberating Women's History: Theoretical & Critical Essays*, ed. Bernice A. Carroll (Urbana: University of Illinois Press, 1976), 120–121.
48. Doherty and Summers, *West Virginia University*, 42.
49. *The Revised Statutes of South Carolina Vol. 1 Containing The Civil Statutes, Approved by the General Assembly of 1893* (Columbia, SC: Charles A. Calvo, Jr., 1894), 392.
50. Montgomery, Folmsbee, and Greene, *To Foster Knowledge*, 104. See also Ralph D. Christy and Lionel Williamson, eds., *A Century of Service: Land-Grant Colleges and Universities, 1890–1990* (New Brunswick and London: Transaction Publishers, 1992), 4–5, 53–68.
51. George W. Summers, *The Mountain State: A Description of the Natural Resources of West Virginia, Prepared for Distribution at the World's Columbian Exposition* (Charleston, WV: Moses W. Donnally, 1893), 65.
52. Montgomery, Folmsbee, and Greene, *To Foster Knowledge*, 88, and James A. Raffel, *Historical Dictionary of School Desegregation: The American Experience* (Westport, CT: Greenwood Press, 1998), xxv, 139. The Universities of Tennessee and Mississippi were desegregated in 1952 and 1962 respectively.
53. Joseph Thompson, *The Owens College: Its Foundation and Growth; and its Connection with the Victoria University, Manchester* (Manchester: J. E. Cornish, 1886), 387–388, 405, 416–417, and William Jack, "The New English University," in *Macmillan's Magazine Vol. XLIII November 1880, to April 1881* (London: Macmillan and Co., 1881), 112–113.
54. H. B. Charlton, *Portrait of a University, 1851–1951: To Commemorate the Centenary of Manchester University* (Manchester: Manchester University Press, 1951), 138.
55. "Report from The Owens College, Manchester," in *Education Department Reports from University Colleges 1899* (London: Wyman and Sons, 1899), 259, and Christina Sinclair Bremner, *Education of Girls and Women in Great Britain* (London: Swan Sonnenschein & Co., 1897), 149.
56. Thompson, *The Owens College*, 493.
57. Bremner, *Education of Girls and Women in Great Britain*, 149.
58. *Essays and Addresses, by Professors and Lecturers of the Owens College, Manchester* (London: Macmillan and Co., 1874), vii. A. S. Wilkins became a professor of classical literature there in 1903 and later pro-vice chancellor of Victoria University and was, among other accomplishments, a contributor to *Encyclopedia Britannica.* For more see *Who's Who, 1904: An Annual Biographical Dictionary, Fifty-Sixth Year of Issue* (London: Adam and Charles Black and New York: The Macmillan Company, 1904), 1642.
59. Thompson, *The Owens College*, 492.
60. Mayo W. Hazeltine, *British and American Education: The Universities of the Two Countries Compared* (New York: Harper & Brothers, 1880), 129; Thompson, *The Owens College*, 511, 517, 533, 536–550; and J. E. G. De Montmorency, *The Progress of Education in England: A Sketch of the Development of English Educational Organization from Early Times to the Year 1904* (London: Knight & Co., 1904), 266.
61. "Report from The Owens College, Manchester," 286, and Barnes, "Crossing the Invisible Line," 44.
62. Stephen Gwynn, *The Famous Cities of Ireland* (Dublin and London: Maunsel & Co. and New York: The Macmillan Company, 1915), 106.
63. Constant Reader, "Female Education," *The Nation*, April 24, 1847, 11.
64. Ibid.
65. Parkes and Harford, "Women and Higher Education in Ireland," 105.
66. Ibid., 106.

67. "Female Students in Ireland," *Freeman's Journal*, May 18, 1877, 6. William Francis Cowper-Temple would become Baron Mount-Temple in 1880; Mount-Temple was in county Sligo in Ireland. For more see *The London Gazette*, May 25, 1880, 3173.
68. Harford, *The Opening of University Education to Women in Ireland*, 77, 86–87, 94.
69. "Irish Education," *Anglo-Celt*, June 8, 1867, 2, and Harford, *The Opening of University Education to Women in Ireland*, 80.
70. Eibhlin Breathnach, "Women and Higher Education in Ireland (1879–1914)," in *The Irish Women's History Reader*, ed. Alan Hayes and Diane Urquhart (London: Routledge, 2001), 46–48. Originally called the Catholic University of Ireland, the institution was renamed University College Dublin in 1881.
71. Parkes and Harford, "Women and Higher Education in Ireland," 109–110, and Harford, *The Opening of University Education to Women in Ireland*, 87.
72. Parkes and Harford, "Women and Higher Education in Ireland," 107–108, 111, and Harford, *The Opening of University Education to Women in Ireland*, 77–78.
73. Parkes and Harford, "Women and Higher Education in Ireland," 111.
74. Breathnach, "Women and Higher Education in Ireland (1879–1914)," in *The Irish Women's History Reader*, 46.
75. The Lord Advocate (Mr. J. B. Balfour), Secretary Sir William Harcourt, and Mr. Solicitor General for Scotland brought forward draft bills in 1883, 1884, and 1885. The Lord Advocate/Solicitor General for Scotland (Rt. Hon. J. H. A. Macdonald) did so in 1887, and the Lord Advocate (Mr. J. P. B. Robertson), Mr. Chancellor of the Exchequer (Rt. Hon. G. J. Goschen), Mr. Solicitor General for Scotland (Mr. Moir T. Stormonth Darling), and Sir Herbert Maxwell in 1889. Macdonald and Darling were each the representative for Edinburgh and St. Andrews Universities; this makes their presence most understandable. Harcourt was busy putting forward legislation for Oxford and Cambridge at the same time; this presumably shows higher education to be a particular interest of his. See also Christina Struthers, *The Admission of Women to Scottish Universities* (Aberdeen: John Rae Smith, 1883), and "University education for women in Scotland," *Ladies' Edinburgh Magazine* (November 5, 1879), 517.
76. Moore, *Bajanellas and Semilinas*, 31. The only oversight in this book is the 1874 "Bill to Remove Doubts as to the Powers of the Universities of Scotland to Admit Women as Students and to Grant Degrees to Women" which is not included in Moore's study, but which is held at the Glasgow University Archives.
77. Alexander, *First Ladies of Medicine*, 5.
78. Scottish Universities Commission, *General Report of the Commissioners under the Universities (Scotland) Act, 1889. With an Appendix containing Ordinances, Minutes, Correspondence, Evidence, and other documents* (Edinburgh: Mill & Co., 1900), xxi. The Act and its ordinances were published in the *Edinburgh Gazette* on March 8, 15, 22, and 29, 1892. Copies of the *Gazette*, which was an official government publication, could be purchased by the public at a cost of 9p. See also J. N. Morton, *An Analysis of the Universities (Scotland) Act, 1889, with the Act Itself and the Act of 1858, and an Index* (Edinburgh and London: William Blackwood and Sons, 1889), 15.
79. "Appendix," *The Edinburgh University Calendar 1892–1893* (Edinburgh: James Thin, 1892), 51–54.
80. Morton, *An Analysis of the Universities (Scotland) Act*, 12. Morton also made a parenthetical reference for readers to "See Jex Blake *v.* University of Edinburgh, 1873, 11 Macpherson's Reports, 784" for more information on the disappointment expressed by supporters of women's higher education.
81. *The Glasgow University Calendar for the year 1901–2* (Glasgow: James Maclehose and Sons, 1901), 135.
82. Coutts, *A History of the University of Glasgow*, 458.
83. Moore, *Bajanellas and Semilinas*, 39. See also Emily Davies, "Women in the Universities of England and Scotland," in *The Educators: Female Education*, ed. Marie Mulvey Roberts and Tamae Mizuta (London: Routledge/Thoemmes, 1995), 183.
84. Morton, *An Analysis of the Universities (Scotland) Act*, 11–12.
85. *The Glasgow University Calendar for the year 1901–2* (Glasgow: James Maclehose and Sons, 1901), 243, and Appendix, *The Edinburgh University Calendar 1895–1896*, 17–21.
86. John Stuart Mill, *On Liberty: The Subjection of Women* (New York: Henry Holt and Company, 1882), 360. For more see Ann P. Robson and John M. Robson, eds., *Sexual Equality: Writings by John Stuart Mill, Harriet Taylor Mill and Helen Taylor* (Toronto: University of Toronto Press, 1994).
87. GUA, *Glasgow University Magazine* 3, 2 (December 10, 1890), 28.

88. Harford, *The Opening of University Education to Women in Ireland*, 92.
89. William Cadwaladr Davies and William Lewis Jones, *The University of Wales and Its Constituent Colleges* (London: F. E. Robinson & Co., 1905), 24, and W. J. Wallis-Jones, "The University College of Wales," *WALES: A National Magazine for the English Speaking Parts of Wales* 3, 26 (June 1896), 247. For more see Gareth Rees and David Istance, "Higher Education in Wales: The (Re-)emergence of a National System?" *Higher Education Quarterly* 51, 1 (January 1997), 51–53.
90. "The University," *The West Virginia School Journal* 15, 2 (February 1896), 571.
91. Doherty and Summers, *West Virginia University*, 36, 44–45. Lyon's daughters were not the only female students Willey taught in the fall of 1883 as others from Morgantown were also invited. For more see "Co-Education at West Virginia University," *WVU Women: The First Century* (Morgantown: WVU Women's Centenary Project, West Virginia University, 1989).
92. A. R. Whitehill, *History of Education in West Virginia* (Washington, DC: Government Printing Office, 1902), 128.
93. Franklin L. Riley, *School History of Mississippi for use in Public and Private Schools* (Richmond, VA: B. F. Johnson Publishing Company, 1915), 331.
94. "Punch's Essence of Parliament," *Punch* 70 (July 15, 1876), 16.
95. *Public Acts of the Legislature of the State of Michigan Passed at the Regular Session of 1899 with an Appendix Containing Joint and Concurrent Resolutions, Amendments to the Constitution, and the State Treasurer's Report for the Year Ending June 30, 1899* (Lansing, MI: Robert Smith Printing Co., 1899), 281–282.
96. Campbell, "The Rise of the Higher Education of Women Movement," 134.
97. Charlotte Wronker, "Co-Education in the 'Varsity," *The Missouri Alumni Quarterly* (December 1905), 27, 32. Read Hall would later be converted into a student union building. For more on Read Hall see Albert Ross Hill, "Advantages and Disadvantages of Residential Halls for Women in Co-Educational Universities," in *Transactions and Proceedings of the National Association of State Universities in the United States of America,* No. 8, 1910 (Hamilton, OH: Republican Publishing Company, 1910), 88–92, and Olson, *The University of Missouri*, 42–43, 69.
98. UWA, *Minutes of the Board of Regents* (1866 through 1876), 194–198, 201. Pages 196–197 of the *Regents' Minutes* detail the firing of President Twombly due to "incompetency...he possessing neither the learning to teach, the capacity to govern, or the wisdom to direct." They subsequently amended and softened this statement greatly.
99. David V. Mollenhoff, *Madison: A History of the Formative Years* (Dubuque, IA: Kendall/Hunt Pub. Co., 1982), 343; Allan G. Bogue and Robert Taylor, eds., *The University of Wisconsin: One Hundred and Twenty-Five Years* (Madison: University of Wisconsin Press, 1975), 3–35; and Reuben Gold Thwaites, ed., *The University of Wisconsin: Its History and Its Alumni* (Madison, WI: J. N. Purcell, 1900), 360. See also Florence Bascom, "The University in 1874–1887," *Wisconsin Magazine of History* 8 (March 1925), 303.
100. Frederick Rudolph, *The American College and University: A History* (New York: Knopf, 1962), 321.
101. Clement L. Martzolff, "Ohio University—The Historic College of the Old Northwest," *Ohio Archaeological and Historical Quarterly* 19, 2 (April 1910), 437.
102. Thomas N. Hoover, *The History of Ohio University* (Athens: Ohio University Press, 1954), 139, and Wm. Raimond Baird, *Betas of Achievement: Being Brief Biographical Records of Members of the Beta Theta Pi Who Have Achieved Distinction in Various Fields of Endeavor* (New York: The Beta Publishing Co., 1914), 17–18, 51. William Harvey Glenn Adney was valedictorian of the Ohio University Class of 1860 and would become a professor of natural science in 1872 before leaving for a job at Washington and Jefferson College in 1873.
103. Martzolff, "Ohio University—The Historic College of the Old Northwest," 437–438. See also Betty Hollow, *Ohio University, 1804–2004: The Spirit of a Singular Place* (Athens: Ohio University Press, 2003), 71.
104. William Kimok, email message to author, September 24, 2009. According to the archival records held at Ohio University, Margaret's other siblings included John who was a physician; Jane, Kate, and Lucy who were all teachers; and Fanny who was a housewife. Hugh Boyd was a minister in Ohio at the time he assisted Margaret in her quest for admission and later became a classics professor in Iowa. William worked as an attorney in Cincinnati.
105. *The West Virginia School Journal* 7, 6 (June 1888), 1, and Doherty and Summers, *West Virginia University*, 45.

106. Franklin Lyon was the nephew of Mary Lyon, founder of Mt. Holyoke College, so the belief in women's higher education was clearly strong in the family. For more see A. B. Lyon and G. W. A. Lyon, eds., *Lyon Memorial: Massachusetts Families, Including Descendants of the Immigrants William Lyon, of Roxbury, Peter Lyon, of Dorchester, George Lyon, of Dorchester, with Introduction Treating of the English Ancestry of the American Families* (Detroit: Wm. Graham Printing Co., 1905), 274–275, 279–280.
107. Doherty and Summers, *West Virginia University*, 36, 44.
108. *West Virginia University, Morgantown, Catalogue 1889–90: Announcements for 1890–91* (Charleston, WV: Moses W. Donnally, 1890), 10, and James Morton Callahan, *Semi-Centennial History of West Virginia* ([Charleston]: Semi-Centennial Commission of West Virginia, 1913), 232. For more see Doherty and Summers, *West Virginia University*, 45, 64.
109. Thomas Waverly Palmer, comp., *A Register of the Officers and Students of the University of Alabama 1831–1901* (Tuscaloosa: The University of Alabama, 1901), 31.
110. Palmer, *A Register of the Officers and Students of the University of Alabama*, 24, 380. See also James B. Sellers, *History of the University of Alabama* (Tuscaloosa: University of Alabama Press, 1953), 478. William Asa Parker was professor of Greek (1871–1872) and modern languages from 1871 to 1898 when he began teaching only German.
111. Palmer, *A Register of the Officers and Students of the University of Alabama*, 115, 139. Another brother, John Marshall Parker, also attended the University of Alabama.
112. Sellers, *History of the University of Alabama*, 467.
113. Palmer, *A Register of the Officers and Students of the University of Alabama*, 314, 331, 338, 376, 393, 412, 431. Bessie's siblings who attended Alabama were William Clayton Parker (entered in 1884), Osborne Parker (entered in 1886), Graham Parker (entered in 1892), John Scott Parker (entered in 1887), Thornton Parker (entered in 1895), Allen Parker (entered in 1897), and Mary Parker (entered in 1901).
114. "Medical Items and News: Medical Women in Ireland," *The Medical Record: A Weekly Journal of Medicine and Surgery* (November 25, 1876), 774.
115. Tadhg Foley, ed., *From Queen's College to National University: Essays on the Academic History of QCG/UCG/NUI, Galway* (Dublin: Four Courts Press, 1999), 74, 390–395.
116. "Queen's Institute," *Freeman's Journal*, April 12, 1877, 7.
117. Harford, *The Opening of University Education to Women in Ireland*, 78.
118. Parkes and Harford, "Women and Higher Education in Ireland," 111.
119. "Editorials," *The University Courant* 4, 4 (April 1890), 39.
120. *Report of the Board of Curators of the State University of the State of Missouri to the XXXIst General Assembly* (Jefferson City, MO: Tribune Printing Company, 1881), 14, and Edward Fiddes, "The University Movement in Manchester (1851–1903)," in *Historical Essays in Honour of James Tait*, ed. J. G. Edwards, V. H. Galbraith, and E. F. Jacob (Manchester: Printed for the subscribers, 1933), 106.
121. Bremner, *Education of Girls and Women in Great Britain*, 152.
122. "Girl Graduates," *The Students' Journal and Hospital Gazette* (August 30, 1879), 207.

Four: Academic Student Life

1. "The Status of Women Students in the 1890s," *WVU Women: the First Century* (Morgantown: WVU Women's Centenary Project, West Virginia University, 1989). For more see Lillian J. Waugh and Judith G. Stitzel, "'Anything but Cordial': Coeducation and West Virginia University's Early Women," *West Virginia History* 49 (1990), 69–80.
2. "Jottings," *The Durham University Journal* 5, 6 (December 16, 1882), 70.
3. For more on the history of curricular changes at U.S. colleges and universities see Frederick Rudolph, *Curriculum: A History of the American Undergraduate Course of Study Since 1636* (San Francisco: Jossey-Bass Publishers, 1977).
4. Johanna Geyer-Kordesch and Rona Ferguson, *Blue Stockings, Black gowns, White Coats: A Brief History of Women Entering the Medical Profession in Scotland in Celebration of One Hundred Years of Women Graduates at the University of Glasgow* (Glasgow: Wellcome Unit for the History of Medicine, 1994), 14, 39–40.
5. Lewis Campbell, *On the Nationalisation of the Old English Universities* (London: Chapman and Hall, 1901), 232.

6. University of London, *The Calendar for the Year 1870* (London: Taylor and Francis, 1870), 444, and University of London, *The Calendar for the Year 1871* (London: Taylor and Francis, 1871), 465.
7. "Home Notes: The Higher Education of Women," *The Sunday Magazine: For Family Reading* (London: Daldy, Isbister, & Co., 1878), 356.
8. Susan M. Parkes and Judith Harford, "Women and Higher Education in Ireland," in *Female Education in Ireland 1700–1900: Minerva or Madonna*, ed. Deirdre Raftery and Susan M. Parkes (Dublin and Portland, OR: Irish Academic Press, 2007), 106–107.
9. David Woodside, *The Life of Henry Calderwood, LL.D., F.R.S.E.* (London: Hodder and Stoughton, 1900), 233. Professor Calderwood was a longtime supporter of the admission of women to the University of Edinburgh. See also W. N. Boog Watson, "LA of Edinburgh University," *University of Edinburgh Journal* 25 (1971–72), 215–219; A. Logan Turner, ed. *History of the University of Edinburgh 1883–1933* (Edinburgh: Oliver and Boyd, 1933); and Beatrice Welsh, *After the Dawn. A Record of the Pioneer Work in Edinburgh for the Higher Education of Women* (Edinburgh: Oliver and Boyd, 1939).
10. *The Edinburgh University Calendar 1878–1879* (Edinburgh: James Thin, 1878), 57. The organization would change its name to the Edinburgh Association for the University Education of Women. See Appendix, *The Edinburgh University Calendar 1882–1883* (Edinburgh: James Thin, 1882), 15, and *British Universities: Notes and Summaries Contributed to the Welsh University Discussion by Members of the Senate of the University College of North Wales* (Manchester: J. E. Cornish, 1892), 69. The subjects included in this program in 1878 were Biblical criticism, botany, chemistry, English literature, experimental physics, geology, Greek, Latin, logic and mental philosophy, mathematics, moral philosophy, physiology, political economy, theory of education, and zoology.
11. *The Edinburgh University Calendar 1878–1879*, 60.
12. Ibid., 60, 120.
13. Ibid., 123, and Appendix, *The Edinburgh University Calendar 1882–1883*, 15, 20.
14. GUA, Glasgow Association for the Higher Education of Women, Draft petition to Glasgow University Senate for a degree in Arts for women (21 October 1882). See also Geyer-Kordesch and Ferguson, *Blue Stockings, Black Gowns, White Coats*, 41. At times the term "Licentiate" was used in place of "Literate." See also GUA, Janet Galloway, Letter to Principal Caird, Principal of Glasgow University, relating to progress of the College (29 April 1889); Frances H. Melville, "Queen Margaret College," *Pass It On* 15, 1 (November 1935). 1; and Carol Dyhouse, *No Distinction of Sex? Women in British Universities 1870–1939* (London: UCL Press, 1995), 12.
15. *The St Andrews University Calendar for the year 1899–1900* (Edinburgh: William Blackwood and Sons, 1899), 115.
16. J. G. Fitch, "Women and the Universities," *The Contemporary Review* 58 (London: Isbister and Company, 1890), 253.
17. For more on the efforts being made to establish a university in the Scottish Highlands see Christine D. Myers, "A Plea for the Highlands of Scotland: University Reform in the Early 20th Century," in *Contemporary Issues in Education*, ed. David Seth Preston (Amsterdam and New York: Rodopi, 2004), 141–158.
18. GUA, Glasgow Association for the Higher Education of Women, Draft petition to Glasgow University Senate for a degree in Arts for women (21 October 1882).
19. GUA, Glasgow Association for the Higher Education of Women, Petition to the Senate of Glasgow University for a University title for women, plus copies of Mrs. Lindsay's earlier draft, and suggestions of possible alterations (1883).
20. GUA, Lecturers' Committee, Report to Queen Margaret College Executive Council of the affiliation of the College to Glasgow University (21 April 1890).
21. UWA, *Minutes of the Board of Regents* (1866 through 1876), 113. For more see Merle Curti and Vernon Carstensen, *The University of Wisconsin, 1848–1925, 2 vols.* (Madison: University of Wisconsin Press, 1949), and Allan G. Bogue and Robert Taylor, eds., *The University of Wisconsin: One Hundred and Twenty-Five Years* (Madison: University of Wisconsin Press, 1975), 3–35.
22. OUMC, Margaret Boyd Diary (1873), 170, in Ohio Memory: An Online Scrapbook of Ohio History, www.ohiomemory.org/index.html (accessed June 21, 2009). The University of Mississippi also made a shift from writing their diplomas in Latin to English in 1884. For more see David G. Sansing, *The University of Mississippi: A Sesquicentennial History* (Jackson: University Press of Mississippi, 1999), 141, and James Allen Cabaniss, *A History of the University of Mississippi* (University: University of Mississippi, 1949), 106.

23. Ibid., 171.
24. James W. H. Trail, "Natural Science in the Aberdeen Universities," in *Studies in the Development of the University, Aberdeen University Studies: No. 19*, ed. P. J. Anderson (Aberdeen: Aberdeen University Press, 1906), 184.
25. Emily Janes, *The Englishwoman's Year Book and Directory 1900* (London: Adam and Charles Black, 1900), 9.
26. Ibid., 8.
27. Annie McMillan, "Queen Margaret College in the Middle Ages," in *The Book of the Jubilee: In Commemoration of the Ninth Jubilee of the University of Glasgow, 1451–1901*, ed. the Students' Jubilee Celebrations Committee (Glasgow: J. Maclehose and Sons, 1901), 141.
28. GUA, Copy correspondence between John Caird and Mrs. Elder, 9. See also Dyhouse, *No Distinction of Sex?* 41–44, 48.
29. GUA, Correspondence between Sir Richard Lodge, Professor of History, and Secretary of Court, concerning teaching a separate course of lectures for women (27 January 1896).
30. GUA, Glasgow University Court, Excerpt minute from meeting concerning outcome controversy of Mrs. Elder's complaint about the ineffectual treatment of her proposal of equal teaching of women at the College (15 March 1897).
31. Janes, *The Englishwoman's Year Book and Directory 1900*, 6 and *The Calendar of King's College, London for 1896–97* (London: Published by the College, 1896), 136, 275–276, 314. Wood carving was taught by A. J. Bull or W. H. Howard, and women were admitted by special permission.
32. *The Calendar of King's College, London for 1896–97*, 283–284.
33. Ibid., 308–309.
34. Alice Zimmern, *The Renaissance of Girls' Education in England: A Record of Fifty Years' Progress* (London: A. D. Innes & Company, 1898), 135.
35. *The Calendar of King's College, London for 1896–97*, 6, 9, 12, 13, 15, 41.
36. Ibid., lii.
37. Isabel Maddison, ed., *Handbook of Courses Open to Women in British, Continental and Canadian Universities* (New York: The Macmillan Company, 1896), 100.
38. Frederick O'Dwyer, *The Architecture of Deane & Woodward* (Cork: Cork University Press, 1997), 52–53.
39. *Queen's College, Galway, Calendar for 1900–1901* (Dublin: The University Press, 1901), 106, 108, 131, and Royal University of Ireland, *The Calendar for the Year 1908* (Dublin: Alex. Thom & Co., 1908), 266, 442.
40. Judith Harford, *The Opening of University Education to Women in Ireland* (Dublin and Portland, OR: Irish Academic Press, 2008), 80.
41. MEWC, Warren H. Lewis, ed., *Memoirs of the Lewis Family 1850–1930, Volume One: From October 17th, 1850 to September 23rd, 1881* (Oxford: Leeborough Press, 1933), iii, 309, 313, and A. N. Wilson, *C. S. Lewis: A Biography* (New York: Norton, 2002), 2–3.
42. MEWC, Lewis, *Memoirs of the Lewis Family 1850–1930*, 312.
43. Harriet A. McElwain, "Ladies' Department," in *Report of the Pennsylvania State College, for the year 1888* (Harrisburg, PA: Edwin K. Meyers, 1889), 46–47.
44. Michael Bezilla, *Penn State: An Illustrated History* (University Park and London: The Pennsylvania State University Press, 1985), 23.
45. McElwain, "Ladies' Department," in *Report of the Pennsylvania State College, for the year 1888*, 47.
46. *Report of the Pennsylvania State College, for the year 1888* (Harrisburg, PA: Edwin K. Meyers, 1889), 4, and Bezilla, *Penn State*, 60.
47. "The Pennsylvania State College," *The Free Lance* 11, 4 (October 1897), back cover, and "The Pennsylvania State College," *The Free Lance* 14, 6 (February 1901), back cover.
48. *Annual Report of the Regents of the University of Wisconsin, for the Fiscal Year ending September 30, 1869* (Madison, WI: Published by the Board of Regents, 1869), 28–29.
49. Barry Teicher and John W. Jenkins, *A History of Housing at the University of Wisconsin* (Madison, WI: UW History Project, 1987), 7.
50. *Catalogue of the Officers and Students of the University of Wisconsin for the year ending June 21, 1871* (Madison, WI: Atwood & Rublee, 1871), 45.
51. UWA, *Minutes of the Board of Regents* (1866 through 1876), 103, 118, 163, 180. Chadbourne did offer to remain president until a replacement could be found, so the board continued with their usual business.

52. *Catalogue of the Officers and Students of the University of Wisconsin for the year ending June 21, 1871*, 49.
53. *Catalogue of the Officers and Students of the University of Wisconsin for the year ending June 19, 1872* (Madison, WI: Atwood & Rublee, 1872), 41, and *Annual Report of the Regents of the University of Wisconsin for the Fiscal Year Ending September 30, 1877* (Madison, WI: David Atwood, 1877), 12.
54. *West Virginia University, Morgantown, Catalogue 1889–90: Announcements for 1890–91* (Charleston, WV: Moses W. Donnally, 1890), 11.
55. Ibid., 10–11.
56. *Catalogue of West Virginia University, Morgantown, For the Year 1891–92* (Charleston, WV: Moses W. Donnally, 1892), 9.
57. James Albert Woodburn, *Higher Education in Indiana* (Washington, DC: Government Printing Office, 1891), 86.
58. Burton Dorr Myers, *History of Indiana University Volume II: The Bryan Administration* (Bloomington: Published by Indiana University, 1952), 551.
59. *Annual Catalogue of the Indiana University for the Sixty-Seventh College Year, 1890–91* (Indianapolis, IN: Wm. B. Burford, 1891), 24.
60. *Annual Catalogue of the Ohio University 1875* (Athens, OH: Published by the University, 1876), 27.
61. Thomas N. Hoover, *The History of Ohio University* (Athens: Ohio University Press, 1954), 142.
62. OUMC, Margaret Boyd Diary (1873), 17, 19, 21, 26, 53, 114, 160–161.
63. Albert Shaw, ed., *The Review of Reviews* 11, 60 (January 1895), 8. This was followed by the phrase "Send for one" without the more avid appeals made by some other colleges and universities.
64. *Calendar of the University of Michigan for 1880–1881* (Ann Arbor, MI: The Courier Steam Printing House, 1881), 93, 108.
65. Ibid., 72, 93, and *Calendar of the University of Michigan for 1894–95* (Ann Arbor, MI: The Register Publishing Company, 1895), 138, 176.
66. GUA, Volume of Presscuttings of Queen Margaret College, 1891–1894, *Glasgow Herald*, May 5, 1893.
67. GUA, Volume of Presscuttings of Queen Margaret College, 1891–1894, 82.
68. Janes, *The Englishwoman's Year Book and Directory 1900*, 6.
69. *The Department of Education in the University of Manchester 1890–1911* (Manchester: Manchester University Press, 1911), 58. The year 1894 marked another change for women at Manchester because that is the year that most of the college's scientific laboratories were opened to them after much complaint from the female students. For more see Sarah V. Barnes, "Crossing the Invisible Line: Establishing Co-education at the University of Manchester and Northwestern University," *History of Education* 23, 1 (1994), 47.
70. Barnes, "Crossing the Invisible Line," 41.
71. "The Affiliated Colleges and the Prizes of the University," *The Durham University Journal* 5, 3 (May 27, 1882), 26.
72. Lilian Daly, "Women and the University Question," *The New Ireland Review* 17 (March 1902 to August 1902), 74–75.
73. *Reports from Commissioners, Inspectors, and Others: Thirty-Four Volumes, 21. Wales and Monmouthshire, Session 11 February 1896–14 August 1896, Vol. XXXV* (1896), 466; and *The Parliamentary Debates (Authorised Edition), Fourth Series: Commencing with the Fifth Session of the Twenty-sixth Parliament of the United Kingdom of Great Britain and Ireland. 62 Victoriae. Volume LXXIII, Comprising the period from the Twentieth Day of June to the Fifth Day of July 1899* (London: Wyman and Sons, 1899), 374.
74. W. J. Wallis-Jones, "The University College of Wales," *WALES: A National Magazine for the English Speaking Parts of Wales* 3, 26 (June 1896), 248.
75. Harriet A. McElwain, "Ladies' Department," in *Annual Report of the Pennsylvania State College, for the year 1894* ([Harrisburg, PA]: Clarence M. Busch, 1895), 148.
76. William T. Doherty, Jr. and Festus P. Summers, *West Virginia University: Symbol of Unity in a Sectionalized State* (Morgantown: West Virginia University Press, 1982), 65.
77. GUA, Queen Margaret College Letterbook, 1878–1883 (Correspondence Courses), 18 and 23 September 1878, 13, 15. There was also an Edinburgh School of Cookery at the same time offering similar courses for women in that city. For more see Christine D. Myers, "The Glasgow Association for the Higher Education of Women, 1878–1883," *Historian* 63, 2

(Winter 2001), 357–371, and Tom Begg, *The Excellent Women: The Origins and History of Queen Margaret College* (Edinburgh: John Donald Publishers, 1994).

78. Frank K. Prochaska, *Women and Philanthropy in Nineteenth-Century England* (Oxford: Clarendon Press, 1980), 2–3. See also Emily Davies, "Some account of a proposed new College for women," in *The Educators: Female Education*, ed. Marie Mulvey Roberts and Tamae Mizuta (London: Routledge/Thoemmes Press, 1995), 97–98. This paper was read at the Annual Meeting of the National Association for the Promotion of Social Science in 1868 and was subsequently printed in 1872.
79. *Report of the Pennsylvania State College, for the year 1888* (Harrisburg, PA: Edwin K. Meyers, 1889), 3–4, and Bezilla, *Penn State*, 45.
80. "Should University Degrees be given to Women?" in *The Westminster Review Vol. CXV January-April, 1881, American Edition* (New York: Leonard Scott Publishing Company, 1881), 239.
81. Doherty and Summers, *West Virginia University*, 64, and *The Volunteer Vol. I Published Annually by The Students of the University of Tennessee* (Knoxville, TN: S. B. Newman & Co., 1897), 121. The class in domestic science at Tennessee was pictured in the 1897 yearbook and included eighteen students and one teacher.
82. "Domestic Science in the Agricultural Colleges," *The American Kitchen Magazine* 7, 6 (September 1897), 219.
83. Charles W. Dabney, "Report of the President," *University of Tennessee Record* 2 (February 1901), 24.
84. "Horticulture and Forestry," *University of Tennessee Record* 2 (February 1901), 110, 199.
85. *Calendar of the University of Michigan for 1880–1881*, 125.
86. *Catalogue and Announcements of the University of Mississippi, University P. O., (Near Oxford, Miss.) Forty-Fifth Session 1896–'97* (Yazoo City, MS: Mott Printing Company, 1897), 69.
87. *Historical and Current Catalogue of the Officers and Students of the University of Mississippi, Forty-Second Session, 1893–'94* (Oxford, MS: Published by the University, 1894), 197.
88. Cabaniss, *A History of the University of Mississippi*, 122.
89. *The Garnet and Black 1899, Published by the Students Volume I* (Columbia, SC: The Bryan Printing Co., 1899), 23. South Carolina also had a Normal Course that grew in popularity over time.
90. *The Garnet and Black, Published by the Students of the South Carolina College, Nineteen Hundred and One* (Columbia, SC: The Bryan Printing Co., 1901), 120.
91. *The St Andrews University Calendar for the year 1899–1900*, 518–519. Interestingly the male studentships were £25 per annum and the women's were only £20.
92. *Annual Catalogue of the Indiana University for the Sixty-Eighth College Year, 1891–92* (Indianapolis, IN: Wm. B. Burford, 1892), 83–84.
93. *The Indiana University Catalogue, Seventh-Fourth College Year 1897–98* (Bloomington, IN: Published by the University, 1898), 19, and Edward Mussey Hartwell, "Physical Training," in *The Report of the Commissioner of Education for 1897–98*, United States Bureau of Education (Washington, DC: Government Printing Office, 1899), 550.
94. *The Indiana University Catalogue, Seventh-Fourth College Year 1897–98*, 94–95.
95. *The Calendar of King's College, London for 1896–97* (London: Published by the College, 1896), 310–311.
96. For more see Maresi Nerad, *The Academic Kitchen: A Social History of the Gender Stratification at the University of California, Berkeley* (Albany: State University of New York Press, 1988).
97. Angie Warren Perkins, "Report of the Acting Dean, Woman's Department," *University of Tennessee Record* 8 (January 1899), 24–25.
98. M. Carey Thomas, "The Future of Woman's Higher Education," in *Mount Holyoke College: The Seventy-fifth Anniversary* (South Hadley, MA, 1913), 100–104. For more see Rosalind Rosenberg, "The Limits of Access: The History of Coeducation in America," in *Women and Higher Education in American History*, ed. John M. Faragher and Florence Howe (New York: Norton, 1988), 124; Joan N. Burstyn, "Historical Perspectives on Women in Educational Leadership," in *Women and Educational Leadership*, ed. Sari Knopp Biklen and Marilyn B. Brannigan (Lexington, MA: Lexington Books, 1980), 65–75; and Susan B. Carter, "Academic Women Revisited: An Empirical Study of Changing Patterns in Women's Employment as College and University Faculty, 1890–1963," *Journal of Social History* 14 (1981), 675–699.
99. UWA, *Minutes of the Board of Regents* (1866 through 1876), 147, 163, 207, 217.
100. Ibid., 138.

101. *Annual Report of the Regents of the University of Wisconsin for the Fiscal Year Ending September 30, 1877* (Madison, WI: David Atwood, 1877), 37.
102. Reuben Gold Thwaites, ed., *The University of Wisconsin: Its History and its Alumni.* (Madison, WI: J. N. Purcell, 1900), 766–775. Frederick Jackson Turner, John M. Olin, and (university president) Charles R. Van Hise are three examples of instructors who went on to become prominent Wisconsin professors.
103. *Catalogue of the University of Wisconsin for the academic year 1891–92* (Madison, WI: Published by the University, 1891), 129. For more on Miss Frisby's life and career see Thwaites, *The University of Wisconsin*, 458.
104. Carol Sonenklar, *We Are a Strong, Articulate Voice: A History of Women at Penn State* (University Park: Pennsylvania State University Press, 2006), 26.
105. Bezilla, *Penn State*, 22.
106. Ibid.
107. *Calendar of the University of Michigan for 1894–95* (Ann Arbor, MI: The Register Publishing Company, 1895), 16, 18.
108. *Biennial Report of the Board of Curators of the University of Missouri to the 36th General Assembly for the Two Years Ending December 31, 1890* (Jefferson City, MO: Tribune Printing Company, 1891), 6.
109. Frances E. Willard, *Occupations for Women: A Book of Practical Suggestions for the Material Advancement, the Mental and Physical Development, and the Moral and Spiritual Uplift of Women* (Cooper Union, NY: Success Company, 1897), 276, and *Report of the Board of Curators of the State of Missouri to the XXXIst General Assembly* (Jefferson City, MO: Tribune Printing Company, 1881), 8.
110. Sansing, *The University of Mississippi*, 3–4, 136, 138–139. Thomas Dudley Isom is credited with giving Oxford, Mississippi, its name. For more see Cabaniss, *A History of the University of Mississippi*, 7–8.
111. *Catalogue of the University of Mississippi at University P. O., Near Oxford, Miss.: Prepared this Year with Special Reference to the Schools of English and Belles Lettres. Thirty-Ninth Session 1890–'91* (Oxford, MS: Published by the University, 1890), 53–54. Elocution was not a required course of study at Mississippi.
112. Sansing, *The University of Mississippi*, 139.
113. *Historical and Current Catalogue of the Officers and Students of the University of Mississippi, Forty-Second Session, 1893–'94* (Oxford, MS: Published by the University, 1894), 195.
114. James Maitland Anderson, ed., *The Matriculation Roll of the University of St. Andrews 1747–1897* (Edinburgh and London: William Blackwood and Sons, 1905), lxxxiv.
115. Joseph Thompson, *The Owens College: Its Foundation and Growth; and its Connection with the Victoria University, Manchester* (Manchester: J. E. Cornish, 1886), 498, 501–502, and *The Victoria University of Manchester: Register of Graduates up to July 1st, 1908* (Manchester: Manchester University Press, 1908), 392. See also Barnes, "Crossing the Invisible Line," 43.
116. Wallis-Jones, "The University College of Wales," 244–245.
117. Ibid., 243–245, and A. Wallis Myers, "Women Students in Wales," *The Ludgate Illustrated Magazine Vol. VIII* (London: F. V. White & Co., 1899), 137.
118. Wallis-Jones, "The University College of Wales," 244–245.
119. Janes, *The Englishwoman's Year Book and Directory 1900*, 5, and Zimmern, *The Renaissance of Girls' Education in England*, 134.
120. M. Montgomery Campbell, "The Central Conference of Women Workers," in *The Monthly Packet*, ed. Christabel R. Coleridge and Arthur Innes (London: A. D. Innes and Co., 1894), 751.
121. McElwain, "Ladies' Department," in *Report of the Pennsylvania State College, for the year 1888*, 45.
122. *University of Tennessee Register for 1897–'98 and Announcement for 1898–'99* (Knoxville: The University of Tennessee Press, 1898), 3.
123. *Eleventh Annual Report of the Agricultural Experiment Station of the University of Tennessee to the Governor 1898* (Knoxville: The University Press, 1899), 22. and *The Volunteer Published by the Students Vol. II* (Knoxville, TN: Bean, Warters & Gaut, 1898), 13, 15.
124. *The Volunteer Volume V 1901 Published Annually by the Students' Association, University of Tennessee* (Knoxville, TN: Ogden Bros. & Co., 1901), 251.
125. James B. Sellers, *History of the University of Alabama* (Tuscaloosa: University of Alabama Press, 1953), 413–415. Mrs. Gorgas was the daughter of former Alabama governor John Gayle and

the mother of future U.S. surgeon general William C. Gorgas. The library at the University of Alabama is named in her honor. For more see *Memorial Services Held in Honor of Major General William Crawford Gorgas by the Southern Society of Washington, D.C.* (Washington, DC: Government Printing Office, 1921), 8–11, and Thomas McAdory Owen, *History of Alabama and Dictionary of Alabama Biography Volume II* (Chicago: S. J. Clarke Publishing Company, 1921), 897, 1363, 1364.

126. Thomas Waverly Palmer, comp., *A Register of the Officers and Students of the University of Alabama 1831–1901* (Tuscaloosa: The University of Alabama, 1901), 17, 21, 23, 25–26, 28–30, and Sellers, *History of the University of Alabama*, 414.
127. *The Corolla '96* (Tuscaloosa. AL: W. H. Ferguson, 1896), 2, 6.
128. Ibid., 113. See also Sellers, *History of the University of Alabama*, 463. Mrs. Gorgas became matron of the infirmary in 1879 while her husband was still president.
129. Sansing, *The University of Mississippi*, 154.
130. Doherty and Summers, *West Virginia University*, 65.
131. Parkes and Harford, "Women and Higher Education in Ireland," 112.
132. "Queen's College Morality," *The Nation*, February 20, 1864, 11.
133. W. Le Conte Stevens, *The Admission of Women to Universities* (New York: Press of S. W. Green's Son, 1883), 11.
134. H. B. Hutchins, "The University and Co-Education," *The Michigan Alumnus* 17, 160 (January 1911), 182.
135. UMC, *The Savitar—1899; by Students of the Junior Class* (Columbia, MO: E. W. Stephens, 1899), 188, 190. In both pictures several of the women can be seen wearing hats that were in style at the end of the nineteenth century. See also James and Vera Olson, *The University of Missouri: An Illustrated History* (Columbia: University of Missouri Press, 1988), 35, and *The Savitar—1900; Published by the Junior Class of the University of Missouri 1900* (Columbia, MO: E. W. Stephens, 1901), 10, 13.
136. Wallis-Jones, "The University College of Wales," 243; Harford, *The Opening of University Education to Women in Ireland*, 79; and *The University of Glasgow Through Five Centuries* ([Glasgow]: Published by the University in commemoration of the Fifth Centenary, 1951).
137. Sellers, *History of the University of Alabama*, 481.
138. Christina Sinclair Bremner, *Education of Girls and Women in Great Britain* (London: Swan Sonnenschein & Co., 1897), 142.
139. J. T. Fowler, *Durham University: Earlier Foundations and Present Colleges* (London: F. E. Robinson, 1904), 137–139, 140.
140. Helen M. Nimmo, "Some Recent Notes and Recollections of Queen Margaret College Life," in *The Book of the Jubilee: In Commemoration of the Ninth Jubilee of the University of Glasgow*, 149–150.
141. Lawrence Hutton, *Literary Landmarks of the Scottish Universities* (New York and London: G. P. Putnam's Sons, 1904), 118–119.
142. John Malcolm Bulloch, ed., *College Carols* (Aberdeen: D. Wylie and Son, 1894), 3, 35.
143. L. K. Sabine, "The Romance of a Freshman," *The Inlander* 11, 8 (May 1901), 318.
144. "Edwin Hammett's Sigh for Telepathy," in *The Savitar—1895; by Junior Class of 1894* (Columbia, MO: E.W. Stephens Printing Company, 1895), 109.
145. *The Volunteer Vol. I*, 139.
146. Emma O. Lundberg, "Women in the University of Wisconsin," *The Wisconsin Alumni Magazine* (April 1908), 64.
147. "U. W. One of First to Admit Women," *The Wisconsin State Journal*, December 31, 1919.
148. Henrietta Wood Kessenich, "'Twas Long, Long Ago," *The Wisconsin Alumnus* (1938), 307.
149. Wallis-Jones, "The University College of Wales," 245. Other descriptions of the location also emphasize the attractiveness of the setting, with all of the lecture rooms overlooking Cardigan Bay. For more see T. Levi, "Welsh Education," *The Cambrian Volume Twenty-Two* (Utica, NY: Thomas J. Griffiths, 1902), 12.
150. Olson, *The University of Missouri*, 15–16.
151. A basic difference between British and U.S. universities is that those in the United States focus on graduation rates, while in the nineteenth century most British universities focused on compiling matriculation data.
152. Amy Thompson McCandless, "Maintaining the Spirit and Tone of Robust Manliness: The Battle against Coeducation at Southern Colleges and Universities, 1890–1940," *NWSA Journal* 2, 2 (Spring 1990), 211–212.

153. Andrew Cunningham McLaughlin, *History of Higher Education in Michigan* (Washington, DC: Government Printing Office, 1891), 68.
154. McCandless, "Maintaining the Spirit and Tone of Robust Manliness," 203.
155. Rheta Childe Dorr, "Breaking Into the Human Race," *Hampton's Magazine* 27, 3 (September 1911), 328.
156. *The Garnet and Black 1899, Published by the Students Volume I* (Columbia, SC: The Bryan Printing Co., 1899), 106.
157. Ibid., 112.
158. *Glasgow News*, April 4, 1877, in GUA, Glasgow Association for the Higher Education of Women. General committee meeting minutes with presscuttings re: inaugural meeting.
159. Stevens, *The Admission of Women to Universities*, 10–12.
160. Sellers, *History of the University of Alabama*, 477–478.
161. *The Corolla of Ninety-Four* (Tuscaloosa: Published by the Students of the University of Alabama, 1894), 162.
162. Sellers, *History of the University of Alabama*, 481.
163. John William Abercrombie, "Address of Welcome: For the University," *1831–1906 University of Alabama Bulletin Commemoration Number* (November 1906), 20.
164. Doherty and Summers, *West Virginia University*, 44.
165. James Riley Montgomery, Stanley J. Folmsbee, and Lee Seifert Greene, *To Foster Knowledge: A History of The University of Tennessee 1794–1970* (Knoxville: The University of Tennessee Press, 1984), 148.
166. *Phi Beta Kappa: Catalogue of the Alpha of Missouri, 1901–1909* (Columbia, MO: E. W. Stephens Publishing Company, 1909), 56, 62, 65, 68, 71, 73, 75–76, 79–80, 82. Other University of Missouri undergraduate women who became members of Phi Beta Kappa during the Victorian Era were Zannie May Denny and Lucy Gentry in 1885; Ulie B. Denny in 1889; Leila Ruth Britt in 1891; Jean Augusta Shaefer in 1894; Cora Alice Eitzen in 1896; Minnie Katherine Organ and Ethel Swearingen in 1897; Jessie Alice Blair in 1898; Ida Moore Edwards and Mary Basset Potter in 1899; Meta Therese Eitzen, Rosalie Gerig, and Emily Guitar in 1900; and Talitha Jennie Green in 1901.
167. "The First Co-Ed to Graduate from the University," *The Missouri Alumni Quarterly* (September 1905), 15.
168. *The Addresses and Journal of Proceedings of the National Educational Association, Session of the Year 1874, at Detroit, Michigan* (Worcester, MA: Published by the Association, 1874), 134.
169. Wallis-Jones, "The University College of Wales," 243.
170. Ibid., 246.
171. *Catalogue of the Western University of Pennsylvania for the year ending 1897 with detailed statements of the courses of instruction* ([Pittsburgh]: Western University of Pennsylvania, 1897), 213.
172. *Catalogue of the Western University of Pennsylvania for the year ending 1899 with detailed statements of the courses of instruction* ([Pittsburgh]: Western University of Pennsylvania, 1899), 212.
173. *The Western University of Pennsylvania, 75th Annual Commencement, Carnegie Music Hall, Pittsburgh, Pa. June ninth, Eighteen Hundred and Ninety-eight.* (Pittsburgh: University of Pittsburgh, Digital Research Library, 2006), and Robert C. Alberts, *Pitt: The Story of the University of Pittsburgh, 1787–1987* (Pittsburgh: University of Pittsburgh Press, 1986), 43. The sisters graduated with identical grades in 1898.
174. *The Western University of Pennsylvania 1878–1900, Annual Commencement of the Collegiate, Engineering and Legal Departments. Carnegie Music Hall, Pittsburgh, Pennsylvania, June 14th, 1900, 8:15 P. M.* (Pittsburgh: University of Pittsburgh, Digital Research Library, 2006).
175. Alberts, *Pitt*, 48.
176. *Report of the Chancellor to the Board of Trustees of Western University of Pennsylvania in Annual Session June 3rd, 1907* ([Pittsburgh]: Western University of Pennsylvania, 1906), 4, 17. Previously, Samuel B. McCormick noted that there were only one or two female students in any of the graduating classes, and only fourteen total alumnae in the first decade of coeducation. In 1906–1907, however, nineteen women were taking courses at the university.
177. *Queen's College, Galway, Calendar for 1898–1899* (Dublin: The University Press, 1899), 100, and *Queen's College, Galway, Calendar for 1900–1901* (Dublin: The University Press, 1901), 104. Both Clarke and Daly would earn their degrees in 1904. Clarke went on to teach at Galway from 1910 to 1942. See The Royal University of Ireland, *The Calendar for the Year 1908*, 281, 288, and Parkes and Harford, "Women and Higher Education in Ireland," 112.

178. *Queen's College, Galway, Calendar for 1900–1901*, 104; "Women and Their Work We Ought to Know About," *The Woman's Medical Journal* 17, 4 (April 1907), 226; and The Royal University of Ireland, *The Calendar for the Year 1908*, 398, 448–449, 497. Alice Perry earned First Class Honours in civil engineering and was appointed to replace her father as interim county surveyor in Galway after his death in 1906. For more on Alice Perry's life after graduation see "Collected Poems of First Ever Woman Engineer Are Presented to UCG," *City Tribune*, December 13, 1996, 6.
179. Harford, *The Opening of University Education to Women in Ireland*, 77.
180. The Royal University of Ireland, *The Calendar for the Year 1908*, 173, 246.
181. *Queen's College, Galway, Calendar for 1900–1901*, 160.
182. "The Calendar of the Royal University," *The Nation*, March 24, 1888, 4.
183. *Dickens's Dictionary of London, 1879: An Unconventional Handbook* (London: Charles Dickens and Evans, 1879), 267.
184. Katharine Lake, ed., *Memorials of William Charles Lake, Dean of Durham 1869–1894* (London: Edward Arnold, 1901), 129.
185. Fowler, *Durham University*, 62. See also *Durham Calendar with Almanack 1910–1911* (Durham: Thomas Caldcleugh & Son and London: Whittaker & Co., 1910), 118–119, 289.
186. *The St Andrews University Calendar for the year 1899–1900*, 223.
187. GUA, Examination results, notes on bursaries and qualified medical students, 1895–96, and List of Bursaries from GAHEW for Glasgow University Local Exams (1883).
188. Recent historiography questions the democratic nature of Scottish education, especially the pervasive belief in the accessibility of the university system that became an integral part of Scotland's national identity. For more see R. D. Anderson, "The Scottish University Tradition: Past and Present," in *Scottish Universities: Distinctiveness and Diversity*, ed. Jennifer J. Carter and Donald J. Witherington (Edinburgh: Edinburgh University Press, 1992), 72, and "In Search of the 'Lad of Parts': The Mythical History of Scottish Education," *History Workshop Journal* 19 (Spring 1985), 82–104; George Elder Davie, *The Democratic Intellect: Scotland and her Universities in the Nineteenth Century* (Edinburgh: Edinburgh University Press, 1961), and *The Crisis of the Democratic Intellect: The Problem of Generalism and Specialisation in Twentieth-Century Scotland* (Edinburgh: Polygon, 1986).
189. GUA, G. G. Henderson, Letter to Janet Galloway, reporting on the Committee of Lecturers' recommendations to the Council of Queen Margaret College concerning prizes (4 February 1892).
190. OUMC, Margaret Boyd Diary (1873), 17, 21, 38, 81–86, 93, 147.
191. Ibid., 147.
192. Ibid., 63.
193. Ibid., 52.
194. *Notes and Materials for the History of University College, London: Faculties of Arts and Science* (London: H. K. Lewis, 1898), 17, 22.
195. Emily Davies, "Women in the Universities of England and Scotland," in *The Educators: Female Education*, ed. Marie Mulvey Roberts and Tamae Mizuta (London: Routledge/Thoemmes Press, 1995), 158. See also Eileen Breathnach, "Women and Higher Education in Ireland (1879–1914)," *Crane Bag* 4, 1 (1980), 50.
196. "Educational News Items," *The Southern Educational Journal* 13, 1 (November 1899), 9.
197. Charlotte Wronker, "Co-Education in the 'Varsity," *The Missouri Alumni Quarterly* (December 1905), 29.
198. Hutchins, "The University and Co-Education," 181.

Five: Facilitating Coeducation

1. UWA, *Catalogue of the Officers and Students of the University of Wisconsin, For the Year 1872–73 and the First Term of 73–74* (Madison, WI: Atwood & Culver, 1873), frontispiece.
2. Jim Feldman, in his work on *The Buildings of the University of Wisconsin* (Madison, WI: University Archives, 1997), remarked that Bascom Hill "may be the longest, steepest, stepless incline in American higher education."
3. "Items of Interest," *The School World* 6, 64 (April 1904), 153.
4. Henrietta Wood Kessenich, "'Twas Long, Long Ago," *The Wisconsin Alumnus* (1938), 306.

5. "Co-Education of the Sexes in Colleges," *Indiana School Journal* 25, 8 (August 1880), 421–422.
6. Emily Davies, "Women in the Universities of England and Scotland," in *The Educators: Female Education*, ed. Marie Mulvey Roberts and Tamae Mizuta (London: Routledge/Thoemmes Press, 1995), 193, 196.
7. *The Calendar of the University College of Wales, Aberystwyth, Fourteenth Session, 1885–6* (Manchester: J. E. Cornish, 1885), 35.
8. W. J. Wallis-Jones, "The University College of Wales," *WALES: A National Magazine for the English Speaking Parts of Wales* 3, 26 (June 1896), 243.
9. A. Wallis Myers, "Women Students in Wales," *The Ludgate Illustrated Magazine Vol. VIII* (London: F. V. White & Co., 1899), 137.
10. Ibid., 141.
11. Wallis-Jones, "The University College of Wales," 244.
12. "Boards of Visitors: Report for the Year 1883–84" in *Biennial Report of the Board of Regents of the University of Wisconsin, for the Two Years Ending September 30, 1884* (Madison, WI: Democrat Printing Co., 1883), 53.
13. Helen Lefkowitz Horowitz, *Alma Mater: Design and Experience in the Women's Colleges from Their Nineteenth-Century Beginnings to the 1930s* (Amherst: University of Massachusetts Press, 1993), xv.
14. *Annual Report of the Regents of the University of Wisconsin, for the year ending September 30, 1857* (Madison, WI: Calkins & Webb Printers, 1857), 10, and UWA, *Minutes of the Board of Regents* (1866 through 1876), 124, 163.
15. Barry Teicher and John W. Jenkins, *A History of Housing at the University of Wisconsin* (Madison, WI: UW History Project), 6.
16. M. Montgomery Campbell, "The Central Conference of Women Workers," in *The Monthly Packet*, Vol. 88, ed. Christabel R. Coleridge and Arthur Innes (London: A. D. Innes and Co., 1894), 750–751. See also Foster Watson, *The Encyclopaedia and Dictionary of Education* (London, Bath, Melbourne, Toronto, and New York: Sir Isaac Pitman & Sons, 1921), 1563.
17. GUA, Letters (8) from Miss Galloway; Masson Hall, Edinburgh; University Hall St. Andrews; and students to Mrs. Riddoch relating to the Hall regulations concerning visitors (8 March 1902).
18. Elizabeth C. Wallace, "Queen Margaret Hall," *Pass It On* 15, 1 (November 1935), 11.
19. Robert C. Alberts, *Pitt: The Story of the University of Pittsburgh, 1787–1987* (Pittsburgh, PA: University of Pittsburgh Press, 1986), 45, 47.
20. Agnes Lynn Starrett, *Through One Hundred and Fifty Years: The University of Pittsburgh* (Pittsburgh, PA: University of Pittsburgh Press, 1937), 204.
21. *Catalogue of the Western University of Pennsylvania for the year ending 1901 with detailed statements of the courses of instruction* ([Pittsburgh]: Western University of Pennsylvania, 1901), 215.
22. "Famous Firsts for WVU Women," *WVU Women: The First Century* (Morgantown: WVU Women's Centenary Project, West Virginia University, 1989).
23. *Biennial Report of the Board of Curators of the University of Missouri to the 36th General Assembly for the Two Years Ending December 31, 1890* (Jefferson City, MO: Tribune Printing Company, 1891), 20, and John E. Swanger, comp., *Official Manual of the State of Missouri for the Years 1907–1908* (Jefferson City, MO: The Hugh Stephens Printing Company, 1907), 231.
24. Calvin M. Woodward, "Acceptance of the Buildings," in *The Order of Exercises and the Addresses at the Dedication of Academic Hall and the New Department Buildings, on Tuesday, June the Fourth, A. D. One Thousand, Eight Hundred and Ninety-Five* (Columbia, MO: Printed by the University, 1895), 13. See also Charlotte Wronker, "Co-Education in the 'Varsity," *The Missouri Alumni Quarterly* (December 1905), 28, and James and Vera Olson, *The University of Missouri: An Illustrated History* (Columbia: University of Missouri Press, 1988), 32–33.
25. *The Calendar of King's College, London for 1896–97* (London: Published by the College, 1896), 274.
26. J. M. Horsburgh, "Report of University College, London," in *Reports from University Colleges 1899, Presented to both Houses of Parliament by Command of Her Majesty* (London: Wyman and Sons, 1899), 210.
27. Judith Harford, *The Opening of University Education to Women in Ireland* (Dublin and Portland, OR: Irish Academic Press, 2008), 79, and NUIG, "Ladies' Notes," *QCG: a record of college life in the city of the tribes* 2, 3 (May 1904), 21.

28. Florence V. Skeffington, "Report of the Dean of the Woman's Department," *University of Tennessee Record* 2 (February 1901), 42.
29. *The Corolla of Ninety-Five, Volume III* (Tuscaloosa: Published by the Students of the University of Alabama, 1895), 177.
30. *Catalogue of the Officers and Students of the University of Mississippi, at Oxford, Mississippi, Twenty-Seventh Session* (Jackson, MS: The Clarion Steam Printing Establishment, 1879), 65.
31. *Catalogue of the University of Mississippi at University P. O., Near Oxford, Miss.: Prepared this Year with Special Reference to the Schools of English and Belles Lettres. Thirty-Ninth Session 1890–'91* (Oxford, MS: Published by the University, 1890), 92, and David G. Sansing, *The University of Mississippi: A Sesquicentennial History* (Jackson: University Press of Mississippi, 1999), 53, 158–159.
32. *Catalogue and Announcements of the University of Mississippi at University P. O., Forty-Third Session 1894–'95* (Vicksburg, MS: Vicksburg Printing & Publishing Co., 1895), 76.
33. GUA, Professor Latta, Letter to Miss Galloway concerning the use of the University Library by women students (26 October 1908).
34. Norman Fraser, *Student Life at Edinburgh University* (Paisley: J. and R. Parlane, 1884), 73.
35. University of Aberdeen, *Handbook to the City and University* (Aberdeen: Printed for the University, 1906), 84.
36. Wallis-Jones, "The University College of Wales," 245–246, and Iwan Morgan, *The College by the Sea (A Record and a Review): "Nid Byd Byd Heb Wybodaeth"* (Aberystwyth: Published by the Students' Representative Council in Collaboration with the College Council, 1928), 231.
37. James Albert Woodburn, *Higher Education in Indiana* (Washington, DC: Government Printing Office, 1891), 84, 89–90. More importantly, university officials guaranteed that they were "thoroughly fireproof."
38. *Annual Report of the Indiana University including the Catalogue for the Academical Year, 1882–1883* (Indianapolis, IN: Wm. B. Burford, 1883), 44.
39. Theophilus A. Wylie, *Indiana University, Its History from 1820, When Founded, to 1890, with Biographical Sketches of Its Presidents, Professors and Graduates, and a List of Its Students from 1820 to 1887* (Indianapolis, IN: Wm. B. Burford, 1890), 84.
40. "Indiana University," *The Educator-Journal* 2, 1 (September 1901), 42, and Mary Caroline Crawford, *The College Girl of American and the Institutions Which Make Her What She Is* (Boston: L.C. Page & Company, 1905), 260.
41. Fraser, *Student Life at Edinburgh University*, 11, 19.
42. *The St Andrews University Calendar for the year 1899–1900* (Edinburgh: William Blackwood and Sons, 1899), 26.
43. Ibid., 514.
44. F. S. Dumaresq de Carteret-Bisson, *Our Schools and Colleges, Vol. II: For Girls* (London: Simpkin, Marshall & Co., 1884), 177.
45. Ibid.
46. Marion Gilchrist, "Some Early Recollection of the Queen Margaret Medical School," *Surgo* (March 1948), 81.
47. W. Le Conte Stevens, *The Admission of Women to Universities* (New York: Press of S. W. Green's Son, 1883), 10–11.
48. Woodburn, *Higher Education in Indiana*, 90, and *Annual Catalogue of the Indiana University for the Sixty-Seventh College Year, 1890–91* (Indianapolis, IN: Wm. B. Burford, 1891), 71.
49. OUMC, Margaret Boyd Diary (1873), 26, in Ohio Memory: An Online Scrapbook of Ohio History, www.ohiomemory.org/index.html (accessed June 21, 2009).
50. Thomas N. Hoover, *The History of Ohio University* (Athens: Ohio University Press, 1954), 186–187. The first women's residence hall built on campus, Boyd Hall, was not built until 1906–1907. The university also decided to purchase Women's Hall in 1908, increasing the number of female students that could be housed in official university lodgings.
51. "Ohio University," *Journal of Pedagogy* 19, 1 (September 1906), 89. The *Journal of Pedagogy* was first published at Ohio University. See also "Beta—Ohio University," *The Rainbow of Delta Tau Delta* 11, 1 (January 1888), 82.
52. *Catalogue of the South Carolina College 1904–1905* (Columbia, SC: The R. L. Bryan Company, 1905), 67.
53. *Catalogue of the University of Mississippi, Thirty-Ninth Session 1890–91*, 27.
54. *The Corolla of Ninety-Five*, 177.

55. Amy Thompson McCandless, "Maintaining the Spirit and Tone of Robust Manliness: The Battle against Coeducation at Southern Colleges and Universities, 1890–1940," *NWSA Journal* 2, 2 (Spring 1990), 203.
56. *Bulletin of the University of Mississippi: Announcements and Catalogue of the University of Mississippi, University P. O., (Near Oxford), Fifty-Second Session, (Fifty-Fifth Year), 1903–1904* 3, 1 (April 1904), 44. The women's residence hall was named for "Mrs. Fanny J. Ricks of Yazoo City."
57. "The Queen's University," *The Nation*, October 16, 1875, 12.
58. "The University Question," *The Nation*, September 18, 1880, 8.
59. "Co-Education at West Virginia University," *WVU Women: The First Century* (Morgantown: WVU Women's Centenary Project, West Virginia University, 1989).
60. "Women's Housing, 1889–1918," *WVU Women: The First Century* (Morgantown: WVU Women's Centenary Project, West Virginia University, 1989).
61. GUA, Volume of Presscuttings of Queen Margaret College, 1884–1890, dated June 22, 1888, no source. The source is unknown, though it is likely to be the *Evening Times*.
62. *Biennial Report of the Board of Curators of the University of Missouri to the 36th General Assembly*, 8.
63. Olson, *The University of Missouri: An Illustrated History*, 42, and "University Dictionary," in *The Savitar—1898; by the Junior Class of the University of Missouri 1897* (Columbia, MO: E. W. Stephens, 1898), 166.
64. Alberts, *Pitt*, 33, 36, 49.
65. "Editorials," *The Western University Courant* 11, 3 (November 1895), 1. See also "Pittsburgh as a Site for Universities," *The University Courant* 5, 6 (October 1891), 51–54.
66. "Editorial," *The Western University Courant* 10, 2 (November 1894), 6.
67. *Annual Report of the Regents of the University for the year ending September 30, 1857*, 10. See also Horowitz, *Alma Mater*, 32.
68. Emily Janes, *The Englishwoman's Year Book and Directory 1900* (London: Adam and Charles Black, 1900), 5, and Helene Lange, *Higher Education of Women in Europe* (New York: D. Appleton and Company, 1897), 40.
69. Alice Zimmern, *The Renaissance of Girls' Education in England: A Record of Fifty Years' Progress* (London: A. D. Innes & Company, 1898), 134.
70. Skeffington, "Report of the Dean of the Woman's Department," 43.
71. James Maitland Anderson, ed., *The Matriculation Roll of the University of St. Andrews 1747–1897* (Edinburgh and London: William Blackwood and Sons, 1905), xlvii.
72. *The St Andrews University Calendar for the year 1899–1900* (Edinburgh: William Blackwood and Sons, 1899), 27.
73. GUA, Volume of Presscuttings of Queen Margaret College, 1884–1890, dated October 1894, and *Glasgow Herald*, November 5, 1894. See also GUA, Volume of Presscuttings of Queen Margaret College, 1884–1890, June 22 and 28 and October 22,1888, and Prospectus (printed) of Queen Margaret Hall with terms of board and general regulations, n.d.
74. GUA, Samples of advertisement styles of other University Halls plus Queen Margaret Hall, n.d. At Edinburgh and London the advertisements for women's residence halls did not stipulate rates, whereas at the University College in Cardiff rooms were listed at £30 or £40. Cardiff also noted details of academic courses they offered in their notice.
75. Sheila Hamilton, "Women and the Scottish Universities circa 1869–1939: A Social History" (Ph.D. thesis, University of Edinburgh, 1987), 317, 326–332.
76. J. T. Fowler, *Durham University: Earlier Foundations and Present Colleges* (London: F. E. Robinson, 1904), 47, 62, and Maria G. Grey, "The Women's Educational Movement," in *The Woman Question in Europe: A Series of Original Essays*, ed. Theodore Stanton (New York, London, and Paris: G. P. Putnam's Sons, 1884), 57. Women who lived off campus even after the Women's Hostel opened were still called "home students," and they greatly outnumbered those that lived on campus. For more see *Durham Calendar with Almanack 1910–1911* (Durham: Thomas Caldcleugh & Son and London: Whittaker & Co., 1910), 328–330.
77. Fowler, *Durham University*, 47, 62.
78. Ibid., 281.
79. *Durham Calendar with Almanack 1910–1911*, 244.
80. Isabel Maddison, ed., *Handbook of Courses Open to Women in British, Continental and Canadian Universities: Supplement for 1897* (New York: The Macmillan Company, 1897), 42.
81. Ibid.
82. *The Edinburgh University Calendar 1899–1900* (Edinburgh: James Thin, 1899), 879.

83. Ibid.
84. Janes, *The Englishwoman's Year Book and Directory 1900*, 8–9.
85. Appendix, in *The Edinburgh University Calendar 1895–1896* (Edinburgh: James Thin, 1895), 16.
86. Maddison, *Handbook of Courses Open to Women in British, Continental and Canadian Universities*, 100, and Christina Sinclair Bremner, *Education of Girls and Women in Great Britain* (London: Swan Sonnenschein & Co., 1897), 143.
87. A. Robertson and Walter Smith, "Report from King's College, London," in *Reports from University Colleges 1899, Presented to both Houses of Parliament by Command of Her Majesty* (London: Wyman and Sons, 1899), 180.
88. *The Calendar of King's College, London for 1896–97* (London: Published by the College, 1896), 273.
89. Ibid., 402.
90. Edward Fiddes, "The University Movement in Manchester (1851–1903)," in *Historical Essays in Honour of James Tait*, ed. J. G. Edwards, V. H. Galbraith, and E. F. Jacob (Manchester: Printed for the subscribers, 1933), 106. Ashburne House later became Ashburne Hall. See also Edward Fiddes, *Chapters in the History of Owens College and of Manchester University, 1851–1914* (Manchester: Manchester University Press, 1937), 119.
91. M. Winifred Jones, "What Women Are Doing in Manchester," *Womanhood* 3, 18 (May 1900), 427, and Janes, *The Englishwoman's Year Book and Directory 1900*, 7.
92. "Report from The Owens College, Manchester," in *Education Department Reports from University Colleges 1899* (London: Wyman and Sons, 1899), 286.
93. *The Victoria University of Manchester Medical School* (Manchester: Manchester University Press, 1908), 40. Not surprisingly the Victoria Church Hostel was run by the Church of England. It was located near the Royal Infirmary for the benefit of women who were studying medicine. The men at Manchester lived in Dalton Hall, a Society of Friends hall that opened in 1876, or Hulme Hall, a Church of England hall that opened in 1881. For more see Fiddes, "The University Movement in Manchester (1851–1903)," 106, and "Manchester," *The University Review* 2 (October 1905–March 1906), 422.
94. Annabel Wharton, "Gender, Architecture, and Institutional Self-Preservation: The Case of Duke University," *South Atlantic Quarterly* 90, 1 (Winter 1991), 198. See also Horowitz, *Alma Mater*, xv.
95. Michael Bezilla, *Penn State: An Illustrated History* (University Park and London: The Pennsylvania State University Press, 1985), 23.
96. Harriet A. McElwain, "Ladies' Department," in *Report of the Pennsylvania State College, for the year 1888* (Harrisburg, PA: Edwin K. Meyers, 1889), 46.
97. Ibid., 47.
98. "Editorials," *The Free Lance* 4, 2 (May 1890), 20, and Bezilla, *Penn State*, 35, 45.
99. Harriet A. McElwain, "Ladies' Department," in *Annual Report of the Pennsylvania State College, for the year 1894* ([Harrisburg, PA]: Clarence M. Busch, 1895), 149–150.
100. *The St Andrews University Calendar for the year 1899–1900* (Edinburgh: William Blackwood and Sons, 1899), 27. See also William Alexander Newman Dorland, *The Sum of Feminine Achievement: A Critical and Analytical Study of Woman's Contribution to the Intellectual Progress of the World* (Boston: The Stratford Company, 1917), 111.
101. Janes, *The Englishwoman's Year Book and Directory 1900*, 9.
102. "Miss Lumsden and University Hall," *The Journal of Education* 22 (September 1900), 598.
103. Ibid.
104. Zimmern, *The Renaissance of Girls' Education in England*, 145.
105. William Cadwaladr Davies and William Lewis Jones, *The University of Wales and Its Constituent Colleges* (London: F. E. Robinson & Co., 1905), 131, and David Barnes, *The Companion Guide to Wales* (Woodbridge, Suffolk: Companion Guides, 1970), 9. The new Victorian buildings located at the end of Marine Terrace would result in that portion of the road being referred to as Victoria Terrace on more recent maps of Aberystwyth.
106. Davies and Jones, *The University of Wales and Its Constituent Colleges*, 131.
107. Myers, "Women Students in Wales," 140.
108. T. Mortimer Green, Advertisement for the University College of South Wales, Aberystwyth, *Educational Review* 18, 13 (London: Educational Review, 1900), 479.
109. *University of Tennessee Record, July 1898 Volume V* (Knoxville: The University of Tennessee Press, 1898), 285–286, 295. The acting dean was also referred to in advertisements as Mrs. Charles

A. Perkins, for she was married to a member of the physics and electrical engineering faculty, whom she had met when they were both working at Lawrence University in the early 1880s. See also Bethenia McLemore Oldham, *Tennessee... and Tennesseans* (Clarksville, TN: W. P. Titus, 1903), 192.

110. T. W. Jordan, "Report of the Dean of the College," in *University of Tennessee Register for 1901–1902 and Announcement for 1902–1903* (Knoxville: The University of Tennessee Press, 1902), 196.
111. Florence Skeffington, "Report of the Dean of the Woman's Department," in *University of Tennessee Register for 1901–1902 and Announcement for 1902–1903* (Knoxville: The University of Tennessee Press, 1902), 197.
112. *The Michigan Book* (Ann Arbor, MI: Edwin H. Humphrey, 1898), 306, and Barbara Miller Solomon, *In the Company of Educated Women* (New Haven, CT, and London: Yale University Press, 1985), 107.
113. Solomon, *In the Company of Educated Women*, 107.
114. Burton Dorr Myers, *History of Indiana University Volume II: The Bryan Administration* (Bloomington: Published by Indiana University, 1952), 181. The other Victorian sororities at Indiana were Alpha Zeta Beta (1892) and Pi Beta Phi (1893). See also "Indiana University," *Beta Theta Pi* 21, 1 (October 1893), 440, and Diana B. Turk, *Bound by a Mighty Vow: Sisterhood and Women's Fraternities, 1870–1920* (New York and London: New York University Press, 2004), 48–49.
115. "Woman at Wisconsin: A Chronology," *The Wisconsin Magazine,* (March 1916), frontispiece.
116. Teicher and Jenkins, *A History of Housing at the University of Wisconsin*, 12, and "Co-Education in the 'Varsity," 30.
117. Myers, *History of Indiana University Volume II*, 181.
118. Lynn D. Gordon, *Gender and Higher Education in the Progressive Era* (New Haven, CT, and London: Yale University Press, 1990), 105. At Chicago this worry resulted in the prohibition of national "Greek" organizations at the university.
119. Sansing, *The University of Mississippi*, 177. The sororities at Mississippi were Chi Omega (1899) and Delta Delta Delta (1904). For more see Ida Shaw Martin, *The Sorority Handbook* (Menasha, WI: George Banta Publishing Company, 1918), 106.
120. *Catalogue of the University of Mississippi, Thirty-Ninth Session 1890–91*, 27.
121. *Catalogue of the Officers and Students of the University of Alabama with a Statement of the Courses of Instruction in the Various Departments 1879–80* (Tuscaloosa, AL: Burton's Book and Job Printing Office, 1880), 28. For more on the military discipline at Alabama see Sellers, *History of the University of Alabama*, 486–513.
122. Stevens, *The Admission of Women to Universities*, 11.
123. UWA, *Minutes of the Board of Regents*, 231.
124. Ibid., 189.
125. Ronald W. Hogeland, "Coeducation of the Sexes at Oberlin College: A Study of Social Ideas in Mid-Nineteenth Century America," *Journal of Social History* 6 (1972), 160. He wrote that it was "not surprising therefore that the school consciously attempted to provide these future evangelical ministers with suitable wives."
126. Ibid., 166, 168.
127. *La Vie '92 published by the Junior Class* (State College, PA: Published by the University, 1892), 143, and Bezilla, *Penn State*, 45.
128. "Locals," *The Free Lance* 3, 3 (June 1889), 192.
129. "Customs at State," in *La Vie '98 published by the Junior Class of the Pennsylvania State College Vol. IX* (State College, PA.: Published by the University, 1898), 190. See also *La Vie '99 published by the Junior Class of the Pennsylvania State College Volume X* (Chicago: A. L. Swift & Co., 1899), 177.
130. GUA, Frances H. Simson, Letter from Masson Hall, University of Edinburgh, to Mrs. Riddoch (13 February 1902).
131. GUA, Frances H. Melville, Letter from University Hall at St. Andrews (16 February 1902).
132. GUA, Frances H. Simson, Letter to Mrs. Riddoch.
133. OUMC, Margaret Boyd Diary, 14.
134. "Annual Examinations: Report of the Board of Visitors to the Board of Regents," in *Annual Report of the Regents of the University of Wisconsin for the Fiscal Year Ending September 30, 1877* (Madison, WI: David Atwood, 1877), 44–46. See also Helen R. Olin, *The Women of a State*

University: An Illustration of the Working of Coeducation in the Middle West (New York and London: G. P. Putman, 1909), 81–82.

135. Ibid. See also Carol Smith-Rosenberg and Charles Rosenberg, "The Female Animal: Medical and Biological Views of Woman and Her Role in Nineteenth-Century America," *Journal of American History* 60 (1973), 341–342, and Olin, *The Women of a State University*, 84–86.
136. John Bascom, "University Colleges: The Report of the President of the University to the Board of Regents," in *Annual Report of the Regents of the University of Wisconsin for the Fiscal Year Ending September 30, 1877*, 35–39, and Olin, *The Women of a State University*, 89–93. The report continued with a detailed examination of the absences from ill health reported for both men and women.
137. "Annual Examinations: Report of the Board of Visitors to the Board of Regents," in *Annual Report of the Regents of the University of Wisconsin for the Fiscal Year Ending September 30, 1878* (Madison, WI: David Atwood, 1878), 32, 37, 39. See also Olin, *The Women of a State University*, 94–95.
138. *Annual Report of the Regents of the University of Wisconsin for the Fiscal Year Ending September 30, 1877*, 7. See also Horowitz, *Alma Mater*, xv, and Olin, *The Women of a State University*, 82–83. Olin discusses the lack of ventilation in Ladies' Hall and adds "All private rooms of students were heated by air-tight stoves and lighted by kerosene lamps."
139. *The President's Report to the Board of Regents for the Year Ending June 30, 1879* (Ann Arbor, MI: Ann Arbor Printing and Publishing Company, 1879), 7.
140. Stevens, *The Admission of Women to Universities*, 3, 10–11.
141. Campbell, "The Central Conference of Women Workers," 750–751.
142. Wallis-Jones, "The University College of Wales," 243.
143. Myers, "Women Students in Wales," 137.
144. T. Levi, "Welsh Education," *The Cambrian Volume Twenty-Two* (Utica, NY: Thomas J. Griffiths, 1902), 12.
145. Myers, "Women Students in Wales," 140.
146. George W. Summers, *The Mountain State: A Description of the Natural Resources of West Virginia, Prepared for Distribution at the World's Columbian Exposition* (Charleston, WV: Moses W. Donnally, 1893), 64.
147. *The Corolla of Ninety-Five*, 177. See also *The Corolla '96* (Tuscaloosa, AL: W. H. Ferguson, 1896), 161, and *The Corolla* (Cleveland, OH: The Cleveland Printing and Publishing Co., 1893), 182.
148. *Catalogue of the South Carolina College 1904–1905* (Columbia, SC: The R. L. Bryan Company, 1905), 89–91, and Edwin L. Green, ed., *A History of the University of South Carolina* (Columbia, SC: The State Company, 1916), 304.
149. *Catalogue of the South Carolina College 1905–1906* (Columbia, SC: The R. L. Bryan Company, 1906), 65, 85. The men did have to pay a "room fee" as well to cover their electric use and "room service."
150. *Catalogue of the South Carolina College 1904–1905*, 90.
151. *Catalogue of the University of Mississippi, Thirty-Ninth Session 1890–91*, 90. This was before electricity and steam heat were available on campus. Similar expenses would be charged to women once they had a dormitory on campus in the early 1900s. For more see *Bulletin of the University of Mississippi: Announcements for Session of 1903–1904*, Series 2, No. 3 (August 1903), 34–35, and Sansing, *The University of Mississippi*, 158.
152. *Catalogue of the University of Mississippi, Thirty-Ninth Session 1890–91*, 91.
153. Woodburn, *Higher Education in Indiana*, 90, and *Annual Catalogue of the Indiana University for the Sixty-Seventh College Year, 1890–91*, 71.
154. Calvin Dill Wilson, *Working One's Way Through College and University: A Guide to Paths and Opportunities to Earn an Education at American Colleges and Universities* (Chicago: A. C. McClurg & Co., 1912), 72.
155. A. L. Brown and Michael Moss, *The University of Glasgow: 1451–1996* (Edinburgh: Edinburgh University Press, 1996), 5.
156. Gilchrist, "Some Early Recollection of the Queen Margaret Medical School," 81.
157. "Report from The Owens College, Manchester," 287.
158. GUA, P. Geddes, Letter from University College Dundee, to Miss Galloway concerning rules for residents of Queen Margaret Hall (10 May 1894), and Annie G. Philip, Letter regarding the regulations for the Hall Committee (4 March 1902).

159. Roy M. Pinkerton, "Of Chambers and Communities: Student Residence at the University of Edinburgh 1583–1983," in *Four Centuries: Edinburgh University Life, 1583–1983*, ed. Gordon Donaldson (Edinburgh: Edinburgh University Press, 1983), 116–130. See also Hamilton, "Women and the Scottish Universities circa 1869–1939," 289–303, 315–317.
160. Anderson, *The Matriculation Roll of the University of St. Andrews 1747–1897*, xlvii, and Janes, *The Englishwoman's Year Book and Directory 1900*, 9.
161. *The St Andrews University Calendar for the year 1899–1900* (Edinburgh: William Blackwood and Sons, 1899), 27. The costs also included light and coal, and were therefore considerably less during the summer when less of each was required.
162. Katharine Lake, ed., *Memorials of William Charles Lake, Dean of Durham 1869–1894* (London: Edward Arnold, 1901), 128.
163. Crawford, *The College Girl of America and the Institutions Which Make Her What She Is*, 258–259. For an extended discussion of this issue see Amy Hague, " 'Give Us a Little Time to Find Our Places': University of Wisconsin Alumnae, Classes of 1875–1900" (M.A. thesis, University of Wisconsin-Madison, 1983), and Rosalind Rosenberg, "The Limits of Access: The History of Coeducation in America," in *Women and Higher Education in American History*, ed. John M. Faragher and Florence Howe (New York: Norton, 1988), 117–118.
164. Teicher and Jenkins, *A History of Housing at the University of Wisconsin*, 4–6.
165. Janes, *The Englishwoman's Year Book and Directory 1900*, 5, and "Notes on the Work," *Oxford University Extension Gazette* 3, 32 (May 1893), 103.
166. Wendy Alexander, *First Ladies of Medicine: The Origins, Education and Destination of Early Women Medical Graduates of Glasgow University* (Glasgow: Wellcome Unit for the History of Medicine, 1987), 63. See also Frances H. Melville, *University Education for Women in Scotland: Its Effects on Social and Intellectual Life. A paper read at the conference of the National Union of Women Workers of Great Britain and Ireland, Edinburgh, October 1902* (St. Andrews, 1902), 5, and James Coutts, *A History of the University of Glasgow: From its Foundation in 1451 to 1909* (Glasgow: J. Maclehose and Sons, 1909), 472–473.
167. GUA, Article from *The Buteman*, Rothesay, regarding the establishment of two bursaries for ladies taking the Glasgow University Local Exams (Saturday April 20, 1878).
168. Catherine Mary Kendall, "The Queen Margaret Settlement 1897–1914: Glasgow Women Pioneers in Social Work" (M.A. thesis, University of Glasgow, 1993), 29. See also R. D. Anderson, *Education and Opportunity in Victorian Scotland: Schools and Universities* (Oxford: Clarendon Press, 1983). For details of Glasgow University bursaries see GUA, Programme of Classes and List of Bursaries Available, 1897–1912.
169. Lange, *Higher Education of Women in Europe*, 40.
170. "Oxford and Science," *Nature* 69, 1783 (December 31, 1903), 208.
171. Horowitz, *Alma Mater*, 38, 54.

Six: Extracurricular Student Life

1. "The Academic Revival, 1864–1914," in *Interamna Borealis: Being Memories and Portraits from an old University Town between the Don and the Dee*, ed. W. Keith Leask (Aberdeen: The Rosemount Press, 1917), 11.
2. Lynn D. Gordon, *Gender and Higher Education in the Progressive Era* (New Haven, CT, and London: Yale University Press, 1990), 3.
3. Mary Caroline Crawford, *The College Girl of America and the Institutions Which Make Her What She Is* (Boston: L. C. Page & Company, 1905), 258. See also Christine A. Ogren, "Where Coeds Were Coeducated: Normal Schools in Wisconsin, 1870–1920," *History of Education Quarterly* 25, 1 (Spring 1995), 17–21.
4. *The Savitar—1891; by Students of the University of Missouri* (Columbia, MO: Published by the University, 1891), 7, 39, 51, 63.
5. Ibid., 7, 82, 94. The ladies of the Philalethean Society rotated the duties of the officers each term. See also Charlotte Wronker, "Co-Education in the 'Varsity," *The Missouri Alumni Quarterly* (December 1905), 29.
6. *The St Andrews University Calendar for the year 1899–1900* (Edinburgh: William Blackwood and Sons, 1899), 511–515.
7. *The Edinburgh University Calendar 1882–1883* (Edinburgh: James Thin, 1882), 41.

8. *The St Andrews University Calendar for the year 1899–1900*, 34.
9. Appendix, in *The Edinburgh University Calendar 1878–1879* (Edinburgh: James Thin, 1878), 94–98, and Appendix, in *The Edinburgh University Calendar 1882–1883*, 30–37.
10. W. J. Wallis-Jones, "The University College of Wales," *WALES: A National Magazine for the English Speaking Parts of Wales* 3, 26 (June 1896), 246 and *The Calendar of the University College of Wales, Aberystwyth, Fourteenth Session, 1885–6* (Manchester: J. E. Cornish, 1885), 83–84.
11. Iwan Morgan, *The College by the Sea (A Record and a Review): "Nid Byd Byd Heb Wybodaeth"* (Aberystwyth: Published by the Students' Representative Council in Collaboration with the College Council, 1928), 229.
12. *The Calendar of King's College, London for 1896–97* (London: Published by the College, 1896), 274.
13. "The Universities," *The University Review* 5, 24 (April 1907), 60, and Emily Janes, *The Englishwoman's Year Book and Directory 1900* (London: Adam and Charles Black, 1900), 6.
14. "Ireland," *The Lancet* (February 22, 1890), 440.
15. "Royal College of Surgeons of England," *The British Medical Journal* 1 (January 21, 1899), 178.
16. "The Royal College of Surgeons of England," *The Lancet* (January 21, 1899), 185.
17. *The Indiana University Catalogue for the Seventy-First College Year, 1894–95* (Bloomington, IN: Published by the University, 1895), 73.
18. P.C.M., "The Evolution of Sex," review of *The Evolution of Sex*, by Professor Patrick Geddes and J. Arthur Thomson, *Nature*, April 10, 1890, 531–532. It should be noted that Thomson had not been hired yet by the University of Aberdeen at the time the book was first published. Geddes and Thompson's conclusions were used against the women's movement because they felt the science of Darwinism proved that too much education for women would damage their health. For more see Margaret Birney Vickery, *Buildings for Bluestockings: The Architecture and Social History of Women's Colleges in Late Victorian England* (Newark: University of Delaware Press, 1999), 153–154.
19. Alfred Russel Wallace, *Darwinism: An Exposition of the Theory of Natural Selection with Some of Its Applications* (London: Macmillan and Co., 1912), iii–iv.
20. *Fifty-Seventh Annual Catalogue of the Indiana University for the Academic Year 1886–1887* (Indianapolis, IN: Wm. B. Burford, 1887), 63, and *Annual Catalogue of the Indiana University for the Sixty-Seventh College Year, 1890–91* (Indianapolis, IN: Wm. B. Burford, 1891), 77.
21. Charles Woodward Hutson, "Letters of a Student of the Late 50's," in *A History of the University of South Carolina*, ed. Edwin L. Green (Columbia, SC: The State Company, 1916), 352.
22. "W.U.P. Dramatic Club Entertainment," *The University Courant* 5, 2–3 (February–March 1891), 21.
23. James Riley Montgomery, Stanley J. Folmsbee, and Lee Seifert Greene, *To Foster Knowledge: A History of The University of Tennessee 1794–1970* (Knoxville: The University of Tennessee Press, 1984), 371, and Florence Skeffington, "Report of the Dean of the Woman's Department," in *University of Tennessee Register for 1901–1902 and Announcement for 1902–1903* (Knoxville: The University of Tennessee Press, 1902), 196–197.
24. Florence V. Skeffington, "Report of the Dean of the Woman's Department," *University of Tennessee Record* 2 (February 1901), 43.
25. *The Volunteer Published by The Students of the University of Tennessee Vol. III* (Knoxville, TN: Bean, Warters & Gaut, 1899), 130–131, and *The Volunteer Published by The Students of the University of Tennessee Vol. IV* (Knoxville, TN: Bean, Warters & Gaut, 1900), 140.
26. *The Garnet and Black 1900, Published by the Students of the South Carolina College* (Columbia, SC: The Bryan Printing Co., 1900), 135.
27. *The St Andrews University Calendar for the year 1899–1900*, 511–515.
28. *The St. Andrews University Calendar for the Year 1904–1905* (Edinburgh: William Blackwood and Sons, 1904), 636.
29. Wallis-Jones, "The University College of Wales," 246, and *The Calendar of the University College of Wales, Aberystwyth, Fourteenth Session, 1885–6*, 83–84.
30. *The Edinburgh University Calendar 1897–1898* (Edinburgh: James Thin, 1897), 474; *The Edinburgh University Calendar 1899–1900* (Edinburgh: James Thin, 1899), 877; and *The St Andrews University Calendar for the year 1899–1900*, 511–515. John R. Elder, "Music in the University since 1898," *The Aberdeen University Review* 1, 3 (June 1914), 237.
31. *The Student's Guide to the University of Durham; with information respecting Expenses, Scholarships, Examinations, and Degrees* (Durham: The "Advertiser" Office, 1880), 28–29.

32. *The Savitar—1891*, 115, and *The Garnet and Black 1900*, 140.
33. *The Monticola Volume II . Published by the Class of 1900* (Morgantown, WV: The Acme Publishing Company, 1899), 166, 169–170.
34. Wilfred B. Shaw, *A Short History of the University of Michigan* (Ann Arbor, MI: George Wahr, 1937), 122–123.
35. Ibid., 123.
36. "College Notes," *The Free Lance* 13, 1 (April 1899), 25.
37. "College Notes," *The Free Lance* 13, 2 (May 1899), 55.
38. Michael Bezilla, *Penn State: An Illustrated History* (University Park and London: The Pennsylvania State University Press, 1985), 45.
39. James B. Sellers, *History of the University of Alabama* (Tuscaloosa: University of Alabama Press, 1953), 472.
40. Walter Penfield, "A History of the Junior Hop," *The Inlander* 11, 5 (February 1901), 184.
41. *The Michigan Book* (Ann Arbor, MI: Edwin H. Humphrey, 1898), 65, 68.
42. Walter Penfield, "A History of the Junior Hop," 188.
43. Caroline Dall, " 'The Opening at the Gates,' in *The College, the Market and the Court, or Women's Relations to Education, Labor and Law* (Boston: Lee and Shepherd, 1867)," and in *American Feminism: Key Source Documents 1848–1920 Volume II: Work and Education*, ed. Janet Beer, Anne-Marie Ford, and Katherine Joslin (London: Routledge, 2002), 79.
44. *Trochos*, 81 (Madison: Junior Class of the University of Wisconsin, 1885).
45. H. B. Lathrop, "The Progress of Half a Century," in *The Badger for Nineteen Hundred and Five* (Madison: Badger Board of the Junior Class of the University of Wisconsin, 1905), 18–19. See also Helen Lefkowitz Horowitz, *Alma Mater: Design and Experience in the Women's Colleges from Their Nineteenth-Century Beginnings to the 1930s* (Amherst: University of Massachusetts Press, 1993), 63.
46. *Trochos*. "Fred J. Turner," more widely known as Frederick Jackson Turner, later became a professor at the university, as well as a noted historian. For more see Ray Allen Billington, *Frederick Jackson Turner: Historian, Scholar, Teacher* (New York: Oxford University Press, 1973).
47. *Catalogue of the South Carolina College 1904–1905* (Columbia, SC: The R. L. Bryan Company, 1905), 86.
48. *The Garnet and Black 1899, Published by the Students Volume I* (Columbia, SC: The Bryan Printing Co., 1899), 29, 128–129.
49. *Catalogue of the South Carolina College 1904–1905*, 47, 50.
50. Amy Thompson McCandless, "Maintaining the Spirit and Tone of Robust Manliness: The Battle against Coeducation at Southern Colleges and Universities, 1890–1940," *NWSA Journal* 2, 2 (Spring 1990), 203.
51. *The Garnet and Black, Published by the Students of the South Carolina College, Nineteen Hundred and One* (Columbia, SC: The Bryan Printing Co., 1901), 82–83, 186.
52. Ibid., 89, 93, 187.
53. Ibid., 189.
54. *The St Andrews University Calendar for the year 1899–1900*, 511, 514, NUIG, "Debating Society," *QCG: a record of college life in the city of the tribes* 1, 1 (November 1902), 12; "Ladies' Notes," *QCG: a record of college life in the city of the tribes* 1, 2 (February 1903), 48; and "The Union Society," *The Durham University Journal* 5, 12 (December 17, 1883), 142.
55. J. T. Fowler, *Durham University; Earlier Foundations and Present Colleges* (London: F. E. Robinson, 1904), 57. The reading materials were made available to students in the Union Society Library that also included books on theology, novels, "and other light reading." See also *The Student's Guide to the University of Durham*, 26.
56. GUA, Queen Margaret College Literary and Debating Society, Minutes of Meetings, January 1899 to May 1905.
57. Norman Fraser, *Student Life at Edinburgh University* (Paisley: J. and R. Parlane, 1884), 48.
58. GUA, Queen Margaret College Literary and Debating Society, Minutes of Meetings, January 1899 to May 1905.
59. Appendix, in *The Edinburgh University Calendar 1894–1895* (Edinburgh: James Thin, 1894), 15.
60. *The Edinburgh University Calendar 1901–1902* (Edinburgh: James Thin, 1901), 890.
61. Sheila Hamilton, "Women and the Scottish Universities circa 1869–1939: A Social History" (Ph.D. thesis, University of Edinburgh, 1987), 354–361.
62. GUA, Queen Margaret College Literary and Debating Society, Minutes of Meetings, January 1899 to May 1905.

63. Helen M. Nimmo, "Some Recent Notes and Recollections of Queen Margaret College Life," in *The Book of the Jubilee: In Commemoration of the Ninth Jubilee of the University of Glasgow, 1451–1901*, ed. the Students' Jubilee Celebrations Committee (Glasgow: J. Maclehose and Sons, 1901), 153.
64. GUA, Queen Margaret College Literary and Debating Society, Minutes of Meetings, January 1899 to May 1905.
65. Ibid.
66. *Catalogue of West Virginia University Morgantown 1887–8: Announcements for 1888–9* (Charleston, WV: M. W. Donnally, 1888), 18.
67. *Fifty-Seventh Annual Catalogue of the Indiana University for the Academic Year 1886–1887*, 63, and *The Indiana University Catalogue, Seventh-Fourth College Year 1897–98* (Bloomington, IN: Published by the University, 1898), 98.
68. *Annual Report of the Indiana University, including the Catalogue for the Academical Year 1881–1882* (Indianapolis. IN: Wm. B. Burford, 1882), 36.
69. *Catalogue of the Western University of Pennsylvania for the year ending 1895 with detailed statements of the courses of instruction* ([Pittsburgh]: Western University of Pennsylvania, 1895), 15. The students had collected a substantial number of books to create their own library, which they had donated to the university in 1875.
70. Agnes Lynn Starrett, *Through One Hundred and Fifty Years: The University of Pittsburgh* (Pittsburgh: University of Pittsburgh Press, 1937), 156–157.
71. James Allen Cabaniss, *A History of the University of Mississippi* (University: University of Mississippi, 1949), 104.
72. *Catalogue and Announcements of the University of Mississippi at University P. O., Forty-Third Session 1894–'95* (Vicksburg, MS: Vicksburg Printing & Publishing Co., 1895), 76, and David G. Sansing, *The University of Mississippi: A Sesquicentennial History* (Jackson: University Press of Mississippi, 1999), 53–54. For much of the Victorian Era each had their own hall in the Chapel Building, which was more of an assembly hall than a religious structure.
73. Montgomery, Folmsbee, and Greene, *To Foster Knowledge*, 63–64.
74. Ibid., 36.
75. *The Volunteer Published by the Students Vol. II* (Knoxville, TN: Bean, Warters & Gaut, 1898), 102; *The Volunteer Published by The Students of the University of Tennessee Vol. III*, 89–90; and *The Volunteer Published by The Students of the University of Tennessee Vol. IV*, 101.
76. *The Calendar of the University College of Wales, Aberystwyth, Fourteenth Session, 1885–6*, 82–83.
77. Betty Hollow, *Ohio University: The Spirit of a Singular Place, 1804–2004* (Athens: Ohio University Press, 2003), 18.
78. OUMC, Margaret Boyd Diary, 178, in Ohio Memory: An Online Scrapbook of Ohio History, www.ohiomemory.org/index.html (accessed June 21, 2009).
79. James and Vera Olson, *The University of Missouri: An Illustrated History* (Columbia: University of Missouri Press, 1988), 8, 38.
80. Horowitz, *Alma Mater*, 281.
81. *Report of the Scottish Institution for the Education of Young Ladies with an Appendix containing separate reports, by the different teachers, of the course of instruction, and the system pursued, in their respective classes* (Edinburgh: Oliver & Boyd, 1835), 25.
82. Barbara Miller Solomon, *In the Company of Educated Women* (New Haven, CT, and London: Yale University Press, 1985), 103.
83. Ibid., 104.
84. V. Sturge, "The Physical Education of Women," in *The Education Papers: Women's Quest for Equality in Britain, 1850–1912*, ed. Dale Spender (New York and London: Routledge & Kegan Paul, 1987), 284–294. There is no date listed for this article, but Spender's positioning of it chronologically places it in the late 1880s or 1890s.
85. A. Lapthorn Smith, "Higher Education of Women and Race Suicide," *Popular Science Monthly* 66 (March 1905), 470–471. See also Rosalind Rosenberg, "The Limits of Access: The History of Coeducation in America," in *Women and Higher Education in American History*, ed. John M. Faragher and Florence Howe (New York: Norton, 1988), 117–118.
86. "Editorial," *The Western University Courant* 10, 2 (November 1894), 6–7. For more on the connection between military training and university athletics see Michael Pearlman, "To Make the University Safe for Morality: Higher Education, Football and Military Training from the 1890s through the 1920s," *The Canadian Review of American Studies* 12, 1 (Spring 1981), 44–47.

87. Sellers, *History of the University of Alabama*, 464, 523. Sellers states that the students "begged for, and got, a gymnasium and a physical education department; and they launched a program of competitive sports which, before the turn of the century, had expanded into intercollegiate athletics."
88. Olson, *The University of Missouri*, 20.
89. Ibid., 54.
90. "Our Share in Athletics," in *The Amulet: Published Annually in Their Junior Year by the Ladies of Eighty-Four* (Ann Arbor, MI: Register Printing and Publishing Company, 1882), 14.
91 *Calendar of the University of Michigan for 1894–95* (Ann Arbor, MI: The Register Publishing Company, 1895), 33; and *The President's Report to the Board of Regents for the Year Ending Sept. 30, 1895* (Ann Arbor, MI: The Inland Press, 1895), 20.
92. Shaw, *A Short History of the University of Michigan*, 128.
93. H. B. Hutchins, "The University and Co-Education," *The Michigan Alumnus* 17, 160 (January 1911), 184; *The President's Report to the Board of Regents for the Year Ending Sept. 30, 1895* (Ann Arbor, MI: The Inland Press, 1895), 21; and "Event and Comment: The Michigan Ideal," *The Michigan Alumnus* 10, 90 (January 1904), 167–168.
94. "Letter to the Editor," *The Free Lance* 5, 7 (January 1892), 139.
95. "Editorials," *The Free Lance* 4, 8 (February 1891), 123–124.
96. "We Regret," in *La Vie '98 published by the Junior Class of the Pennsylvania State College Vol. IX* (State College, PA: Published by the University, 1898), 166.
97. Burton Dorr Myers, *History of Indiana University Volume II: The Bryan Administration* (Bloomington: Published by Indiana University, 1952), 185–186.
98. Ibid., 171, 187.
99. *Annual Catalogue of the Indiana University for the Sixty-Seventh College Year, 1890–91*, 77. Mrs. Saunderson was married to George W. Saunderson, who was a professor of rhetoric and oratory. They both moved on to the University of Wisconsin in 1893. For more see *Catalogue of the University of Wisconsin for 1895–96* (Madison, WI: Published by the University, 1896), 16, 18.
100. F. S. Dumaresq de Carteret-Bisson, *Our Schools and Colleges, Vol. II: For Girls* (London: Simpkin, Marshall & Co., 1884), 181–182. See also M. Pointon, "Factors Influencing the Participation of Women and Girls in Physical Education, Physical Recreation and Sport in Great Britain During the Period 1850–1920," *History of Education Society Bulletin* 24 (1979), 46–56.
101. *The Calendar of King's College, London for 1896–97* (London: Published by the College, 1896), 274, and Mary Frances Billington, "Alexandra House," in *The Woman's World: Volume III*, ed. Oscar Wilde (London, Paris, and Melbourne: Cassell & Company, 1890 and London: Source Book Press, 1970), 154–157.
102. "Editorials," *The Free Lance* 4, 8 (February 1891), 124.
103. Skeffington, "Report of the Dean of the Woman's Department," *University of Tennessee Record* 2, 44.
104. *Catalogue of the Western University of Pennsylvania for the year ending 1895 with detailed statements of the courses of instruction* ([Pittsburgh]: Western University of Pennsylvania, 1895), 15.
105. "Editorials," *The Western University Courant* 10, 3 (December 1894), 17.
106. "Wanted," *The University Courant* 5, 10 (February 1892), 113.
107. "Durham: University News and Notes," *The University Review* 1, 4 (August 1905), 424.
108. *The Student's Guide to the University of Durham*, 6, 28.
109. Ibid., 27. There was also a combined "University Boat Club" that was "an association of the College and Hall clubs for various purposes, such as the selection of a boat or boats to represent the University in the Durham Regatta." See also Scott A.G.M. Crawford, ed., *'Serious Sport': J. A. Mangan's Contribution to the History of Sport* (London: Taylor & Francis, 2004), 30.
110. OCLA, Women's Boating Club in the 1890s; Appendix, in *The Edinburgh University Calendar 1878–1879*, 94–98; Appendix, in *The Edinburgh University Calendar 1882–1883*, 30–37; and Wallis-Jones, "The University College of Wales," 244–245.
111. "Locals," *The Free Lance* 9, 7 (January 1896), 163, and "College Notes," *The Free Lance* 14, 6 (February 1901), 164. These trips were an "old custom" on the campus, with students from each class making the drive on a different day of the week after classes let out. Typically they would leave at four o'clock and return late at night, having dined in Bellefonte.
112. Olson, *The University of Missouri*, 15–16.

113. *The St Andrews University Calendar for the year 1899–1900*, 512–513, 515; Appendix, in *The Edinburgh University Calendar 1878–1879*, 94–98; and Appendix, in *The Edinburgh University Calendar 1882–1883*, 30–37. The University Golf Club at Edinburgh held their meetings at the course at Musselburgh, which hosted the Open Championship in 1874, 1877, 1880, 1883, 1886, and 1889. For more see Horace G. Hutchinson, *Golf* (London: Longmans, Green, and Co., 1895), 463.
114. Olson, *The University of Missouri*, 41.
115. "Early Women's Organizations," *WVU Women: The First Century* (Morgantown: WVU Women's Centenary Project, West Virginia University, 1989), and *The Monticola Volume III, 1901* (Morgantown, WV: The Acme Publishing Company, 1900), 157.
116. Skeffington, "Report of the Dean of the Woman's Department," *University of Tennessee Record* 2, 44.
117. *The Student's Guide to the University of Durham*, 28–29, *Catalogue of the Western University of Pennsylvania for the year ending 1895*, 15; *The Calendar of King's College, London for 1896–97* (London: Published by the College, 1896), 274; Billington, "Alexandra House," 154–157; and Skeffington, "Report of the Dean of the Woman's Department," *University of Tennessee Record* 2, 44.
118. "Locals," *The Free Lance* 9, 7 (January 1896), 163, and "College Notes," *The Free Lance* 14. 6 (February 1901), 164.
119. *The St Andrews University Calendar for the year 1899–1900*, 512–513, 515, and *The Student's Guide to the University of Durham*, 28–29.
120. William T. Doherty Jr. and Festus P. Summers, *West Virginia University: Symbol of Unity In a Sectionalized State* (Morgantown: West Virginia University Press, 1982), 54.
121. Wallis-Jones, "The University College of Wales," 244–245.
122. *The Garnet and Black 1900*, 127, and Skeffington, "Report of the Dean of the Woman's Department," *University of Tennessee Record* 2, 44.
123. Frances H. Melville, *University Education for Women in Scotland: Its Effects on Social and Intellectual Life. A paper read at the Conference of the National Union of Women Workers of Great Britain and Ireland* (St. Andrews, 1902), 6.
124. NUIG, "Debating Society," *QCG: a record of college life in the city of the tribes* 1, 1 (November 1902), 12; "Ladies' Notes," *QCG: a record of college life in the city of the tribes* 1, 2 (February 1903). 48; and Wallis-Jones, "The University College of Wales," 244–245.
125. Isabel Maddison, ed., *Handbook of Courses Open to Women in British, Continental and Canadian Universities* (New York: The Macmillan Company, 1896), 100, and Edith Thompson, *Hockey as a Game for Women* (London: Edward Arnold, 1905), 4.
126. Foster Watson, ed., *The Encyclopaedia and Dictionary of Education in Four Volumes, Volume I* (London, Bath, Melbourne, Toronto, and New York: Sir Isaac Pitman & Sons, 1921), 299.
127. NUIG, "Ladies' Notes," *QCG: a record of college life in the city of the tribes* 1, 2 (February 1903), 48.
128. Horowitz, *Alma Mater*, 58.
129. For the full text of the ordinance, see *The Universities (Scotland) Act, 1889 together with Ordinances of the Commissioners under the said Act and of University Court Ordinances* (Glasgow: James MacLehose and Sons, 1915), 188.
130. GUA, *Glasgow University Magazine* 6, 10 (February 7, 1894), 153. The QMC SRC was first mentioned by the male students on Gilmorehill in February 1894: "Q.M., we are informed, has set up a Representative Council of her very own. Whether it have attained, or be likely to attain, legal status, we know not, but *Prosit!*"
131. GUA, Volume of Presscuttings of Queen Margaret College, 1891–1894, January 15, 1894.
132. GUA, *Glasgow University Magazine* 41, 2 (November 6, 1929): 45–46.
133. *The St Andrews University Calendar for the year 1899–1900*, 217, 510–511.
134. P. J. Anderson, ed., *Record of the Celebration of the Quatercentenary of the University of Aberdeen* (Aberdeen: Aberdeen University Press, 1907), 332.
135. Wallis-Jones, "The University College of Wales," 243.
136. Morgan, *The College by the Sea*, photo facing page 242.
137. Royal Commission on University Education in Wales, *Minutes of Evidence taken before the Royal Commissioners appointed to inquire into the organisation and work of the University and its three constituent Colleges, and into the elations of the University to those Colleges and to other institutions in Wales providing education of a post-secondary nature, and to consider in what respects the present*

organisation of University Education in Wales can be improved and what changes, if any, are desirable in the constitution, functions and powers of the University and its three colleges (Eleventh Day, Friday, 1st December, 1916), 103.

138. Grace Smith, "Indiana University," *Kappa Alpha Theta* 17, 2 (January 1903), 101, 105; Doherty and Summers, *West Virginia University*, 64; *The Volunteer Published by the Students Vol. II*, 110–111; and Hutchins, "The University and Co-Education," 184.
139. Skeffington, "Report of the Dean of the Woman's Department," *University of Tennessee Record* 2, 42.
140. *The President's Report to the Board of Regents for the Year Ending Sept. 30, 1891* (Ann Arbor, MI: The Register Publishing Company, 1891), 11.
141. *Calendar for the University of Michigan for 1891–92* (Ann Arbor, MI: J. S Cushing & Company, 1892), 29.
142. "Event and Comment: Co-Educational Problems," *The Michigan Alumnus* 17, 1 (October 1910), 5.
143. *The Monticola Volume II*, 148–153.
144. UWA, "A Half Century of Progress, A Future of Promise," WSGA Commemorative Booklet, 1897–1947, 5.
145. Ibid. For more on the Women's Self-Government Association see Christine D. Myers, "Gendering the 'Wisconsin Idea': The Women's Self-Government Association, c. 1898–1948," in *Gender, Politics and the Experience of Education: An International Perspective*, ed. Jane Martin and Jayne Goodman (London: Woburn Press, 2002), 148–172.
146. GUA, Volume of Presscuttings of Queen Margaret College, 1891–1894, *Glasgow Herald*, November 11, 1893.
147. Gordon, *Gender and Higher Education in the Progressive Era*, 1.
148. Sir Alexander Grant, *The Story of the University of Edinburgh in its First Three Hundred Years* (London: Longmans, Green, and Co., 1884), 237. Sir Alexander Grant was the principal of the University of Edinburgh at the time. For more see Fraser, *Student Life at Edinburgh University*, 35.
149. Robert Sangster Rait, *The Universities of Aberdeen: A History* (Aberdeen: James Gordon Bisset, 1895), 356–357. See also J. N. Morton, *An Analysis of the Universities (Scotland) Act, 1889, with the Act Itself and the Act of 1858, and an Index* (Edinburgh and London: William Blackwood and Sons, 1889), and Campbell Fox Lloyd, "Relationships between Scottish Universities and Their Communities c. 1858–1914" (Ph.D. thesis, University of Glasgow, 1993).
150. *Aurora Borealis Academia: Aberdeen University Appreciations 1860–1889* (Aberdeen: The University Printers, 1899), 8–9.
151. Lindy Moore, *Bajanellas and Semilinas: Aberdeen University and the Education of Women* (Aberdeen: Aberdeen University Press, 1991), 95.
152. There was a small amount of concern at QMC that the women would not be included in the Rectorial election, but the decision was made quite easily in the spring of 1893 for them to be allowed to participate. For more see GUA, Marion Buchanan, Letter regarding Q. M. Students and Rectorial Election (28 March 1893), and Report of Committee on Women Students and Rectorial Election (22 April 1893).
153. GUA, Volume of Presscuttings of Queen Margaret College, 1891–1894, "The Rectorial Contest," *Glasgow Herald*, November 16, 1893.
154. There are numerous examples of student misbehavior at the time of Rectorial elections, including duels, street fights, mud fights, and so on.
155. Sellers, *History of the University of Alabama*, 462, and Joseph Thompson, *The Owens College: Its Foundation and Growth; and its Connection with the Victoria University, Manchester* (Manchester: J. E. Cornish, 1886), 159.
156. Doherty and Summers, *West Virginia University*, 52–53.
157. OUMC, Margaret Boyd Diary, 97, 100, 156.
158. Appendix, in *The Edinburgh University Calendar 1895–1896* (Edinburgh: James Thin, 1895), 9.
159. Hollow, *Ohio University*, 81; Sellers, *History of the University of Alabama*, 462; *The Indiana University Catalogue, Seventy-Fourth College Year 1897–98* (Bloomington, IN: Published by the University, 1898), 100; *The Michigan Book*, 90; Sansing, *The University of Mississippi*, 165–166; *The Savitar—1891*, 75, 81; Bezilla, *Penn State*, 39; "Y.M.C.A.," *The University Courant* 2, 8 (October 1888), 92; Montgomery, Folmsbee, and Greene, *To Foster Knowledge*, 350; *The*

Monticola Volume II, 154–156; *Catalogue of the South Carolina College 1904–1905*, 93; and UWA, *Students' Hand-Book*, Presented by the University Young Men's and Young Women's Christian Associations, 1894–1895 (Milwaukee: Press of The Evening Wisconsin Company, 1894).

160. Montgomery, Folmsbee, and Greene, *To Foster Knowledge*, 350, and *The Volunteer Published by the Students Vol. II*, 110–111.
161. Montgomery, Folmsbee, and Greene, *To Foster Knowledge*, 377.
162. *The Michigan Book*, 90. A new branch of the Y.M.C.A. was started by students in 1895.
163. Ibid., 89.
164. *Calendar of the University of Michigan for 1894–95* (Ann Arbor, MI: The Register Publishing Company, 1895), 32.
165. Ibid.
166. *Catalogue of the University of Mississippi at University P. O., Near Oxford, Miss.: Prepared this Year with Special Reference to the Schools of English and Belles Lettres. Thirty-Ninth Session 1890–'91* (Oxford, MS: Published by the University, 1890), 26; Dumaresq de Carteret-Bisson, *Our Schools and Colleges*, 180; and James Heywood, "The Owens College, Manchester, and a Northern University," *Journal of the Statistical Society* 41, Part 3 (September 1878), 544.
167. *The Corolla '96* (Tuscaloosa, AL: W. H. Ferguson, 1896), 115, and Sellers, *History of the University of Alabama*, 460–461.
168. Bezilla, *Penn State*, 25.
169. Doherty and Summers, *West Virginia University*, 52.
170. "College Notes," *The Free Lance* 12, 1 (April 1898), 28.
171. Wallis-Jones, "The University College of Wales," 246; *The Calendar of the University College of Wales, Aberystwyth, Fourteenth Session, 1885–6*, 83–84; *The St Andrews University Calendar for the year 1899–1900*, 511–515; Appendix, in *The Edinburgh University Calendar 1878–1879*, 94–98; and Appendix, in *The Edinburgh University Calendar 1882–1883*, 30–37.
172. *Fifty-Fifth Annual Report of the Indiana University, Including the Catalogue for the Academical Year 1884–1885* (Indianapolis, IN: Wm. B. Burford, 1885), 62.
173. Gordon, *Gender and Higher Education in the Progressive Era*, 4.
174. GUSC, Edith Oakley, "The Formation of Character," essay and Prof. Bowman's notes, dated 6 February 1929. Sir Henry Jones was professor of moral philosophy at Glasgow from 1894 to 1922 and has been described as the "leading British idealist of his generation." See A. L. Brown and Michael Moss, *The University of Glasgow: 1451–1996* (Edinburgh: Edinburgh University Press, 1996), 94. William Smart was Glasgow's first professor of political economy from 1896 to 1915.
175. The Right Hon. Sir John Gorst, MP, " 'Settlements' in England and America," in *The Universities and the Social Problem: An Account of the University Settlements in East London*, ed. John M. Knapp (London: Rivington, Percival & Co., 1895), 1–30.
176. Catherine Mary Kendall, "The Queen Margaret Settlement 1897–1914: Glasgow Women Pioneers in Social Work" (M.A. thesis, University of Glasgow, 1993), 32–34.
177. Wendy Alexander, *First Ladies of Medicine: The Origins, Education and Destination of Early Women Medical Graduates of Glasgow University* (Glasgow: Wellcome Unit for the History of Medicine, University of Glasgow, 1987), 64.
178. Cyril Jackson, "The Children's Country Holidays Fund, and the Settlements," in *The Universities and the Social Problem: An Account of the University Settlements in East London*, ed. John M. Knapp (London: Rivington, Percival & Co., 1895), 89.
179. W. Reason, ed., *University and Social Settlements* (London: Methuen & Co., 1898), 188; C. R. Henderson, *Social Settlements* (New York: Lentilhon & Company, 1899), 33–34; and John Palmer Gavit, comp., *Bibliography of College, Social and University Settlements* (Cambridge, MA: Co-operative Press, 1897), 62.
180. GUA, QMC Christian Union, Constitution, office bearers and branches, 1898–99 (printed).
181. *The Edinburgh University Calendar 1897–1898*, 474, and *The Edinburgh University Calendar 1899–1900*, 877.
182. UWA, *Students' Hand-Book*, 9–11. See also Pearlman, "To Make the University Safe for Morality," 44.
183. Gordon, *Gender and Higher Education in the Progressive Era*, 95, 107.
184. George W. Truett, "The Making of a Life," in *University of Alabama Bulletin; Centennial Celebration 1831–1931*, No. 90 (June 1931), 3.
185. Pearlman, "To Make the University Safe for Morality," 38, 43–44.
186. Horowitz, *Alma Mater*, 178.

Seven: Student Publications

1. *Trochos* (Madison, WI: Junior Class of the University of Wisconsin, 1885), 11–12.
2. *The Student's Guide to the University of Durham; with information respecting Expenses, Scholarships, Examinations, and Degrees* (Durham: The "Advertiser" Office, 1880), 34.
3. University of London, *The Calendar for the Year 1870* (London: Taylor and Francis, 1870), 135–139.
4. Norman Fraser, *Student Life at Edinburgh University* (Paisley: J. and R. Parlane, 1884), 72.
5. Ibid.
6. "University Studies: Scottish and English," in *The British and Foreign Evangelical Review Vol. XVI* (London: James Nisbet & Co. and Edinburgh: Oliver & Boyd, 1867), 27.
7. GUA, *Glasgow University Students' Handbook*, 1893.
8. *Students' Hand-Book, presented by the University Young Men's and Women's Christian Associations, 1894–1895* (Milwaukee: Press of The Evening Wisconsin Company, 1894), 6.
9. Ibid., 27.
10. UWA, Notes on Other Publications.
11. Burton Dorr Myers, *History of Indiana University Volume II: The Bryan Administration* (Bloomington: Published by Indiana University, 1952), 620–621, and *Fifty-Seventh Annual Catalogue of the Indiana University for the Academic Year 1886–1887* (Indianapolis, IN: Wm. B. Burford, 1887), 63.
12. Michael Bezilla, *Penn State: An Illustrated History* (University Park and London: Pennsylvania State University Press, 1985), 41.
13. UWA, *Cardinal* Articles, 1892–1912. See also WHS, *Daily Cardinal*, beginning with April 4, 1892 issue, and *Daily Cardinal*, Miscellaneous publications.
14. Jennifer L. Stein, "The History of the Daily Cardinal from 1892–1991: A Look at the University of Wisconsin-Madison's Oldest Student Daily" (Senior thesis, University of Wisconsin, Madison, 1991), 9.
15. UWA, *Minutes of the Board of Regents* (1866 through 1876), July 15, 1895 and April 21, 1896. Complimentary copies were also sent to state high schools to encourage interest among the students in attending the university.
16. Stein, "The History of the Daily Cardinal from 1892–1991," 15–16. The paper also received financial assistance from Mary Adams, wife of University of Wisconsin president Charles Adams, who subscribed and paid for 50 copies of the *Cardinal* in advance to help defray costs.
17. H., "The Dangers of Socialism," *The Free Lance* 8, 8 (February 1895), 106–107.
18. Bezilla, *Penn State*, 33.
19. Mabel Tylecote, *The Education of Women at Manchester University 1883 to 1933* (Manchester: Manchester University Press, 1941), 38.
20. Ibid., 39.
21. Ibid., 38–40.
22. *The St Andrews University Calendar for the year 1899–1900* (Edinburgh: William Blackwood and Sons, 1899), 515. *College Echoes* was first published by the St. Andrews' SRC in 1889. It was initially eight pages in length and grew in size in the twentieth century.
23. Henry Sell, *Sell's Dictionary of the World's Press* (London: Sell's Advertising Offices, 1886), 218.
24. "The Social Life of the Student," *The Students' Journal and Hospital Gazette* (September 13, 1879), 232.
25. GUA, *Glasgow University Magazine* 1, 1 (February 5, 1889), 1.
26. Madge Wildfire was an alias used by a prolific male writer and illustrator, or possibly several different students. The name came from a "half-crazed gypsy girl" in Sir Walter Scott's "The Heart of Midlothian" which was first published in 1818. For more see James F. Hunnewell, *The Lands of Scott* (Boston: Houghton, Osgood and Company and Cambridge: The Riverside Press, 1880), 265, 271, and GUA, *Glasgow University Magazine* 3, 10 (February 18, 1891), 109.
27. GUA, *Glasgow University Magazine* 3, 1 (December 3, 1890), 3, and Daniel Walker Hollis, *University of South Carolina Volume II. College to University* (Columbia: University of South Carolina Press, 1956), 187.
28. GUA, *Glasgow University Magazine* 3, 3 (December 17, 1890), front cover, and GUA, *Glasgow University Magazine* 3, 8 (February 4, 1891), front cover.

29. Wilfred B. Shaw, *A Short History of the University of Michigan* (Ann Arbor, MI: George Wahr, 1937), 121–122.
30. Ibid. Shaw also noted that *The Inlander,* which "existed precariously for some fifteen years, was finally discontinued, but the *Daily* was a successful enterprise from the first."
31. Ibid.
32. Howard H. Peckham, *The Making of The University of Michigan 1817–1992*, ed. and upd. Margaret L. Steneck and Nicholas H. Steneck (Ann Arbor: The University of Michigan Press, 1967, 1994), 120, and Shaw, *A Short History of the University of Michigan*, 188–189.
33. "Beta—Ohio University," *The Rainbow of Delta Tau Delta* 11, 1 (January 1888), 82; C. H. Fouts, "Ohio University," *The Beta Theta Pi* 15, 3 (April 1888), 207; and "Exchange Notes," *The Vassar Miscellany* 16, 3 (December 1886), 115.
34. *Catalogue of the Western University of Pennsylvania for the year ending 1895 with detailed statements of the courses of instruction* ([Pittsburgh]: Western University of Pennsylvania, 1895), 15.
35. "Exchanges," *The University Courant* 2, 7 (September 1888), 84.
36. "The Exchange Editor's Table," *Pennsylvania Western* 5, 2 (February 1886), 44–45, and "The Exchange Editor's Table," *Pennsylvania Western* 5, 4 (April 1886). 98–99.
37. George Eyre Evans, *Aberystwyth and its Court Leet* (Aberystwyth: *Welsh Gazette*, 1902), 193. See also "Literary and Art Notes of the Month, &c.," in *The Red Dragon: The National Magazine of Wales, Vol.* IV—*July to December, 1883*, ed. Charles Wilkins (Cardiff: Daniel Owen and Company, 1883), 473–477, and Iwan Morgan, *The College by the Sea (A Record and a Review): "Nid Byd Byd Heb Wybodaeth"* (Aberystwyth: Published by the Students' Representative Council in Collaboration with the College Council, 1928), 280.
38. *The Calendar of the University College of Wales, Aberystwyth, Fourteenth Session, 1885–6* (Manchester: J. E. Cornish, 1885), 14, 83.
39. Morgan, *The College by the Sea*, 231, 294–295. The name change took place in 1903 when the magazine was taken over by the Students' Representative Council. The magazine was also "altered in shape and size" at the same time.
40. Ibid., 290–295, and W. J. Wallis-Jones, "The University College of Wales," *WALES: A National Magazine for the English Speaking Parts of Wales* 3, 26 (June 1896), 244–245.
41. *Trochos*, 11–12.
42. Shaw, *A Short History of the University of Michigan*, 121.
43. Grace Smith, "Indiana University," *Kappa Alpha Theta* 17, 2 (January 1903), 103.
44. Bezilla, *Penn State*, 43.
45. David G. Sansing, *The University of Mississippi: A Sesquicentennial History* (Jackson: University Press of Mississippi, 1999), 137, and James Allen Cabaniss, *A History of the University of Mississippi* (University: University of Mississippi, 1949), 129.
46. James Riley Montgomery, Stanley J. Folmsbee, and Lee Seifert Greene, *To Foster Knowledge: A History of The University of Tennessee 1794–1970* (Knoxville: The University of Tennessee Press, 1984), 350–351.
47. *The Corolla '96* (Tuscaloosa, AL: W. H. Ferguson, 1896), 115.
48. Sixteen was not the youngest age admitted to the University of Alabama. James Sellers notes that in "1887 two fourteen-year-olds, P. W. White and John Parker, were admitted." John Parker was one of Bessie Parker's brothers, and the success of the family surely contributed to his being allowed to enter at such a young age. By the end of the century young ladies of sixteen were admitted if they lived at home with their parents. For more see James B. Sellers, *History of the University of Alabama* (Tuscaloosa: University of Alabama Press, 1953), 421, 483.
49. *The Corolla '96*, 115.
50. *The Garnet and Black 1899, Published by the Students Volume I* (Columbia, SC: The Bryan Printing Co., 1899), 230.
51. UWA, Official Ballot, January 1900.
52. "Jottings," *The Durham University Journal* 5, 9 (May 26, 1883), 112. Ascension Day, or Holy Thursday, was May 3 in 1883. See also C. W. Leffingwell and Arthur Seymour, eds., *The Living Church Annual and Almanac and Calendar, for the Year of Our Lord, 1883* (New York: E. & J. B. Young & Company, 1881), 21.
53. *The Durham University Calendar, with Almanack, MDCCCLXXXII* (Durham: Andrews and Co., and London: Whittaker and Co., 1882), ciii, and *The Student's Guide to the University of Durham*, 13.
54. GUA, *Glasgow University Magazine* 3, 2 (December 10, 1890), 28.

55. GUA, *Glasgow University Magazine* 3, 13 (March 11, 1891), 154.
56. Agnes Lynn Starrett, *Through One Hundred and Fifty Years: The University of Pittsburgh* (Pittsburgh: University of Pittsburgh Press, 1937), 202–203.
57. "The Exchange Editor's Table," *Pennsylvania Western* 5, 2 (February 1886), 45.
58. "The Exchange Editor's Table," *Pennsylvania Western* 5, 4 (April 1886), 99.
59. William Johnston, ed., *Roll of the Graduates of the University of Aberdeen 1860–1900, Aberdeen University Studies: No. 18* (Aberdeen: Aberdeen University Press, 1906), 62.
60. John Malcolm Bulloch, ed., *College Carols* (Aberdeen: D. Wylie and Son, 1894), 25, Lines 1–4.
61. Ibid., Lines 5–8.
62. GUA, *Glasgow University Magazine* 4, 7 (January, 13 1892), 1.
63. Helen M. Nimmo, "Some Recent Notes and Recollections of Queen Margaret College Life," in *The Book of the Jubilee: In Commemoration of the Ninth Jubilee of the University of Glasgow, 1451–1901*, ed. the Students' Jubilee Celebrations Committee (Glasgow: J. Maclehose and Sons, 1901), 146.
64. Mrs. Campbell, "The Rise of the Higher Education of Women Movement," in *The Book of the Jubilee: In Commemoration of the Ninth Jubilee of the University of Glasgow*, 136. The stipulation was made by Mrs. Elder who had given the buildings and grounds to the college to use and would then give them to the university permanently, along with an endowment of £25,482. The majority of the endowment funds were raised at a bazaar in the fall of 1892 which was the social event of the season in Glasgow attended by everyone who was anyone. For more see Christine D. Myers, "'The brilliant opening of a stubborn battle': The Queen Margaret College Bazaar and Women's Admission to Higher Education in Scotland, c. 1892," in *Leeds Working Papers in Victorian Studies, Volume 3: Platform-Pulpit-Rhetoric*, ed. Martin Hewitt (Leeds: Leeds Centre for Victorian Studies, 2000), 150–165.
65. GUA, *Glasgow University Magazine* 3, 5 (January 14, 1891).
66. GUA, *Glasgow University Magazine* 3, 6 (January 21, 1891).
67. She is also holding an item that is either a jump rope or a whip, though debates over which it is will probably never cease.
68. UWA, *Cardinal* Articles, 1892–1912.
69. *The Garnet and Black 1900, Published by the Students of the South Carolina College* (Columbia, SC: The Bryan Printing Co., 1900), 33–34.
70. Ibid., 34.
71. Bezilla, *Penn State*, 82.
72. *Wisconsin Wickedness; being some wondrous wailings of western college life* (New York: W. S. Sterling, 1900), 24.
73. Stein, "The History of the Daily Cardinal from 1892–1991," 13, 27.
74. "Book Reviews," *The Inlander* 11, 7 (April 1901), 291.
75. *The Garnet and Black 1899*, 222.
76. Ibid., 225–228.
77. *The Garnet and Black 1900*, 156.
78. "Up on the Hill," *The Western University Courant* 16, 4 (January 1901), 125.
79. The name initially was just *Savitar*, but for consistency I have followed the practice of the University of Missouri-Columbia Archives and referred to it throughout as *The Savitar*.
80. *The Savitar—1891; by Students of the University of Missouri* (Columbia, MO: Published by the University, 1891), 126.
81. Ibid., 122–123.
82. J. D. Derelic, "To My College Girl," in *The Savitar—1899*, 156.
83. GUA, *Glasgow University Magazine* 4, 7 (January 13, 1892).
84. Lindy Moore, *Bajanellas and Semilinas: Aberdeen University and the Education of Women* (Aberdeen: Aberdeen University Press, 1991), 118.
85. GUA, *Glasgow University Magazine* 5, 7 (January 18, 1893), 70–71.
86. GUA, *Glasgow University Magazine* 5, 9 (February 1, 1893), 102.
87. GUA, *Glasgow University Magazine* 5, 6 (January 11, 1893), 64, and *Glasgow University Magazine* 5, 7 (January 18, 1893), 78.
88. *The Corolla '96*, 33.
89. Stein, "The History of the Daily Cardinal from 1892–1991," 13, 27.
90. "Shall the Gentlemen Go?" in *The Amulet: Published Annually in Their Junior Year by the Ladies of Eighty-Four* (Ann Arbor, MI: Register Printing and Publishing Company, 1882), 21.

91. Ibid.
92. Barbara Miller Solomon, *In the Company of Educated Women* (New Haven, CT, and London: Yale University Press, 1985), 129. See also Christine A. Ogren, "Where Coeds Were Coeducated: Normal Schools in Wisconsin, 1870–1920," *History of Education Quarterly* 25, 1 (Spring 1995), 17.
93. *The Garnet and Black 1899*, 6–9.
94. Ibid., 133–136.
95. Amy Thompson McCandless, "Maintaining the Spirit and Tone of Robust Manliness: The Battle against Coeducation at Southern Colleges and Universities, 1890–1940," *NWSA Journal* 2, 2 (Spring 1990), 212.
96. *The Corolla of Ninety-Four, Volume II* (Tuscaloosa: Published by the Students of the University of Alabama, 1894), 121. *The Crimson-White* was "a 4-page, 4-column sheet, devoted to the interests of the students and the University." It was published every Friday. See also Sellers, *History of the University of Alabama*, 479, 521, 548.
97. *The Corolla of Ninety-Five, Volume III* (Tuscaloosa: Published by the Students of the University of Alabama, 1895), 141–144.
98. Sellers, *History of the University of Alabama*, 521–522.
99. *The Volunteer Published by The Students of the University of Tennessee Vol. III* (Knoxville, TN: Bean, Warters & Gaut, 1899), 7, 120, 122–123.
100. *The Monticola Volume II. Published by the Class of 1900* (Morgantown, WV: The Acme Publishing Company, 1899), 6.
101. "General College News," *The American Educational Review* 30, 4 (January 1909), 176.
102. *The Savitar—1891*, 112.
103. J. S. Snoddy, *A Little Book of Missouri Verse: Choice Selections from Missouri Verse-Writers* (Kansas City, MO: Hudson-Kimberly Publishing Co., 1897), 48, 175.
104. "Editorial," *The University Courant* 2, 8 (October 1888), 85.
105. "Locals," *The Western University Courant* 12, 2 (November 1896), 25.
106. Starrett, *Through One Hundred and Fifty Years*, 204.
107. GUA, Queen Margaret College Bazaar scrap book, includes: presscuttings, circular letters, minutes of meeting, programmes, tickets, etc., 1889.
108. X. Y. Z., "Letter to the Editor," *The Free Lance* 10, 5 (November 1896), 100.
109. "Locals," *The Free Lance* 3, 3 (June 1889), 194.
110. "Editorial," *The Free Lance* 14, 1 (April 1900), 20.
111. NUIG, "Ladies' Notes," *QCG: a record of college life in the city of the tribes* 1, 1 (November 1902), 12.
112. "Scotch and Irish Medical Colleges: Cork and Galway," *The Students' Journal and Hospital Gazette* 7 (September, 13, 1879), 225; "Queen's College, Cork," *The Students' Journal and Hospital Gazette* 10 (March 18, 1882), 121; and "Provincial Medical Colleges: Belfast," *The Students' Journal and Hospital Gazette* 10 (September 16, 1882), 866.
113. Stein, "The History of the Daily Cardinal from 1892–1991," 3.
114. Lynn D. Gordon, *Gender and Higher Education in the Progressive Era* (New Haven, CT, and London: Yale University Press, 1990), 9.

Eight: Life After Graduation

1. OUMC, Margaret Boyd Diary (1873), 176 in Ohio Memory: An Online Scrapbook of Ohio History, www.ohiomemory.org/index.html (accessed June 21, 2009). It is unclear from Boyd's diary who was preaching at chapel on the day in question. At various points during the spring and summer of 1873 different members of the community, including some students and faculty from the university, took turns preaching on Sundays and at prayer meetings.
2. Adele Simmons, "Education and Ideology in Nineteenth Century America: The Response of Educational Institutions to the Changing Role of Women," in *Liberating Women's History: Theoretical & Critical Essays*, ed. Bernice A. Carroll (Urbana: University of Illinois Press, 1976), 123.
3. T. Claye Shaw, "The Collegiate Training of Women," in *The Edinburgh Medical Journal Vol. XV*, ed. G. A. Gibson and Alexis Thomson (Edinburgh and London: Young J. Pentland, 1904), 445.

4. B. L. Hutchins, "Higher Education and Marriage," in *The Education Papers: Women's Quest for Equality in Britain, 1850–1912*, ed. Dale Spender (London: Routledge & Kegan Paul, 1987), 333.
5. Glen H. Elder, "Appearance and Education in Marriage Mobility," *American Sociological Review* 34 (1969), 531–532.
6. D. I. Mackay, *Geographical Mobility and the Brain Drain: A Case Study of Aberdeen University Graduates, 1860–1960* (London: George Allen and Unwin, 1969), 171–180.
7. Thomas Waverly Palmer, comp., *A Register of the Officers and Students of the University of Alabama 1831–1901* (Tuscaloosa: The University of Alabama, 1901), 386.
8. Ibid., 391.
9. Ibid., 389, 392.
10. Ibid., 392. They were married in December 1900 and subsequently moved to Decatur.
11. Ibid., 401.
12. Ibid., 406.
13. Ibid., 408.
14. *The Savitar—1891; by Students of the University of Missouri* (Columbia, MO: Published by the university, 1891), 64.
15. *The Monticola Volume II. Published by the Class of 1900* (Morgantown, WV: The Acme Publishing Company, 1899), 46.
16. Hutchins, "Higher Education and Marriage," 331.
17. Robert C. Alberts, *Pitt: The Story of the University of Pittsburgh, 1787–1987* (Pittsburgh: University of Pittsburgh Press, 1986), 48, and Agnes Lynn Starrett, *Through One Hundred and Fifty Years: The University of Pittsburgh* (Pittsburgh: University of Pittsburgh Press, 1937), 203. Margaret Stein was also the first woman to earn her master of arts degree at the Western University of Pennsylvania in 1901. Dr. Fetterman's sister, Valeria, was also a student at the time and graduated in 1900.
18. Ruben Gold Thwaites, ed., *The University of Wisconsin: Its History and its Alumni* (Madison, WI: J. N. Purcell, 1900), 522, 572.
19. Hutchins, "Higher Education and Marriage," 333.
20. Peter J. Schakel, *The Way into Narnia: A Reader's Guide* (Grand Rapids, MI: William B. Eerdmans Publishing Company, 2005), 3.
21. MEWC, Warren H. Lewis, ed., *Memoirs of the Lewis Family 1850–1930, Volume One: From October 17th, 1850 to September 23rd, 1881* (Oxford: Leeborough Press, 1933), 328.
22. Bruce L. Edwards, *C. S. Lewis: Life, Works, and Legacy* (Westport, CT: Praeger, 2007), 150–152.
23. Schakel, *The Way into Narnia*, 3–6.
24. James Coutts, *A History of the University of Glasgow: From its Foundation in 1451 to 1909* (Glasgow: J. Maclehose and Sons, 1909), 459.
25. A. Wallis Myers, "Women Students in Wales," *The Ludgate Illustrated Magazine Vol. VIII* (London: F. V. White & Co., 1899), 141.
26. Theodore W. Koch, *Handbook of the Libraries of the University of Michigan* (Ann Arbor, MI: George Wahr, 1910), 16.
27. Olive San Louis Anderson, *An American Girl and Her Four Years in a Boys' College*, ed. Elisabeth Israels Perry and Jennifer Ann Price (Ann Arbor: University of Michigan Press, 2006), 43–44, 165. One of Anderson's classmates at Michigan was Alice Freeman Palmer who had made the decision not to marry her first love so that she could attend college instead. For more see Ruth Bordin, *Alice Freeman Palmer: The Evolution of a New Woman* (Ann Arbor: University of Michigan Press, 1993), 27–28, and Alice Freeman Palmer, *Why Go to College?* (Boston: T. Y. Crowell & Co., 1897).
28. GUA, Queen Margaret College Literary and Debating Society—Topics of Debate (January 1899–May 1905).
29. *The Monticola Volume III, 1901* (Morgantown, WV: The Acme Publishing Company, 1900), 120.
30. Mercy Grogan, *How Women May Earn a Living* (London, Paris, and New York: Cassell & Company, 1883), iii.
31. Charlotte Wronker, "Co-Education in the 'Varsity," *The Missouri Alumni Quarterly* (December 1905), 27.
32. Martha Vicinus, *Independent Women: Work and Community for Single Women, 1850–1920* (Chicago: University of Chicago Press, 1985), Chapter 6. See also Alice M. Gordon, "The After-careers of University-Educated Women," *Nineteenth Century* 37 (1895), 955–960.

33. Palmer, *A Register of the Officers and Students of the University of Alabama 1831–1901*, 393–394, 398–399.
34. Ibid., 401, 411.
35. Ibid., 394.
36. Ibid., 398, 400–401, 404, 406, 409. Other Alabama graduates who became teachers but did not list a location were Augusta Harrison Cleary, Minnie May Cox, Rosa Lawhon, Laura Scotte McGhee, and Adele Glenn Quarles.
37. Ibid., 380.
38. Ibid., 401, 411.
39. Edward Mayes, *History of Education in Mississippi* (Washington, DC: Government Printing Office, 1899), 178, and James Allen Cabaniss, *A History of the University of Mississippi* (University: University of Mississippi, 1949), 111, 118.
40. David G. Sansing, *The University of Mississippi: A Sesquicentennial History* (Jackson: University Press of Mississippi, 1999), 137–138.
41. *The Monticola Volume II*, 37.
42. Waitman Barbe, ed., *Alumni Record: West Virginia University* ([Morgantown, WV]: Published by the Alumni Association, 1903), 124.
43. "Personal," *The Ohio Educational Monthly; Organ of the Ohio Teachers' Association and The National Teacher* 37 11 (November 1888), 599. See also Thomas N. Hoover, *The History of Ohio University* (Athens: Ohio University Press, 1954), 172.
44. OUMC, Margaret Boyd Diary, 183.
45. Ibid., 241.
46. Palmer, *A Register of the Officers and Students of the University of Alabama 1831–1901*, 398–399.
47. Sarah V. Barnes, "Crossing the Invisible Line: Establishing Co-education at the University of Manchester and Northwestern University," *History of Education* 23, 1 (1994), 47.
48. *The Victoria University of Manchester: Register of Graduates up to July 1st, 1908* (Manchester: Manchester University Press, 1908), 73.
49. Sheila Hamilton, "Women and the Scottish Universities circa 1869–1939: A Social History" (Ph.D. thesis, University of Edinburgh, 1987), 119–125.
50. Wendy Alexander, "Early Glasgow Women Medical Graduates," in *The World is Ill Divided: Women's Work in Scotland in the Nineteenth and Early Twentieth Centuries*, ed. Eleanor Gordon and Esther Breitenbach (Edinburgh: Edinburgh University Press, 1990), 88.
51. *The Calendar of King's College, London for 1896–97* (London: Published by the College, 1896), 320–321, 335–336, 349–350.
52. Mary A. Marshall, "Medicine as a Profession for Women," in *The Woman's World: Volume I*, ed. Oscar Wilde (London, Paris, and Melbourne: Cassell & Company, 1888 and London: Source Book Press, 1970), 106.
53. Lady Frances Balfour, *Dr. Elsie Inglis* (New York: George H. Doran Company, 1919), 67.
54. Ibid., 124, 149–240, and Adelaide Ellsworth, "The Woman Doctor in War," *The Pennsylvania Medical Journal: Official Organ of the Medical Society of the State of Pennsylvania* 22, 1 (October 1918), 24.
55. James Maitland Anderson, ed., *The Matriculation Roll of the University of St. Andrews 1747–1897* (Edinburgh and London: William Blackwood and Sons, 1905), 303, 321.
56. *The Glasgow University Calendar for the year 1902–3* (Glasgow: James Maclehose and Sons, 1902), 389.
57. "The Dermatology Society of Great Britain and Ireland," *The British Journal of Dermatology* (December 1905), 460, and "Pieric Acid and Camphor Cure Ringworm," *The Eclectic-Medical Journal* 74, 5 (May 1914), 258.
58. S. E. P., "British Medical Association. Joint Discussion on the Treatment of Uterine Fibroids," *Medical Science: Abstracts & Reviews* 3, 3 (December 1920), 289.
59. "Brilliant Bandon Doctor, Resignation from Armagh Mental Hospital Post," *Southern Star*, October 17, 1936, 9; "Bandon Brieflets," *Southern Star*, March 5, 1955, 5; and Royal University of Ireland, *The Calendar for the Year 1908* (Dublin: Alex. Thom & Co., 1908), 267.
60. Ja. F. Kellan Johnstone, "Tuesday, 25th September," in *Record of the Celebration of the Quatercentenary of the University of Aberdeen*, ed. P. J. Anderson (Aberdeen: Aberdeen University Press, 1907), 70–71.
61. William Johnston, ed., *Roll of the Graduates of the University of Aberdeen 1860–1900, Aberdeen University Studies: No. 18* (Aberdeen: Aberdeen University Press, 1906), 670, 678. Miss Harrison also received an honorary D.Litt. from the University of Durham in 1897.

62. Ibid., 324–325.
63. Frances E. Willard, *Woman and Temperance: Or, The Work and Workers of The Woman's Christian Temperance Union* (Hartford, CT: James Betts & Co., 1883), 608.
64. Sarah P. Morrison, *Among Ourselves: To A Mother's Memory Vol. I: Out of North Carolina* (Plainfield, IN: Publishing Association of Friends, 1901) and *Among Ourselves To A Mother's Memory Vol. II: Catherine and Her Surroundings* (Plainfield, IN: Publishing Association of Friends, 1902). See also "Literary Notes," in *Friends' Intelligencer and Journal* (Philadelphia: Friends' Intelligencer Association, 1902), 56.
65. *National Normal*, quoted in "A Lady Graduate of the Indiana State University," *Arthur's Home Magazine Vol. XXXIII January to June*, ed. T. S. Arthur and Miss Virginia F. Townsend (Philadelphia: T. S. Arthur, 1869), 301.
66. Judith Barger, *Elizabeth Stirling and the Musical Life of Female Organists in Nineteenth-Century England* (Aldershot, England, and Burlington, VT: Ashgate, 2007), 17, and Derek B. Scott, *The Singing Bourgeois: Songs of the Victorian Drawing Room and Parlour* (Milton Keynes and Philadelphia: Open University Press, 1989), 64.
67. Emily Janes, *The Englishwoman's Year Book and Directory 1900* (London: Adam and Charles Black, 1900), 136.
68. Ibid., 137–138.
69. James D. Brown and Stephen S. Stratton, *British Musical Biography: A Dictionary of Musical Artists, Authors and Composers, born in Britain and its Colonies* (Birmingham: S. S. Stratton, 1897), 197. The composer was Henry Hiles.
70. Frederick H. Martens, Mildred W. Cochran, and W. Dermot Darby, eds., *The Art of Music: Volume Eleven, A Dictionary-Index of Musicians Book I A-L* (New York: The National Society of Music, 1917), 18.
71. Palmer, *A Register of the Officers and Students of the University of Alabama 1831–1901*, 391. One of the few female students from the University of Alabama who did not list teaching as her profession was Clara Virginia Hollomon, a stenographer in Demopolis. A woman at Wisconsin listed "Law stenographer" as her occupation as well. For more see Thwaites, *The University of Wisconsin*, 776–844.
72. Johanna Geyer-Kordesch and Rona Ferguson, *Blue Stockings, Black Gowns, White Coats: A Brief History of Women Entering the Medical Profession in Scotland in Celebration of One Hundred Years of Women Graduates at the University of Glasgow* (Glasgow: University of Glasgow, Wellcome Unit for the History of Medicine, 1994), 36–45.
73. *The President's Report to the Board of Regents for the Year Ending September 30, 1882* (Ann Arbor, MI: Courier Book and Job Printing House, 1882), 4.
74. *The Student Missionary Appeal: Addresses at the Third International Convention of the Student Volunteer Movement for Foreign Missions held at Cleveland, Ohio, February 23–27, 1898* (New York: Student Volunteer Movement for Foreign Missions, 1898), v.
75. Ibid., 541, 543, 545–547.
76. James Earl Russell, "The Extension of University Teaching in England and America: A Study in Practical Pedagogics" (Ph.D. thesis, University of Leipsic, 1895), 152.
77. General Alumni Association, *Alumni directory, University of Pittsburgh, Vol. 2, 1787–1916* (Pittsburgh: Aldine Printing Company; Smith Bros. Col., 1916), 18–19, 102, 184. See also Marion Talbot and Lois Kimball Mathews Rosenberry, *The History of the American Association of University Women, 1881–1931* (Boston: Houghton Mifflin, 1931).
78. *La Vie: The Annual Publication of the Junior Class of the Pennsylvania State College Volume XXXIII* (State College, PA: Published by the Class of Nineteen Twenty-Two, 1921), 47.
79. GUA, Melville, "Presentation Address," On the occasion of the first award of the Frances Melville Medal in Philosophy on the final closure of the College (November 1935), 6. See also Carol Dyhouse, *No Distinction of Sex? Women in British Universities 1870–1939* (London: UCL Press, 1995), 219.
80. John Malcolm Bulloch, ed., *College Carols* (Aberdeen: D. Wylie and Son, 1894), 28, Line 10.
81. Ibid. Lines 25–32. This includes a reference to Alexander Bain who was the author of *The Senses and the Intellect* (London: John W. Parker and Son, 1855) and *The Emotions and the Will* (London: John W. Parker and Son, 1859) among other works.
82. Amos Chiseler, "Cockie Law! Is a Girl a Person?" *Glasgow Evening Citizen*, July 18, 1901, Lines 1–16. The case under consideration that inspired this poem is described in *The Admission of Women to the Legal Profession. Proceedings in the Case of Miss Margaret Hall before the Supreme Court of Scotland* (Dunoon, 1901).

83. Eibhlin Breathnach, "Women and Higher Education in Ireland (1879–1914)," in *The Irish Women's History Reader*, ed. Alan Hayes and Diane Urquhart (London: Routledge, 2001), 48.
84. F. S., "The Irish University Question As Affecting Women," *The Westminster Review* 159 (January to June 1903), 616. Trinity College, Dublin had opened its doors to women in 1904. They would be followed by the newly configured National University of Ireland (NUI), which began operations in 1908. For more see Judith Harford, "An Experiment in the Development of Social Networks for Women: Women's Colleges in Ireland in the Nineteenth Century," *Paedagogica Historica* 43, 3 (June 2007), 374, 380–381, and Breathnach, "Women and Higher Education in Ireland (1879–1914)," 48.
85. "Higher Education, University College, Lady Students' Claims," *Irish Independent*, March 22, 1905, 3. See also " 'Sweet Girl Graduates.' Work of the Irish Association," *Irish Independent*, November 1, 1905, 5.
86. "Women Graduates' Association," *Irish Independent*, February 21, 1905, 7. See also "Women Graduates. Royal University Status," *Irish Independent*, February 20, 1906, 7.
87. W. Le Conte Stevens, *The Admission of Women to Universities* (New York: Press of S. W. Green's Son, 1883), 4.
88. "The Andrew Carnegie Donation," *The Journal of the Iron and Steel Institute Vol. LVIII* (London: E. & F. N. Spon and New York: Spon & Chamberlain, 1900), 6–7. The second woman mentioned, Mrs. Hertha Ayrton, was a graduate of the University of London, having studied at Girton College, Cambridge. She became the first female member of the Institution of Electrical Engineers. For more see James Johnson, "Women Inventors and Discoverers," *Cassier's Magazine: An Engineering Monthly* 36, 6 (October 1909), 548–552.
89. *Proceedings of the Board of Regents of the University of Michigan, from January, 1881, to January, 1886* (Ann Arbor, MI: The Courier Book and Job Printing Establishment, 1886), 385.
90. Virginia D. Young, "Reports from Auxiliary States: South Carolina," in *Proceedings of the Twenty-Seventh Annual Convention of the National-American Woman Suffrage Association held in Atlanta, GA., January 31st to February 5th, 1895*, ed. Harriet Taylor Upton (Warren, OH: Wm. Ritezel & Co., 1895), 88.
91. Mrs. Campbell, "The Rise of the Higher Education of Women Movement," in *The Book of the Jubilee: In Commemoration of the Ninth Jubilee of the University of Glasgow, 1451–1901*, ed. the Students' Jubilee Celebrations Committee (Glasgow: J. Maclehose and Sons, 1901), 128, 138.
92. Alexander, "Early Glasgow Women Medical Graduates," 90.
93. Barnes, "Crossing the Invisible Line," 43.
94. GUA, Queen Margaret College Literary and Debating Society—Topics of Debate (January 1899–May 1905).
95. Barnes, "Crossing the Invisible Line," 46.
96. Campbell, "The Rise of the Higher Education of Women Movement," 128.

Nine: Drawing Conclusions

1. Amy Thompson McCandless, "Maintaining the Spirit and Tone of Robust Manliness: The Battle against Coeducation at Southern Colleges and Universities, 1890–1940," *NWSA Journal* 2, 2 (Spring 1990), 215.
2. Charles F. Richardson and Henry A. Clark, eds., *The College Book* (Boston: Houghton, Osgood and Company, 1878), 353.
3. Louisa Innes Lumsden, "On the Higher Education of Women in Great Britain and Ireland," in *Journal of Social Science, Containing the Transactions of the American Association, Number XX June, 1885, Saratoga Papers of 1884, Part II* (Boston: Cupples, Upham & Co. and New York: G. P. Putnam's Sons, 1885), 55.
4. MEPL, Life Drawing Class at University College, London, 1881, and *The Calendar of King's College, London for 1896–97* (London: Published by the College, 1896), 315.
5. Lucia Gilbert Runkle, "A New Knock at an Old Door," in *Woman and the Higher Education*, ed. Anna C. Brackett (New York: Harper & Brothers Publishers, 1893), 94–95, and GUA, Supplementary Report on Women's University Education: The British Educational Mission to the United States, October–December 1918, 14.
6. Lumsden, "On the Higher Education of Women in Great Britain and Ireland," 55.

7. Carol Dyhouse, *No Distinction of Sex? Women in British Universities 1870–1939* (London: UCL Press, 1995).
8. Edward T. Sanford, *Blount College and the University of Tennessee: An Historical Address Delivered Before the Alumni Association and Members of the University of Tennessee* (Knoxville, TN: Published by the University, 1894), 79.
9. Judith Harford, *The Opening of University Education to Women in Ireland* (Dublin and Portland, OR: Irish Academic Press, 2008), 91.
10. John Malcolm Bulloch, *A History of the University of Aberdeen 1495–1895* (London: Hodder and Stoughton, 1895), 210–211.
11. A. L. Brown and Michael Moss, *The University of Glasgow: 1451–1996* (Edinburgh: Edinburgh University Press, 1996), 71.
12. Ronald Gordon Cant, *The University of St. Andrews: A Short History* (Edinburgh: Scottish Academic Press, 1970), 132, and Mrs. Mina Aitken, "What Women Are Doing in Scotland," *Womanhood* 13, 74 (January 1905), 94. See also Iain Catto, ed., *"No spirits and precious few women": Edinburgh University Union 1889–1989* (Edinburgh: Edinburgh University Union, 1989), 87.
13. W. Le Conte Stevens, *The Admission of Women to Universities* (New York: Press of S. W. Green's Son, 1883), 11.
14. Edith L. Sheffield, "Student Life in the University of Michigan," *The Cosmopolitan* 7, 2 (June 1889), 110.
15. OUMC, Margaret Boyd Diary (1873), 4, in Ohio Memory: An Online Scrapbook of Ohio History, www.ohiomemory.org/index.html (accessed June 21, 2009), 7.
16. "Co-Education of the Sexes in Colleges," *Indiana School Journal* 25, 8 (August 1880), 422.
17. *The President's Report to the Board of Regents for the Academic Year Ending September 30, 1899 and the Report of the Treasurer for the Fiscal Year ending June 30, 1899* (Ann Arbor, MI: The Courier Office, 1899), 5.
18. May Wright Sewall, " 'A Report on the Position of Women in Industry and Education in the State of Indiana,' Indiana Department of the New Orleans Exposition, 1885," in *American Feminism: Key Source Documents 1848–1920 Volume II: Work and Education*, ed. Janet Beer, Anne-Marie Ford and Katherine Joslin (London: Routledge, 2002), 133.
19. *Proceedings of the Board of Regents of the University of Michigan from January, 1876, to January, 1881* (Ann Arbor, MI: Ann Arbor Printing and Publishing Company, 1881), 151–152.
20. "Gov. Hayes: His Address to the Graduating Class of Ohio University," *Chicago Tribune*, June 24, 1876, 6, and William Kimok, email message to author, September 24, 2009. See also Russell H. Conwell, *Life and Public Services of Gov. Rutherford B. Hayes* (Boston: B. B. Russell, 1876), 299. "Ella" Boyd was in fact Jane Elliot Boyd, Margaret's niece. Because Margaret was the youngest of nine children, it is not surprising that she had a niece who was old enough to be in college at the same time as her aunt. Ella went on to marry one of Margaret's classmates, John M. Davis. For more see *Ohio Alumnus* (March 1941), 2.
21. "Domestic Science in the Agricultural Colleges," *The American Kitchen Magazine* 7, 6 (September 1897), 219.
22. *University of Alabama Bulletin; Centennial Celebration 1831–1931* 90 (June 1931), 74.
23. Ibid., 74–75.
24. Harford, *The Opening of University Education to Women in Ireland*, 80.
25. Christina Sinclair Bremner, *Education of Girls and Women in Great Britain* (London: Swan Sonnenschein & Co., 1897), 151.
26. Angie Warren Perkins, "Report of the Acting Dean, Woman's Department," *University of Tennessee Record* 8 (January 1899), 24–25.
27. Cynthia Eagle Russett, *Sexual Science: The Victorian Construction of Womanhood* (Cambridge, MA: Harvard University Press, 1989), 171.
28. George J. Romanes, "Mental Differences between Men and Women" in *The Education Papers: Women's Quest for Equality in Britain, 1850–1912*, ed. Dale Spender (London: Routledge & Kegan Paul, 1987), 30.

BIBLIOGRAPHY

Archival and Unpublished Sources

The Admission of Women to the Legal Profession. Proceedings in the Case of Miss Margaret Hall before the Supreme Court of Scotland. Dunoon, 1901. (GUA, DC 233/2/24/17)

Article from *The Buteman*, Rothesay, regarding the establishment of two bursaries for ladies taking the Glasgow University Local Exams, Saturday April 20, 1878. (GUA 50084)

"A Bill to Remove Doubts As to the Powers of the Universities of Scotland to Admit Women as Students and to Grant Degrees to Women," April 14, 1874. (GUA, DC 233/1/9/1)

Booklet of views of interior, exterior and grounds of Queen Margaret College (6 copies) (some negatives), n.d. (GUA, DC 233/2/22/3/1)

Buchanan, Marion. Letter regarding Q. M. Students and Rectorial Election, 28 March 1893. (GUA 20530)

Cardinal Articles, 1892–1912. (UWA, Series 91/46, Box M12 h1)

Checkland, Olive. "Women in Glasgow University, Queen Margaret's College, Hall, Settlement and Union." Typescript, July 1979. (GUA, DC 233/2/21/11)

Copy correspondence between John Caird and Mrs. Elder concerning separate teaching of women at QMC with reference to teaching of women at Edinburgh University, 1892. (GUA 62398)

Correspondence between Sir Richard Lodge, Professor of History, and Secretary of Court, concerning teaching a separate course of lectures for women, January 1896. (GUA 62413–15)

The Daily Cardinal, beginning with April 4, 1892 issue (Madison, WI). (WHS 98–526 Super over size)

The Daily Cardinal, Miscellaneous publications (Madison, WI). (WHS 90–4726)

Droste, Jean Rasmusen. "Women at Wisconsin." M.A. thesis, University of Wisconsin, 1967.

Examination results, notes on bursaries and qualified medical students, 1895–1896. (GUA, DC 233/2/8/4/1)

Galloway, Janet. "Historical Sketch of the Movement for the Higher Education of Women in Glasgow and Queen Margaret College." On the occasion of the golden wedding anniversary of Mrs. Jean Campbell of Tullichewan, May 1896. (GUA, DC 233/2/21/3)

———. Letter to John Caird, Principal of Glasgow University, relating to progress of the College, 29 April 1889. (GUA, DC 233/2/4/4/13)

Geddes, P. Letter from University College Dundee, to Miss Galloway concerning rules for residents of Queen Margaret Hall, 10 May 1894. (GUA, DC 233/2/13/10/10)

Glasgow Association for the Higher Education of Women. Draft petition to Glasgow University Senate for a degree in Arts for women, 21 October 1882. (GUA, DC 233/1/4/2/5)

———. General Committee meeting minutes with presscuttings re: inaugural meeting, 4 April 1877. (GUA, DC 233/1/1/2)

———. Petition to the Senate of Glasgow University for a University title for women, plus copies of Mrs. Lindsay's earlier draft, and suggestions of possible alterations, 1883. (GUA, DC 233/1/9/3)

Glasgow University Court. Excerpt minute from meeting concerning outcome controversy of Mrs. Elder's complaint about the ineffectual treatment of her proposal of equal teaching of women at the College, 15 March 1897. (GUA, DC 233/2/4/4/53)

Glasgow University Magazine. Volumes 1–43, 1889–1936. (GUA, DC 198/1/1–47)

Glasgow University Students' Handbook, 1893, 1903–1940. (GUA, DC 157/18/1–49)

Hague, Amy. "'Give Us a Little Time to Find Our Places': University of Wisconsin Alumnae, Classes of 1875–1900." M.A. thesis, University of Wisconsin-Madison, 1983.

"A Half Century of Progress, A Future of Promise." WSGA Commemorative Booklet, 1897–1947. (UWA, Series No. 20/2/3/1–2, Box No. 7)

Hamilton, Sheila. "Women and the Scottish Universities circa 1869–1939: A Social History." Ph.D. thesis, University of Edinburgh, 1987.

Henderson, G. G. Letter to Janet Galloway, reporting on the Committee of Lecturers' recommendations to the Council of Queen Margaret College concerning prizes, 4 February 1892. (GUA, DC 233/2/8/5/1)

Holland, William Jacob. *History of the University of Pittsburgh*. Pittsburgh: University of Pittsburgh, Digital Research Library, 2006.

Kendall, Catherine Mary. "The Queen Margaret Settlement 1897–1914: Glasgow Women Pioneers in Social Work." M.A. thesis, University of Glasgow, 1993.

Latta, Professor. Letter to Miss Galloway concerning the use of the University Library by women students, 26 October 1908. (GUA, DC 233/2/4/4/54a)

Lecturers' Committee. Report to Queen Margaret College Executive Council of the affiliation of the College to Glasgow University, 21 April 1890. (GUA, DC 233/2/4/4/16)

Letters (8) from Miss Galloway; Masson Hall, Edinburgh; University Hall St. Andrews; and students to Mrs. Riddoch relating to the Hall regulations concerning visitors, February–March 1902. (GUA, DC 233/2/13/10/9)

Lewis, Warren H., ed. *Memoirs of the Lewis Family 1850–1930, Volume One: From October 17th, 1850 to September 23rd, 1881*. Oxford: Leeborough Press, 1933. (MEWC)

List of Bursaries from GAHEW for Glasgow University Local Exams, 1883. (GUA, DC 233/1/7/3)

Lloyd, Campbell Fox. "Relationships between Scottish Universities and Their Communities c. 1858–1914." Ph.D. thesis, University of Glasgow, 1993.

Margaret Boyd Diary (1873). In Ohio Memory: An Online Scrapbook of Ohio History. http://www.ohiomemory.org/index.html. (OUMC, MSS 015)

Melville, Frances H. Letter from University Hall at St. Andrews, 16 February 1902. (GUA, DC 233/2/13/10/9)

———. "Presentation Address." On the occasion of the first award of the Frances Melville Medal in Philosophy on the final closure of the College, November 1935. (GUA, DC 233/2/21/7)

Mertz, T. J. "'A Peculiar Public Matter': School Politics, Policy and Wisconsin Women, 1885–1921." Paper presented at the History of Education Society Annual Meeting, Chicago, Illinois, October 29–November 1, 1998.

Minutes of the Board of Regents (1866 through 1876). (UWA, Series No. 1/1/1, Vol. No. 3)

Notes on Other Publications. (UWA, Series 91/46, Box M12 h1)

Oakley, Edith. "The Formation of Character." Essay and Prof. Bowman's notes, dated 6 February 1929. (GUSC)

Official Ballot, January 1900. (UWA, Series I-4/13 File 1900)

Philip, Annie G. Letter regarding the regulations for the Hall Committee, 4 March 1902. (GUA, DC 233/2/13/10/4)

Presscuttings book on visits of Queen Victoria and HRH Princess Louise, 1888–1890. (GUA, DC 233/2/18/3)

Programme of Classes and List of Bursaries Available, 1897–1912. (GUA, DC 233/2/8/2/1–14)

Prospectus (printed) of Queen Margaret Hall with terms of board and general regulations, n.d. (GUA, DC 233/2/13/10/3)

QMC Christian Union, Constitution, office bearers and branches, 1898–99 (printed). (GUA)

Queen Margaret College Bazaar scrap book, includes: presscuttings, circular letters, minutes of meeting, programmes, tickets, etc., 1889. (GUA, DC 233/2/15/8)

"Queen Margaret College." *Pass It On: The Magazine of the Women's Educational Union* 15, 1 (November 1935). (GUA, DC 233/2/21/13 or DC 240/5/89)

Queen Margaret College Letterbook, 1878–1883 (Correspondence Courses). (GUA, DC 233/1/4/1/1)

Queen Margaret College Literary and Debating Society, Minutes of Meetings, January 1899 to May 1905. (GUA, DC 233/2/16/4/1)

Report of Committee on Women Students and Rectorial Election, 22 April 1893. (GUA 20562)

Samples of advertisement styles of other University Halls plus Queen Margaret Hall, n.d. (GUA, DC 233/2/13/20)

Simson, Frances H. Letter from Masson Hall, University of Edinburgh to Mrs. Riddoch, 13 February 1902. (GUA, DC 233/2/13/10/9)

Stein, Jennifer L. "The History of the *Daily Cardinal* from 1892–1991: A Look at the University of Wisconsin-Madison's Oldest Student Daily." Senior thesis, University of Wisconsin-Madison, 1991.

Supplementary Report on Women's University Education: The British Educational Mission to the United States, October–December 1918. (GUA, DC 233/2/24/21)

Volume of Presscuttings of Queen Margaret College, 1884–1890. (GUA, DC 233/2/20/1/1)

Volume of Presscuttings of Queen Margaret College, 1891–1894. (GUA, DC 233/2/20/1/2)

The Western University of Pennsylvania, 75th Annual Commencement Carnegie Music Hall, Pittsburgh, Pa. June ninth, Eighteen Hundred and Ninety-eight. Pittsburgh: University of Pittsburgh, Digital Research Library, 2006.

The Western University of Pennsylvania 1878–1900, Annual Commencement of the Collegiate, Engineering and Legal Departments. Carnegie Music Hall, Pittsburgh, Pennsylvania, June 14th, 1900, 8:15 P. M. Pittsburgh: University of Pittsburgh, Digital Research Library, 2006.

"WOMEN. Position American." Extract from an American Supplement of *Encyclopedia Britannica*, 908–913, c. 1889. (GUA, DC 233/2/24/10)

Published Victorian Sources

The Aberdeen University Calendar Part I. Aberdeen: A. King & Co., 1898.

The Addresses and Journal of Proceedings of the National Educational Association, Session of the Year 1874, at Detroit, Michigan. Worcester, MA: Published by the Association, 1874.

"The Affiliated Colleges and the Prizes of the University." *The Durham University Journal* 5, 3 (May 27, 1882): 26.

Allan, J. McGrigor. "On the Real Differences in the Minds of Men and Women." *Anthropological Review* 7 (1869): 195–215.

"American Women: Their Health and Education." *The Westminster Review* 202 (October 1874): 216–235.

The Amulet: Published Annually in Their Junior Year by the Ladies of Eighty-Four. Ann Arbor, MI: Register Printing and Publishing Company, 1882.

Anderson, E. G. "Sex in Mind and Education: A Reply." *Fortnightly Review* 15 (1874): 582–594.

Anderson, Olive San Louis. *An American Girl and Her Four Years in a Boys' College*, edited by Elisabeth Israels Perry and Jennifer Ann Price. Ann Arbor: University of Michigan Press, 2006.

"The Andrew Carnegie Donation." In *The Journal of the Iron and Steel Institute Vol. LVIII*, 6–7. London: E. & F. N. Spon and New York: Spon & Chamberlain, 1900.

Annual Catalogue of the Indiana University for the Sixty-Eighth College Year, 1891–92. Indianapolis, IN: Wm. B. Burford, 1892.

Annual Catalogue of the Indiana University for the Sixty-Seventh College Year, 1890–91. Indianapolis, IN: Wm. B. Burford, 1891.

Annual Catalogue of the Ohio University 1875. Athens, OH: Published by the University, 1876.

Annual Catalogue of the Ohio University 1885. Athens, OH: Published by the University, 1885.

Annual Report of the Indiana University, including the Catalogue for the Academical Year 1881–1882. Indianapolis, IN: Wm. B. Burford, 1882.

Annual Report of the Indiana University including the Catalogue for the Academical Year, 1882–1883. Indianapolis, IN: Wm. B. Burford, 1883.

Annual Report of the Regents of the University for the year ending September 30, 1857. Madison, WI: Calkins & Webb Printers, 1857.

Annual Report of the Regents of the University of Wisconsin, for the Fiscal Year ending September 30, 1869. Madison, WI: Published by the Board of Regents, 1869.

Annual Report of the Regents of the University of Wisconsin for the Fiscal Year Ending September 30, 1877. Madison, WI: David Atwood, 1877.

Annual Report of the Regents of the University of Wisconsin for the Fiscal Year Ending September 30, 1878. Madison, WI: David Atwood, 1878.

Appletons' Annual Cyclopaedia and Register of Important Events of the Year 1883, New Series, Vol. VIII. New York: D. Appleton and Company, 1884.

"Are Men Naturally Cleverer than Women?" *Englishwoman's Journal* 2 (1858): 336.

Atkinson, Geo. W. and Alvaro F. Gibbens. *Prominent Men of West Virginia: Biographical Sketches of Representative Men in Every Honorable Vocation, including Politics, the Law, Theology, Medicine, Education, Finance, Journalism, Trade, Commerce and Agriculture.* Wheeling, WV: W. L. Callin, 1890.

Aurora Borealis Academia: Aberdeen University Appreciations 1860–1889. Aberdeen: The University Printers, 1899.

Baedeker, Karl. *Great Britain: A Handbook for Travellers.* Leipsic: Karl Baedeker, 1901.

Bain, Alexander. *The Emotions and the Will.* London: John W. Parker and Son, 1859.

———. *The Senses and the Intellect.* London: John W. Parker and Son, 1855.

Balfour, Graham. *The Educational Systems of Great Britain and Ireland.* Oxford: The Clarendon Press, 1898.

Barnard, Frederick A. P. *Should American Colleges Be Open to Women as Well as to Men? A Paper Presented to the Twentieth Annual Convocation of the University of the State of New York, at Albany, July 12, 1882.* Albany, NY: Weed, Parsons and Company, 1882.

Baxter, Miss Mary Ann and John Boyd Baxter. *Deed of Endowment & Trust of the University College, Dundee.* Dundee: John Leng & Co., 1882.

"Beta—Ohio University." *The Rainbow of Delta Tau Delta* 11, 1 (January 1888): 82.

Biennial Report of the Board of Curators of the University of Missouri to the 36th General Assembly for the Two Years Ending December 31, 1890. Jefferson City, MO: Tribune Printing Company, 1891.

Biennial Report of the Board of Regents of the University of Wisconsin, for the Two Years Ending September 30, 1884. Madison, WI: Democrat Printing Co., 1883.

Billington, Mary Frances. "Alexandra House." In *The Woman's World: Volume III,* edited by Oscar Wilde, 154–157. London, Paris and Melbourne: Cassell & Company, 1890 and London: Source Book Press, 1970.

Blackburn, Helen, ed. *A Handbook for Women Engaged in Social and Political Work.* Bristol: J.W. Arrowsmith, 1881.

———. *A Handbook for Women Engaged in Social and Political Work, New Enlarged Edition.* Bristol: J.W. Arrowsmith and London: Edward Stanford, 1895.

"Book Reviews." *The Inlander* 11, 7 (April 1901): 290–292.

Brackett, Anna C., ed. *Woman and the Higher Education.* New York: Harper & Brothers Publishers, 1893.

Bremner, Christina Sinclair. *Education of Girls and Women in Great Britain.* London: Swan Sonnenschein & Co., 1897.

British Universities: Notes and Summaries Contributed to the Welsh University Discussion by Members of the Senate of the University College of North Wales. Manchester: J. E. Cornish, 1892.

Brown, James D. and Stephen S. Stratton. *British Musical Biography: A Dictionary of Musical Artists, Authors and Composers, born in Britain and its Colonies.* Birmingham: S. S. Stratton, 1897.

Bulloch, John Malcolm, ed. *College Carols.* Aberdeen: D. Wylie and Son, 1894.

———. *A History of the University of Aberdeen 1495–1895.* London: Hodder and Stoughton, 1895.

———. *University Centenary Ceremonies.* Aberdeen, 1893.

The Calendar of King's College, London for 1896–97. London: Published by the College, 1896.

"The Calendar of the Royal University." *The Nation,* March 24, 1888, 4.

The Calendar of the University College of Wales, Aberystwyth, Fourteenth Session, 1885–86. Manchester: J. E. Cornish, 1885.

Calendar of the University of Michigan for 1880–81. Ann Arbor, MI: The Courier Steam Printing House, 1881.

Calendar for the University of Michigan for 1891–92. Ann Arbor, MI: J. S. Cushing & Company, 1892.

Calendar of the University of Michigan for 1894–95. Ann Arbor, MI: The Register Publishing Company, 1895.

Campbell, Lewis. *On the Nationalisation of the Old English Universities.* London: Chapman and Hall, 1901.

Campbell, M. Montgomery. "The Central Conference of Women Workers." In *The Monthly Packet* 85, edited by Christabel R. Coleridge and Arthur Innes, 750–751. London: A. D. Innes and Co., 1894.

Campbell, Mrs. "The Rise of the Higher Education of Women Movement in Glasgow." In *The Book of the Jubilee: In Commemoration of the Ninth Jubilee of the University of Glasgow, 1451–1901,* edited by the Students' Jubilee Celebrations Committee, 125–138. Glasgow: J. Maclehose and Sons, 1901.

Catalogue and Announcements of the University of Mississippi at University P. O., Forty-Third Session 1894–95. Vicksburg, MS: Vicksburg Printing & Publishing Co., 1895.

Catalogue and Announcements of the University of Mississippi, University P. O., Near Oxford, Miss. Forty-Fifth Session 1896–97. Yazoo City, MS: The Mott Printing Company, 1897.

Catalogue of the Officers and Students of the University of Alabama with a Statement of the Courses of Instruction in the Various Departments 1879–80. Tuscaloosa, AL: Burton's Book and Job Printing Office, 1880.

Catalogue of the Officers and Students of the University of Mississippi, at Oxford, Mississippi, Twenty-Seventh Session. Jackson, MS: The Clarion Steam Printing Establishment, 1879.

Catalogue of the Officers and Students of West Virginia University for the Year 1872–73. Morgantown, WV: Morgan & Hoffman, 1873.

Catalogue of the Officers and Students of the University of Wisconsin for the year ending June 21, 1871. Madison, WI: Atwood & Rublee, 1871.

Catalogue of the Officers and Students of the University of Wisconsin for the year ending June 19, 1872. Madison, WI: Atwood & Rublee, 1872.

Catalogue of the Officers and Students of the University of Wisconsin, For the Year 1872–73 and the First Term of 73–74. Madison, WI: Atwood & Culver, 1873.

Catalogue of the University of Mississippi at University P. O., Near Oxford, Miss.: Prepared This Year with Special Reference to the Schools of English and Belles Lettres. Thirty-Ninth Session 1890–91. Oxford, MS: Published by the University, 1890.

Catalogue of the University of Wisconsin for the academic year 1891–92. Madison, WI: Published by the University, 1891.

Catalogue of the University of Wisconsin for 1895–96. Madison, WI: Published by the University, 1896.

Catalogue of the Western University of Pennsylvania for the year ending 1895 with detailed statements of the courses of instruction. [Pittsburgh]: Western University of Pennsylvania, 1895.

Catalogue of the Western University of Pennsylvania for the year ending 1896 with detailed statements of the courses of instruction. [Pittsburgh]: Western University of Pennsylvania, 1896.

Catalogue of the Western University of Pennsylvania for the year ending 1897 with detailed statements of the courses of instruction. [Pittsburgh]: Western University of Pennsylvania, 1897.

Catalogue of the Western University of Pennsylvania for the year ending 1899 with detailed statements of the courses of instruction. [Pittsburgh]: Western University of Pennsylvania, 1899.

Catalogue of the Western University of Pennsylvania for the year ending 1901 with detailed statements of the courses of instruction. [Pittsburgh]: Western University of Pennsylvania, 1901.

Catalogue of West Virginia University Morgantown 1887–8: Announcements for 1888–9. Charleston, WV: M. W. Donnally, 1888.

Catalogue of West Virginia University, Morgantown, for the Year 1891–92. Charleston, WV: Moses W. Donnally, 1892.

Chiseler, Amos. "Cockie Law! Is a Girl a Person?" *Glasgow Evening Citizen,* July 18, 1901.

Clarke, Edward H. *Sex in Education; or a Fair Chance for Girls.* Boston: James R. Osgood and Co., 1873.

Clouston, T. S. *Female Education from a Medical Point of View.* Edinburgh: Macniver & Wallace, 1882.

Coats, Joseph and John Lindsay Steven, eds. *The Glasgow Medical Journal Vol XLIV.* Glasgow: Alex. Macdougall and London: H. K. Lewis, 1895.

Cocker, W. J. *The Civil Government of Michigan, with Chapters on Political Machinery, and The Government of the United States*. Detroit: The Richmond & Backus Co., 1885.

"Co-Education of the Sexes in Colleges." *Indiana School Journal* 25, 8 (August 1880): 421–422.

"College Notes." *The Free Lance* 12, 1 (April 1898): 28–31.

"College Notes." *The Free Lance* 13, 1 (April 1899): 25–28.

"College Notes." *The Free Lance* 13, 2 (May 1899): 54–59.

"College Notes." *The Free Lance* 14, 6 (February 1901): 163–166.

Constant Reader. "Female Education." *The Nation*, April 24, 1847, 11.

Conwell, Russell H. *Life and Public Services of Gov. Rutherford B. Hayes*. Boston: B. B. Russell, 1876.

The Corolla. Cleveland, OH: The Cleveland Printing and Publishing Co., 1893.

The Corolla '96. Tuscaloosa, AL: W. H. Ferguson, 1896.

The Corolla of Ninety-Five, Volume III. Tuscaloosa: Published by the Students of the University of Alabama, 1895.

The Corolla of Ninety-Four, Volume II. Tuscaloosa: Published by the Students of the University of Alabama, 1894.

"County Items." *The Nation*, November 4, 1876, 2.

Covert, Jennie Muzzy. "At the Dawn of Coeducation." *The Wisconsin Alumni Magazine* 11 (March 1901): 245.

Dabney, Charles W. "Report of the President." *University of Tennessee Record* 2 (February 1901): 15–28.

Dall, Caroline. "'The Opening at the Gates,' *The College, the Market and the Court; or Women's Relations to Education, Labor and Law*. Boston, Lee and Shepherd, 1867." In *American Feminism: Key Source Documents 1848–1920 Volume II: Work and Education*, edited by Janet Beer, Anne-Marie Ford and Katherine Joslin, 75–118. London: Routledge, 2002.

Davies, Emily. "Women in the Universities of England and Scotland." In *The Educators: Female Education*, edited by Marie Mulvey Roberts and Tamae Mizuta, 183–196 (London: Routledge/Thoemmes Press, 1995.

"Day-Students." *The Durham University Journal* 5, 12 (December 17, 1883): 141–142.

Dickens's Dictionary of London, 1879: An Unconventional Handbook. London: Charles Dickens and Evans, 1879.

Distant, W. L. "The Mental Differences between the Sexes." *Journal of the Anthropological Institute* 4 (1875): 78–87.

"Domestic Science in the Agricultural Colleges." *The American Kitchen Magazine* 7, 6 (September 1897): 213–223.

Duffey, Eliza B. *No Sex in Education; or an Equal Chance for both Girls and Boys: Being a Review of Dr. E. H. Clarke's "Sex in Education."* Philadelphia: J.M. Stoddart & Co., 1874.

Dumaresq de Carteret-Bisson, F. S. *Our Schools and Colleges Vol. II: For Girls*. London: Simpkin, Marshall, & Co., 1884.

Dundee, University College. *Calendar for the Second Session 1884–1885*. Dundee: John Leng & Co., 1884.

The Durham University Calendar, with Almanack, MDCCCLXXXII. Durham: Andrews and Co. and London: Whittaker and Co., 1882.

The Edinburgh University Calendar 1871–72. Edinburgh: Edward Ravenscroft, 1871.

The Edinburgh University Calendar 1878–1879. Edinburgh: James Thin, 1878.

The Edinburgh University Calendar 1882–1883. Edinburgh: James Thin, 1882.

The Edinburgh University Calendar 1894–1895. Edinburgh: James Thin, 1894.

The Edinburgh University Calendar 1895–1896. Edinburgh: James Thin, 1895.

The Edinburgh University Calendar 1897–1898. Edinburgh: James Thin, 1897.

The Edinburgh University Calendar 1899–1900. Edinburgh: James Thin, 1899.

The Edinburgh University Calendar 1901–1902. Edinburgh: James Thin, 1901.

"Editorial." *The Free Lance* 14, 1 (April 1900): 19–22.

"Editorial." *The University Courant* 2, 8 (October 1888): 85–86.

"Editorial." *The Western University Courant* 10, 2 (November 1894): 6–7.

"Editorials." *The Free Lance* 4, 2 (May 1890): 19–22.

"Editorials." *The Free Lance* 4, 8 (February 1891): 123–128.
"Editorials." *The University Courant* 4, 4 (April 1890): 39–40.
"Editorials." *The Western University Courant* 10, 3 (December 1894): 16–17.
"Editorials." *The Western University Courant* 11, 1 (September 1895): 1–2.
"Editorials." *The Western University Courant* 11, 3 (November 1895): 1–3.
"Education." *The Critic* 27, 789 (April 3, 1897): 242.
"The Education of Girls: Their Admissibility to Universities." *Westminster Review* 109 (January 1878): 56–90.
"Educational Intelligence." *The Ohio Educational Monthly; Organ of the Ohio Teachers' Association and The National Teacher* 5, 7 (July 1880): 232–235.
"Educational News Items." *The Southern Educational Journal* 13, 1 (November 1899): 7–10.
Eleventh Annual Report of the Agricultural Experiment Station of the University of Tennessee to the Governor 1898. Knoxville, TN: The University Press, 1899.
Elkins, S. B. *Address Delivered before the Literary Societies of the West Virginia University, June 11th, 1888.* New York: Styles & Cash, 1888.
Esler, Robert. *Guide to Belfast, The Giant's Causeway, and the North of Ireland.* Belfast: Wm. Strain & Sons, 1884.
Essays and Addresses, by Professors and Lecturers of the Owens College, Manchester. London: Macmillan and Co., 1874.
"The Exchange Editor's Table." *Pennsylvania Western* 5, 2 (February 1886): 44–45.
"The Exchange Editor's Table." *Pennsylvania Western* 5, 4 (April 1886): 98–99.
"Exchange Notes." *The Vassar Miscellany* 16, 3 (December 1886): 115–118.
"Exchanges." *The University Courant* 2, 7 (September 1888): 84.
Farmer, Lydia Hoyt, ed. *The National Exposition Souvenir: What America Owes to Women.* Buffalo, Chicago, and New York: Charles Wells Moulton, 1893.
Fawcett, M. G. "The Use of Higher Education to Women." *Contemporary Review* (November 1886): 719–728.
"Female Students in Ireland." *Freeman's Journal*, May 18, 1877, 6.
Fifty-Fifth Annual Report of the Indiana University, Including the Catalogue for the Academical Year 1884–1885. Indianapolis, IN: Wm. B. Burford, 1885.
Fifty-Seventh Annual Catalogue of the Indiana University for the Academic Year 1886–1887. Indianapolis, IN: Wm. B. Burford, 1887.
First Annual Report of the American Woman's Educational Association. May, 1853. New York: Kneeland, 1853.
"The First 'Coeds'." In *Indiana University Alumni Quarterly Vol. IX–1922*, 216–218. Indianapolis, IN: C. E. Pauley and Co., 1922.
Fitch, J. G. "Women and the Universities." *The Contemporary Review LVIII*, 240–255. London: Isbister and Company, 1890.
Fouts, C. H. "Ohio University." *The Beta Theta Pi* 15, 3 (April 1888): 207–208.
Fraser, Norman. *Student Life at Edinburgh University.* Paisley: J. and R. Parlane, 1884.
Fulton, John. *Memoirs of Frederick A. P. Barnard, Tenth president of Columbia College in the City of New York.* New York: Macmillan and Co., 1896.
The Garnet and Black 1899, Published by the Students Volume I. Columbia, SC: The Bryan Printing Co., 1899.
The Garnet and Black 1900, Published by the Students of the South Carolina College. Columbia, SC: The Bryan Printing Co., 1900.
The Garnet and Black, Published by the Students of the South Carolina Carolina College, Nineteen Hundred and One. Columbia, SC: The Bryan Printing Co., 1901.
Gaskell, Elizabeth. *Wives and Daughters: A Novel.* New York: Harper & Brothers, 1866.
Gavit, John Palmer, comp. *Bibliography of College, Social and University Settlements.* Cambridge, MA: Co-operative Press, 1897.
General Catalogue of the Ohio University. From the Date of Its Charter in 1804 to 1885. Athens, OH: Published by the University, 1885.
Gilbert, W. S. *Songs of a Savoyard.* London: George Routledge and Sons, 1894.
"Girl Graduates." *The Students' Journal and Hospital Gazette* (August 30, 1879): 207.

The Glasgow University Calendar for the year 1901–2. Glasgow: James Maclehose and Sons, 1901.

Gordon, Alice M. "The After-Careers of University-Educated Women." *Nineteenth Century* 37 (1895): 955–960.

"Gov. Hayes: His Address to the Graduating Class of Ohio University." *Chicago Tribune*, June 24, 1876, 6.

Grant, Sir Alexander. *Happiness and Utility as Promoted by the Higher Education of Women: An Address*. Edinburgh: Edmonston and Douglas, 1872.

———. *The Story of the University of Edinburgh during its First Three Hundred Years*. London: Longmans, Green, and Co., 1884.

Green, T. Mortimer. Advertisement for the University College of South Wales, Aberystwyth. *Educational Review* 18, 13 (London: Educational Review, 1900): 479.

———. "University College of South Wales, Aberystwyth." *Journal of Education* 2, 22 (October 1, 1900): 544.

Grey, Maria G. *On the Education of Women*. London: W. Ridgway, 1871.

———. "The Women's Educational Movement." In *The Woman Question in Europe: A Series of Original Essays*, edited by Theodore Stanton, 30–62. New York, London, and Paris: G. P. Putnam's Sons, 1884.

Grogan, Mercy. *How Women May Earn a Living*. London, Paris and New York: Cassell & Company, 1883.

H. "The Dangers of Socialism." *The Free Lance* 8, 8 (February 1895): 106–107.

Hansard's Parliamentary Debates. 38° Victoriæ, 1875. Volume CCXXII (222). Comprising the Period from the Fifth Day of February 1875, to the Seventeenth Day of March 1875. First Volume of the Session. London, 1875.

Hartwell, Edward Mussey. "Physical Training." In *The Report of the Commissioner of Education for 1897–98*, United States Bureau of Education, 485–589. Washington, DC: Government Printing Office, 1899.

Hazeltine, Mayo W. *British and American Education: The Universities of the Two Countries Compared*. New York: Harper & Brothers, 1880.

Henderson, C. R. *Social Settlements*. New York: Lentilhon & Company, 1899.

Heywood, James. "The Owens College, Manchester, and a Northern University." *Journal of the Statistical Society* 41, 3 (September 1878): 536–547.

Hinsdale, Ellen C. "German Universities." *The Inlander* 11, 8 (May 1901): 312–315.

Historical and Current Catalogue of the Officers and Students of the University of Mississippi, Forty-Second Session, 1893–94. Oxford, MS: Published by the University, 1894.

"Home Notes: The Higher Education of Women." *The Sunday Magazine: For Family Reading*. London: Daldy, Isbister, & Co., 1878.

Horsburgh, J. M. "Report of University College, London." In *Reports from University Colleges 1899, Presented to both Houses of Parliament by Command of Her Majesty*, 207–243. London: Wyman and Sons, 1899.

"Horticulture and Forestry." *University of Tennessee Record* 2 (February 1901): 199–200.

Howe, Julia Ward, ed. *Sex and Education: a Reply to Dr. E. H. Clarke's "Sex in Education."* Boston: Roberts Bros., 1874.

H. R. L. "Co-Education." *The Free Lance* 2, 8 (February 1889): 127–128.

Hunnewell, James F. *The Lands of Scott*. Boston: Houghton, Osgood and Company, and Cambridge: The Riverside Press, 1880.

Hutchinson, Horace G. *Golf*. London: Longmans, Green, and Co., 1895.

"Indiana University." *The Beta Theta Pi* 21, 1 (October 1893): 440–441.

The Indiana University Catalogue for the Seventy-First College Year, 1894–95. Bloomington, IN: Published by the University, 1895.

The Indiana University Catalogue, Seventy-Fourth College Year 1897–98. Bloomington, IN: Published by the University, 1898.

"Indiana University." *The Educator-Journal* 2, 1 (September 1901): 41–42.

"Ireland." *The Lancet* (February 22, 1890): 439–440.

"Irish Education." *Anglo-Celt*, June 8, 1867, 2.

Jack, William. "The New English University." In *Macmillan's Magazine Vol. XLIII November 1880, to April 1881*, 107–113. London: Macmillan and Co., 1881.

Janes, Emily. *The Englishwoman's Year Book and Directory 1900.* London: Adam and Charles Black, 1900.

Jex-Blake, Sophia. *Medical Women: A Thesis and a History.* Edinburgh: Oliphant, Anderson, & Ferrier, 1886.

Jones, M. Winifred. "What Women Are Doing in Manchester." *Womanhood* 3, 18 (May 1900): 427–428.

"Jottings." *The Durham University Journal* 5, 6 (December 16, 1882): 70.

"Jottings." *The Durham University Journal* 5, 9 (May 26, 1883): 112.

Karns, T. C. "The University of Tennessee." In *Higher Education in Tennessee*, edited by Lucius Salisbury Merriam, 63–106. Washington, DC: Government Printing Office, 1894.

Kellogg, Day Otis, ed. *New American Supplement to the Latest Edition of the Encyclopædia Britannica, Volume I.* New York and Chicago: The Werner Company, 1898.

Knapp, John M., ed. *The Universities and the Social Problem: An Account of the University Settlements in East London.* London: Rivington, Percival & Co., 1895.

Knight, George W. and John R. Commons. *The History of Higher Education in Ohio.* Washington, DC: Government Printing Office, 1891.

Lake, Katharine, ed. *Memorials of William Charles Lake, Dean of Durham 1869–1894.* London: Edward Arnold, 1901.

Lange, Helene. *Higher Education of Women in Europe.* New York: D. Appleton and Company, 1901.

La Vie '92 published by the Junior Class. State College, PA: Published by the University, 1892.

La Vie '93 published by the Junior Class. State College, PA: Published by the University, 1893.

La Vie '98 published by the Junior Class of the Pennsylvania State College Vol. IX. State College, PA: Published by the University, 1898.

La Vie '99 published by the Junior Class of the Pennsylvania State College Volume X. Chicago: A. L. Swift & Co., 1899.

Leffingwell, C. W. and Arthur Seymour, eds. *The Living Church Annual and Almanac and Calendar, for the Year of Our Lord, 1883.* New York: E. & J. B. Young & Company, 1881.

"Letter to the Editor." *The Free Lance* 5, 7 (January 1892): 139.

"Localettes." *The University Courant* 6, 3 (June 1892): 148–150.

"Locals." *The Free Lance* 3, 3 (June 1889): 192–195.

"Locals." *The Free Lance* 5, 1 (April 1891): 11–14.

"Locals." *The Free Lance* 9, 7 (January 1896): 162–164.

"Locals." *The Western University Courant* 12, 2 (November 1896): 24–26.

The London Gazette, May 25, 1880, 3173.

Lumsden, Louisa Innes. "On the Higher Education of Women in Great Britain and Ireland." In *Journal of Social Science, Containing the Transactions of the American Association, Number XX June, 1885, Saratoga Papers of 1884, Part II*, 49–60. Boston: Cupples, Upham & Co. and New York: G. P. Putnam's Sons, 1885.

Maddison, Isabel, ed. *Handbook of Courses Open to Women in British, Continental and Canadian Universities.* New York: The Macmillan Company, 1896.

———. *Handbook of Courses Open to Women in British, Continental and Canadian Universities: Supplement for 1897.* New York: The Macmillan Company, 1897.

Marshall, Mary A. "Medicine as a Profession for Women." In *The Woman's World: Volume I*, edited by Oscar Wilde, 105–110. London, Paris, and Melbourne: Cassell & Company, 1888 and London: Source Book Press, 1970.

Maudsley, Henry. "Sex in Mind and in Education." *Fortnightly Review* 15 (1874): 466–483.

———. *Sex in Mind and in Education.* Syracuse, NY: C. W. Bardeen, 1884.

Mayes, Edward. *History of Education in Mississippi.* Washington, DC: Government Printing Office, 1899.

McCarthy, Justin. *A Short History of Our Own Times from the Accession of Queen Victoria to the General Election of 1880 in Two Volumes, Vol. I.* New York: Frederick A. Stokes & Brother, 1888.

McElwain, Harriet A. "Ladies' Department." In *Report of the Pennsylvania State College, for the year 1888*, 45–48. Harrisburg, PA: Edwin K. Meyers, 1889.

McElwain, Harriet A. "Ladies' Department." In *Annual Report of the Pennsylvania State College, for the year 1894*, 147–150. [Harrisburg, PA]: Clarence M. Busch, 1895.

McLaughlin, Andrew Cunningham. *History of Higher Education in Michigan*. Washington, DC: Government Printing Office, 1891.

McMillan, Annie. "Queen Margaret College in the Middle Ages." In *The Book of the Jubilee: In Commemoration of the Ninth Jubilee of the University of Glasgow, 1451–1901*, edited by the Students' Jubilee Celebrations Committee, 139–145. Glasgow: J. Maclehose and Sons, 1901.

"Medical Items and News: Medical Women in Ireland." *The Medical Record: A Weekly Journal of Medicine and Surgery* (November 25, 1876): 774.

"The Medical Society." *The Durham University Journal* 5, 7 (February 17, 1883): 77–78.

The Michigan Book. Ann Arbor, MI: Edwin H. Humphrey, 1898.

Mill, John Stuart. *On Liberty: The Subjection of Women*. New York: Henry Holt and Company, 1882.

"Miss Lumsden and University Hall." *The Journal of Education* 22 (September 1900): 598.

The Monticola Volume II. Published by the Class of 1900. Morgantown, WV: The Acme Publishing Company, 1899.

The Monticola Volume III, 1901. Morgantown, WV: The Acme Publishing Company, 1900.

Morrison, Sarah P. *Among Ourselves: To a Mother's Memory Vol. I: Out of North Carolina*. Plainfield, IN: Publishing Association of Friends, 1901.

Morton, J. N. *An Analysis of the Universities (Scotland) Act, 1889, with the Act Itself and the Act of 1858, and an Index*. Edinburgh and London: William Blackwood and Sons, 1889.

Myers, A. Wallis. "Women Students in Wales." *The Ludgate Illustrated Magazine Vol. VIII*), 136–141. London: F. V. White & Co., 1899.

National Normal. Quoted in "A Lady Graduate of the Indiana State University." *Arthur's Home Magazine Vol. XXXIII January to June*, edited by T. S. Arthur and Miss Virginia F. Townsend (Philadelphia: T. S. Arthur, 1869): 301.

Nimmo, Helen M. "Some Recent Notes and Recollections of Queen Margaret College Life." In *The Book of the Jubilee: In Commemoration of the Ninth Jubilee of the University of Glasgow, 1451–1901*, edited by the Students' Jubilee Celebrations Committee, 146–155. Glasgow: J. Maclehose and Sons, 1901.

Notes and Materials for the History of University College, London: Faculties of Arts and Science. London: H. K. Lewis, 1898.

"Notes on the Work." *Oxford University Extension Gazette* 3, 32 (May 1893): 101–104.

"On the Poetry of the Present Age." In *The London University College Magazine Vol. I*, 140–147. London: H. K. Lewis, 1849.

"Opening of the Central Block of the University College of Wales, Aberystwyth," in *Journal of Education: A Monthly Record and Review, Volume XX*, 695–696. London: William Rice, 1898.

Palmer, Alice Freeman. *Why Go to College?* Boston: T. Y. Crowell & Co., 1897.

Palmer, Thomas Waverly, comp. *A Register of the Officers and Students of the University of Alabama 1831–1901*. Tuscaloosa: The University of Alabama, 1901.

The Parliamentary Debates (Authorised Edition), Fourth Series: Commencing with the Fifth Session of the Twenty-sixth Parliament of the United Kingdom of Great Britain and Ireland. 62 Victoriae. Volume LXXIII, Comprising the period from the Twentieth Day of June to the Fifth Day of July 1899. London: Wyman and Sons, 1899.

The Parliamentary Debates (Authorised Edition). Fourth Series: Commencing with the Seventh Session of the Twenty-Fourth Parliament of the United Kingdom of Great Britain and Ireland. Volumes I, February 1892 (55° Victoriæ) through 36, September 1895 (59° Victoriæ). London: Reuter's Telegram Co., 1892–1895.

P.C.M. "The Evolution of Sex." Review of *The Evolution of Sex*, by Prof. Patrick Geddes and J. Arthur Thomson. *Nature*, April 10, 1890.

Penfield, Walter. "A History of the Junior Hop." *The Inlander* 11, 5 (February 1901): 184–188.

"The Pennsylvania State College." *The Free Lance* 11, 4 (October 1897): back cover.

"The Pennsylvania State College." *The Free Lance* 14, 6 (February 1901): back cover.

Perkins, Angie Warren. "Report of the Acting Dean, Woman's Department." *University of Tennessee Record* 8 (January 1899): 23–25.

"Personal." *The Ohio Educational Monthly; Organ of the Ohio Teachers' Association and The National Teacher* 37, 11 (November 1888): 597–599.

"Pittsburgh as a Site for Universities." *The University Courant* 5, 6 (October 1891): 51–54.

Preceptors, College of. *The Calendar for the Year 1900*. London: Francis Hodgson, 1900.

The President's Report to the Board of Regents for the Academic Year Ending September 30, 1899 and the Report of the Treasurer for the Fiscal Year ending June 30, 1899. Ann Arbor, MI: The Courier Office, 1899.

The President's Report to the Board of Regents for the Fiscal Year Ending June 30, 1870. Ann Arbor, MI: Published by the University, 1870.

The President's Report to the Board of Regents for the Year Ending June 30, 1874. Ann Arbor, MI: Published by the University, 1874.

The President's Report to the Board of Regents for the Year Ending June 30, 1879. Ann Arbor, MI: Ann Arbor Printing and Publishing Company, 1879.

The President's Report to the Board of Regents for the Year Ending Sept. 30, 1887. Ann Arbor, MI: Courier Printing House, 1887.

The President's Report to the Board of Regents for the Year Ending Sept. 30, 1891. Ann Arbor, MI: The Register Publishing Company, 1891.

The President's Report to the Board of Regents for the Year Ending Sept. 30, 1894. Ann Arbor, MI: The Register Publishing Company, 1894.

The President's Report to the Board of Regents for the Year Ending Sept. 30, 1895. Ann Arbor, MI: The Inland Press, 1895.

The President's Report to the Board of Regents for the Year Ending September 30, 1882. Ann Arbor, MI: Courier Book and Job Printing House, 1882.

Proceedings of the Board of Regents of the University of Michigan from January, 1876, to January, 1881. Ann Arbor, MI: Ann Arbor Printing and Publishing Company, 1881.

Proceedings of the Board of Regents of the University of Michigan, from January, 1881, to January, 1886. Ann Arbor, MI: The Courier Book and Job Printing Establishment, 1886.

"Provincial Medical Colleges: Belfast." *The Students' Journal and Hospital Gazette* 10 (September 16, 1882): 866.

Public Acts of the Legislature of the State of Michigan Passed at the Regular Session of 1899 with an Appendix Containing Joint and Concurrent Resolutions, Amendments to the Constitution, and the State Treasurer's Report for the Year Ending June 30, 1899. Lansing, MI: Robert Smith Printing Co., 1899.

The Public General Statutes Passed in the Forty-Fourth and Forty-Fifth Years of the Reign of Her Majesty Queen Victoria, 1881: With a Copious Index, Tables, &c. London: George Edward Eyre and William Spottiswoode, 1881.

"Punch's Essence of Parliament," *Punch* 68 (March 13, 1875): 109–111.

"Punch's Essence of Parliament," *Punch* 70 (July 15, 1876): 14–16.

"Queen's College, Cork." *The Students' Journal and Hospital Gazette* 10 (March 18, 1882): 121.

Queen's College, Galway, Calendar for 1894–1895. Dublin: The University Press, 1895.

Queen's College, Galway, Calendar for 1898–1899. Dublin: The University Press, 1899.

Queen's College, Galway, Calendar for 1900–1901. Dublin: The University Press, 1901.

"Queen's College Morality." *The Nation*, February 20, 1864, 11.

"Queen's Institute." *Freeman's Journal*, April 12, 1877, 7.

"The Queen's University." *Nation*, October 16, 1875, 12.

Rait, Robert Sangster. *The Universities of Aberdeen: A History*. Aberdeen: James Gordon Bisset, 1895.

Reason, W., ed. *University and Social Settlements*. London: Methuen & Co., 1898.

Reavis, L. U. *Saint Louis: The Future Great City of the World with biographical sketches of the representative men and women of St. Louis and Missouri*. St. Louis: C. R. Barns, 1876.

"Recent Removals." In *The United Presbyterian Magazine Vol. II*, 93. Edinburgh: Andrew Elliot, 1885.

Report of the Board of Curators of the State University of the State of Missouri to the XXXIst General Assembly. Jefferson City, MO: Tribune Printing Company, 1881.

Report of the Commissioner of Education for the year 1894–95 Volume 1. Washington, DC: Government Printing Office, 1896.

Report of the Commissioner of Education for the year 1897–98 Volume 2. Containing Parts II and III. Washington, DC: Government Printing Office, 1899.

"Report of the Executive Committee." In *Biennial Report of the Board of Regents of the West Virginia University, for the Years 1877 and 1878*, 21–24. Wheeling, WV: W. J. Johnston, 1878.

"Report from The Owens College, Manchester." In *Education Department Reports from University Colleges 1899*, 245–307. London: Wyman and Sons, 1899.

Report of the Pennsylvania State College, for the year 1888. Harrisburg, PA: Edwin K. Meyers, 1889.

Report of the Scottish Institution for the Education of Young Ladies with an Appendix containing separate reports, by the different teachers, of the course of instruction, and the system pursued, in their respective classes. Edinburgh: Oliver & Boyd, 1835.

Reports from Commissioners, Inspectors, and Others: Thirty-Four Volumes, 21. Wales and Monmouthshire, Session 11 February 1896–14 August 1896, Vol. XXXV (1896).

The Revised Statutes of South Carolina Vol. 1 Containing The Civil Statutes, Approved by the General Assembly of 1893. Columbia, SC: Charles A. Calvo, Jr., 1894.

Richardson, Charles F. and Henry A. Clark, eds. *The College Book.* Boston: Houghton, Osgood and Company, 1878.

Robbie, William. *Aberdeen Its Traditions and History.* Aberdeen: D. Wylie & Son, 1893.

Robertson, A. and Walter Smith. "Report from King's College, London." In *Reports from University Colleges 1899, Presented to both Houses of Parliament by Command of Her Majesty*, 175–206. London: Wyman and Sons, 1899.

Romanes, George J. "Mental Differences between Men and Women," *Nineteenth Century* 21 (1887): 654–672.

Rowold, Katharina, ed. *Gender & Science: Late Nineteenth-Century Debates On the Female Mind and Body.* Bristol: Thoemmes Press, 1996.

"Royal College of Surgeons of England." *The British Medical Journal* 1 (January 21, 1899): 177–178.

"The Royal College of Surgeons of England." *The Lancet* (January 21, 1899): 185.

"The R. U. I. Examinations: Brilliant Success of a Macroom Young Lady Student." *Southern Star*, August 7, 1897, 3.

Russell, James Earl. "The Extension of University Teaching in England and America: A Study in Practical Pedagogics." Ph.D. thesis, University of Leipsic, 1895.

Sabine, L. K. "The Romance of a Freshman." *The Inlander* 11, 8 (May 1901): 317–321.

Sanford, Edward T. *Blount College and the University of Tennessee: An Historical Address Delivered before the Alumni Association and Members of the University of Tennessee.* Knoxville, TN: Published by the University, 1894.

The Savitar—1891; by Students of the University of Missouri. Columbia, MO: Published by the University, 1891.

The Savitar—1895; by Junior Class of 1894. Columbia, MO: E.W. Stephens Printing Company, 1895.

The Savitar—1898; by the Junior Class of the University of Missouri 1897. Columbia, MO: E. W. Stephens, 1898.

The Savitar—1899; by Students of the Junior Class. Columbia, MO: E. W. Stephens, 1899.

The Savitar—1900; Published by the Junior Class of the University of Missouri 1900. Columbia, MO: E. W. Stephens, 1901.

"Scotch and Irish Medical Colleges: Cork and Galway." *The Students' Journal and Hospital Gazette* 7 (September 13, 1879): 225.

Scottish Universities Commission. *General Report of the Commissioners under the Universities (Scotland) Act, 1889. With an Appendix containing Ordinances, Minutes, Correspondence, Evidence, and other documents.* Edinburgh: Mill & Co., 1900.

Seelye, L. Clarke. "The Need of a Collegiate Education for Woman." Paper read before the American Institute of Instruction at North Adams, July 28, 1874.

Sell, Henry. *Sell's Dictionary of the World's Press.* London: Sell's Advertising Offices, 1886.

Sewall, May Wright. "The Education of Woman in the Western States." In *Woman's Work in America*, edited by Annie Nathan Meyer, 54–88. New York: Henry Holt and Company, 1891.

———. "'A Report on the Position of Women in Industry and Education in the State of Indiana,' Indiana Department of the New Orleans Exposition, 1885." In *American Feminism: Key Source Documents 1848–1920 Volume II: Work and Education*, edited by Janet Beer, Anne-Marie Ford and Katherine Joslin, 119–136. London: Routledge, 2002.

Shaw, Albert, ed. *The Review of Reviews* 11, 60 (January 1895): 8.
———. "The Progress of the World." *The Review of Reviews* 16, 5 (November 1897): 515–533.
Sheffield, Edith L. "Student Life in the University of Michigan." *The Cosmopolitan* 7, 2 (June 1889): 105–119.
Sheppard, Nathan. *Before an Audience; Or, The Use of the Will in Public Speaking. Talks to the Students of the University of St. Andrews and the University of Aberdeen.* New York and London: Funk & Wagnalls Company, 1886.
"Should University Degrees be given to Women?" In *The Westminster Review Vol. CXV January–April, 1881, American Edition*, 236–241. New York: The Leonard Scott Publishing Company, 1881.
Skeffington, Florence V. "Report of the Dean of the Woman's Department." *University of Tennessee Record* 2 (February 1901): 42–44.
Skene, Alexander C. J. *Education and Culture as Related to the Health and Diseases of Women.* Detroit: G. S. Davis, 1889.
Smith, Mary Roberts. "Statistics of College and Non-College Women." *American Statistical Association* 49, 50 (March, June 1900): 1–26.
Smith, W. L. *Historical Sketches of Education in Michigan.* Lansing, MI: W. S. George & Co., 1881.
Snoddy, J. S. *A Little Book of Missouri Verse: Choice Selections from Missouri Verse-Writers.* Kansas City, MO: Hudson-Kimberly Publishing Co., 1897.
Snow, Marshall S. *Higher Education in Missouri.* Washington, DC: Government Printing Office, 1898.
"The Social Life of the Student." *The Students' Journal and Hospital Gazette* (September 13, 1879): 232.
"The Song of *Alma Mater.*" *Alma Mater* (January 11, 1899). In *Interamna Borealis: Being Memories and Portraits from an old University Town between the Don and the Dee*, edited by W. Keith Leask, 89–90. Aberdeen: The Rosemount Press, 1917.
Spender, Dale, ed. *The Education Papers: Women's Quest for Equality in Britain, 1850–1912.* London: Routledge & Kegan Paul, 1987.
The St Andrews University Calendar for the year 1899–1900. Edinburgh: William Blackwood and Sons, 1899.
Stanton, Elizabeth Cady, Susan B. Anthony, and Matilda Joslyn Gage, eds. *History of Woman Suffrage Vol. III 1876–1885.* Rochester, NY: Charles Mann, 1887.
Stevens, W. Le Conte. *The Admission of Women to Universities.* New York: Press of S. W. Green's Son, 1883.
Stirling-Maxwell, Sir William. *Miscellaneous Essays and Addresses.* London: John C. Nimmo, 1891.
Struthers, Christina. *The Admission of Women to Scottish Universities.* Aberdeen: John Rae Smith, 1883.
"The State University Troubles." *The Wisconsin State Journal*, January 20, 1874, 1.
The Student Missionary Appeal: Addresses at the Third International Convention of the Student Volunteer Movement for Foreign Missions Held at Cleveland, Ohio, February 23–27, 1898. New York: Student Volunteer Movement for Foreign Missions, 1898.
The Student's Guide to the University of Durham; with Information Respecting Expenses, Scholarships, Examinations, and Degrees. Durham: The "Advertiser" Office, 1880.
Students' Hand-Book. Presented by the University Young Men's and Young Women's Christian Associations, 1894–1895. Milwaukee: Press of The Evening Wisconsin Company, 1894.
Summers, George W. *The Mountain State: A Description of the Natural Resources of West Virginia, Prepared for Distribution at the World's Columbian Exposition.* Charleston, WV: Moses W. Donnally, 1893.
Thomas, Grace Powers. *Where to Educate 1898–1899: A Guide to the Best Private Schools, Higher Institutions of Learning, etc., in the United States.* Boston: Brown and Company, 1898.
Thompson, Joseph. *The Owens College: Its Foundation and Growth; and its Connection with the Victoria University, Manchester.* Manchester: J. E. Cornish, 1886.
Thorburn, John. *Female Education from a Physiological Point of View.* Manchester: Cornish, 1884.
Thwaites, Reuben Gold, ed. *The University of Wisconsin: Its History and its Alumni.* Madison, WI: J. N. Purcell, 1900.

Trochos. Madison, WI: Junior Class of the University of Wisconsin, 1885.

"The Union Society." *The Durham University Journal* 5, 12 (December 17, 1883): 142.

"University education for women in Scotland." *The Ladies' Edinburgh Magazine* (November 5, 1879): 517.

"The University Education of Women." *Chambers's Journal of Popular Literature: Science and Arts* (November 13, 1897): 727–730.

University of London. *The Calendar for the Year 1870*. London: Taylor and Francis, 1870.

University of London. *The Calendar for the Year 1871*. London: Taylor and Francis, 1871.

"The University Question." *The Nation*, September 18, 1880, 8.

"University of St. Andrews. Higher Education for Women, with Title of L.L.A., Equivalent to M.A. for Men." *The Educational Times, and the Journal of the College of Preceptors* 36, 271 (November 1, 1883): 300.

"University Studies: Scottish and English." In *The British and Foreign Evangelical Review Vol. XVI*, 27–42. London: James Nisbet & Co. and Ediburgh: Oliver & Boyd, 1867.

University of Tennessee Register for 1897–98 and Announcement for 1898–99. Knoxville: The University of Tennessee Press, 1898.

University of Tennessee Record, July 1898 Volume V. Knoxville: The University of Tennessee Press, 1898.

"The University." *The West Virginia School Journal* 15, 2 (February 1896): 571.

"Up on the Hill." *The Western University Courant* 16, 4 (January 1901): 124–125.

Venable, W. H. *The Beginnings of Literary Culture in the Ohio Valley: Historical and Biographical Sketches*. Cincinnati: Robert Clarke & Co., 1891.

The Volunteer Published by the Students of the University of Tennessee Vol. III. Knoxville, TN: Bean, Warters & Gaut, 1899.

The Volunteer Published by the Students of the University of Tennessee Vol. IV. Knoxville, TN: Bean, Warters & Gaut, 1900.

The Volunteer Published by the Students Vol. II. Knoxville, TN: Bean, Warters & Gaut, 1898.

The Volunteer Vol. I Published Annually by The Students of the University of Tennessee. Knoxville, TN: S. B. Newman & Co., 1897.

The Volunteer Volume V 1901 Published Annually by the Students' Association, University of Tennessee. Knoxville, TN: Ogden Bros. & Co., 1901.

Wallington, Emma. "The Physical and Intellectual Capacities of Woman Equal to Those of Man." *Anthropologia* 1 (1874): 552–565.

Wallis-Jones, W. J. "The University College of Wales." *WALES: A National Magazine for the English Speaking Parts of Wales* 3, 26 (June 1896): 241–249.

"Wanted." *The University Courant* 5, 10 (February 1892): 113–114.

Watt, William. *A History of Aberdeen and Banff*. Edinburgh and London: William Blackwood and Sons, 1900.

The West Virginia School Journal 7, 6 (June 1888): 1.

West Virginia University, Morgantown, Catalogue 1889–90: Announcements for 1890–91. Charleston, WV: Moses W. Donnally, 1890.

Whedon, Sara. "What Oxford Offers to Women." *The Inlander* 11, 8 (May 1901): 306–311.

Who's Who, 1901: An Annual Biographical Dictionary. London: Adam & Charles Black, 1901.

Wickersham, James Pyle. *A History of Education in Pennsylvania, Private and Public, Elementary and Higher*. Lancaster, PA: Inquirer Publishing Company, 1886.

Wilkins, Charles, ed. *The Red Dragon: The National Magazine of Wales Vol. IV—July to December, 1883*. Cardiff: Daniel Owen and Company, 1883.

Willard, Frances E. *Occupations for Women: A Book of Practical Suggestions for the Material Advancement, the Mental and Physical Development, and the Moral and Spiritual Uplift of Women*. Cooper Union, NY: The Success Company, 1897.

———. *Woman and Temperance: Or, The Work and Workers of The Woman's Christian Temperance Union*. Hartford, CT: James Betts & Co., 1883.

Willey, W. P. "West Virginia's Wrong to Womankind." *The West Virginia School Journal* 7, 6 (June 1888): 6–7.

Wisconsin Wickedness; being some wondrous wailings of western college life. New York: W. S. Sterling, 1900.

Woodburn, James Albert. *Higher Education in Indiana.* Washington, DC: Government Printing Office, 1891.

Woodside, David. *The Life of Henry Calderwood, LL.D., F.R.S.E.* London: Hodder and Stoughton, 1900.

Woodward, Calvin M. "Acceptance of the Buildings." In *The Order of Exercises and the Addresses at the Dedication of Academic Hall and the New Department Buildings, on Tuesday, June the Fourth, A. D. One Thousand, Eight Hundred and Ninety-Five,* 11–17. Columbia, MO: Printed by the University, 1895.

"W.U.P. Dramatic Club Entertainment." *The University Courant* 5, 2–3 (February–March 1891): 21.

Wylie, Theophilus A. *Indiana University, Its History from 1820, When Founded, to 1890, with Biographical Sketches of Its Presidents, Professors and Graduates, and a List of Its Students from 1820 to 1887.* Indianapolis, IN: Wm. B. Burford, 1890.

X. Y. Z. "Letter to the Editor." *The Free Lance* 10, 5 (November 1896): 100–101.

"Y.M.C.A." *The University Courant* 2, 8 (October 1888): 92.

Young, Virginia D. "Reports from Auxiliary States: South Carolina." In *Proceedings of the Twenty-Seventh Annual Convention of the National-American Woman Suffrage Association Held in Atlanta, GA., January 31st to February 5th, 1895,* edited by Harriet Taylor Upton, 86–88. Warren, OH: Wm. Ritezel & Co., 1895.

Zimmern, Alice. *Methods of Education in the United States.* London: Swan Sonnenschein & Co. and New York: Macmillan & Co., 1894.

———. *The Renaissance of Girls' Education in England: A Record of Fifty Years' Progress.* London: A. D. Innes & Company, 1898.

Sources Published Since 1901

Abercrombie, John William. "Address of Welcome: For the University." *1831–1906 University of Alabama Bulletin Commemoration Number* (November 1906): 19–21.

Aberdeen, University of. *Handbook to the City and University.* Aberdeen: Printed for the University, 1906.

"The Academic Revival, 1864–1914." In *Interamna Borealis: Being Memories and Portraits from an Old University Town between the Don and the Dee,* edited by W. Keith Leask, 2–14. Aberdeen: The Rosemount Press, 1917.

Advertisement for the Durham College of Science, *The Journal of Education* 25 (October 1903): 659.

Advertisements for the University of Durham, *The Journal of Education* 25 (October 1903): 659.

Aitken, Mrs. Mina. "What Women Are Doing in Scotland." *Womanhood* 13, 74 (January 1905): 93–95.

Alberts, Robert C. *Pitt: The Story of the University of Pittsburgh, 1787–1987.* Pittsburgh: University of Pittsburgh Press, 1986.

Albisetti, James. "American Women's Colleges through European Eyes, 1865–1914." *History of Education Quarterly* 32, 4 (1992): 439–458.

Alexander, Kern and M. David Alexander. *American Public School Law.* Belmont, CA: Thomson/West, 2005.

Alexander, Wendy. "Early Glasgow Women Medical Graduates." In *The World is Ill Divided: Women's Work in Scotland in the Nineteenth and Early Twentieth Centuries,* edited by Eleanor Gordon and Esther Breitenbach, 70–94. Edinburgh: Edinburgh University Press, 1990.

———. *First Ladies of Medicine: The Origins, Education and Destination of Early Women Medical Graduates of Glasgow University.* Glasgow: Wellcome Unit for the History of Medicine, University of Glasgow, 1987.

Anderson, James Maitland, ed. *The Matriculation Roll of the University of St. Andrews 1747–1897.* Edinburgh and London: William Blackwood and Sons, 1905.

Anderson, P. J., ed., *Record of the Celebration of the Quatercentenary of the University of Aberdeen.* Aberdeen: Aberdeen University Press, 1907.

Anderson, P. J. "The University Library: Past and Present." *The Aberdeen University Review* 1, 2 (February 1914): 123–136.

Anderson, R. D. *Education and Opportunity in Victorian Scotland: Schools and Universities.* Oxford: Clarendon Press, 1983.

Anderson, R. D. *Education and the Scottish People 1750–1918*. Oxford: Clarendon Press, 1995.

———. "In Search of the 'Lad of Parts': the Mythical History of Scottish Education." *History Workshop Journal* 19 (Spring 1985): 82–104.

———. "The Scottish University Tradition: Past and Future." In *Scottish Universities: Distinctiveness and Diversity*, edited by Jennifer J. Carter and Donald J. Witherington, 67–78. Edinburgh: Edinburgh University Press, 1992.

The Badger for Nineteen Hundred and Five. Madison, WI: Badger Board of the Junior Class of the University of Wisconsin, 1905.

Baird, George M. P. "Fragments of University of Pittsburgh Alumni History." *Western Pennsylvania Historical Magazine* 1, 1 (January 1918): 132–138.

Baird, Wm. Raimond. *Betas of Achievement: Being Brief Biographical Records of Members of the Beta Theta Pi Who Have Achieved Distinction in Various Fields of Endeavor*. New York: The Beta Publishing Co., 1914.

Balfour, Lady Frances. *Dr. Elsie Inglis*. New York: George H. Doran Company, 1919.

"Bandon Brieflets." *Southern Star*, March 5, 1955, 5.

Banks, J. A. and Olive. *Feminism and Family Planning among the Victorian Middle Classes*. Liverpool: Liverpool University Press, 1964.

Barbe, Waitman, ed. *Alumni Record: West Virginia University*. [Morgantown, WV]: Published by the Alumni Association, 1903.

Barger, Judith. *Elizabeth Stirling and the Musical Life of Female Organists in Nineteenth-Century England*. Aldershot, England, and Burlington, VT: Ashgate, 2007.

Barnes, David. *The Companion Guide to Wales*. Woodbridge, Suffolk: Companion Guides, 1970.

Barnes, Sarah V. "Crossing the Invisible Line: Establishing Co-Education at the University of Manchester and Northwestern University." *History of Education* 23, 1 (1994): 35–58.

Bascom, Florence. "The University in 1874–1887." *Wisconsin Magazine of History* 8 (March 1925): 303.

Begg, Tom. *The Excellent Women: The Origins and History of Queen Margaret College*. Edinburgh: John Donald Publishers, 1994.

Bellot, Hugh Hale. *University College, London, 1826–1926*. London: University of London Press, 1929.

Bennett, Alice Horlock. *English Medical Women: Glimpses of their Work in War and Peace*. London, Bath, New York, and Melbourne: Sir Isaac Pitman & Sons, 1915.

Bezilla, Michael. *Penn State: An Illustrated History*. University Park and London: The Pennsylvania State University Press, 1985.

Billington, Ray Allen. *Frederick Jackson Turner: Historian, Scholar, Teacher*. New York: Oxford University Press, 1973.

Bogue, Allan G. and Robert Taylor, eds. *The University of Wisconsin: One Hundred and Twenty-Five Years*. Madison: University of Wisconsin Press, 1975.

Boney, F. N. *A Pictorial History of the University of Georgia*. Athens: University of Georgia Press, 2000.

Boog Watson, W. N. "LA of Edinburgh University." *University of Edinburgh Journal* 25 (1971–1972): 215–219.

Bordin, Ruth. *Alice Freeman Palmer: The Evolution of a New Woman*. Ann Arbor: The University of Michigan Press, 1993.

———. *The University of Michigan: A Pictorial History*. Ann Arbor: The University of Michigan Press, 1967.

Boylan, Thomas A. and Timothy P. Foley. *Political Economy and Colonial Ireland: The Propagation and Ideological Function of Economic Discourse in the Nineteenth Century*. London: Routledge, 2002.

Bradley, Ian, ed. *The Complete Annotated Gilbert & Sullivan*. Oxford: Oxford University Press, 2001.

Breathnach, Eibhlin. "Women and Higher Education in Ireland (1879–1914)." In *The Irish Women's History Reader*, edited by Alan Hayes and Diane Urquhart, 44–49. London: Routledge, 2001.

Breathnach, Eileen. "Women and Higher Education in Ireland (1879–1914)." *Crane Bag* 4, 1 (1980): 47–54.

Brickstock, Richard. *Durham Castle: Fortress, Palace, College*. Huddersfield: Jeremy Mills Publishing Company, 2007.

"Brilliant Bandon Doctor, Resignation from Armagh Mental Hospital Post." *Southern Star*, October 17, 1936, 9.

Brown, A. L. and Michael Moss. *The University of Glasgow: 1451–1996*. Edinburgh: Edinburgh University Press, 1996.

Bulletin of the University of Mississippi: Announcements and Catalogue of the University of Mississippi, University P. O., (Near Oxford), Fifty-Second Session, (Fifty-Fifth Year), 1903–1904 Series 3, 1 (April 1904).

Bulletin of the University of Mississippi: Announcements for Session of 1903–1904 Series 2, 3 (August 1903).

Burns, James J. *Educational History of Ohio: A History of its Progress since the Formation of the State Together with the Portraits and Biographies of Past and Present State Officials*. Columbus, OH: Historical Publishing Co., 1905.

Burstyn, Joan N. "Education and Sex: The Medical Case against Higher Education for Women in England, 1870–1900." *Proceedings of the American Philosophical Society* 117 (1973): 79–89.

———. "Historical Perspectives on Women in Educational Leadership." In *Women and Educational Leadership*, edited by Sari Knopp Biklen and Marilyn B. Brannigan, 65–75. Lexington, MA: Lexington Books, 1980.

———. "Religious Arguments against the Higher Education for Women in England 1840–1890." *Women's Studies* 1, 1 (1972): 111–131.

———. *Victorian Education and the Ideal of Womanhood*. London: Croom Helm, 1980.

Cabaniss, James Allen. *A History of the University of Mississippi*. University: University of Mississippi, 1949.

Callahan, James Morton. *Semi-Centennial History of West Virginia*. [Charleston]: Semi-Centennial Commission of West Virginia, 1913.

Cant, Ronald Gordon. *The University of St. Andrews: A Short History*. Edinburgh: Scottish Academic Press, 1970.

Carter, Susan B. "Academic Women Revisited: An Empirical Study of Changing Patterns in Women's Employment as College and University Faculty, 1890–1963." *Journal of Social History* 14 (1981): 675–699.

Catalogue of the South Carolina College 1904–1905. Columbia, SC: The R. L. Bryan Company, 1905.

Catto, Iain, ed. *"No spirits and precious few women": Edinburgh University Union 1889–1989*. Edinburgh: Edinburgh University Union, 1989.

Chamness, Ivy and Burton D. Myers, eds. *Trustees and Officers of Indiana University 1820 to 1950*. Bloomington: Published by Indiana University, 1951.

Chamness, Ivy Leone. "Indiana University." *The Lyre* 25, 3 (April 1922): 259–261.

Charlton, H. B. *Portrait of a University 1851–1951*. Manchester: Manchester University Press, 1951.

Christy, Ralph D. and Lionel Williamson, eds. *A Century of Service: Land-Grant Colleges and Universities, 1890–1990*. New Brunswick and London: Transaction Publishers, 1992.

"Collected Poems of First Ever Woman Engineer Are Presented to UCG." *City Tribune*, December 13, 1996, 6.

"College Hall, London." *Journal of Education* 40, 473 (December 1, 1908): 786.

Coogan, Tim Pat. *Ireland in the Twentieth Century*. New York: Palgrave Macmillan, 2006.

Corr, Helen. "Dominies and Domination: Schoolteachers, Masculinity and Women in 19th Century Scotland." *History Workshop Journal* 40 (Autumn 1995): 151–164.

Coutts, James. *A History of the University of Glasgow: From its Foundation in 1451 to 1909*. Glasgow: J. Maclehose and Sons, 1909.

Crawford, Mary Caroline. *The College Girl of America and the Institutions Which Make Her What She Is*. Boston: L. C. Page & Company, 1905.

Crawford, Scott A.G.M., ed. *'Serious Sport': J. A. Mangan's Contribution to the History of Sport*. London: Taylor & Francis, 2004.

Curti, Merle. *The Social Ideas of American Educators*. Paterson, NJ: Pageant Books, 1959.

Curti, Merle and Vernon Carstensen. *The University of Wisconsin: A History, 1848–1925*, 2 Volumes. Madison: University of Wisconsin Press, 1949.

Daly, Lilian. "Women and the University Question." *The New Ireland Review* 17 (March 1902 to August 1902): 74–80.

Danbom, David B. *Born in the Country: A History of Rural America*. Baltimore, MD: Johns Hopkins University Press, 2006.

Davie, George Elder. *The Crisis of the Democratic Intellect: The Problem of Generalism and Specialisation in Twentieth-Century Scotland*. Edinburgh: Polygon, 1986.

———. *The Democratic Intellect: Scotland and Her Universities in the Nineteenth Century*. Edinburgh: Edinburgh University Press, 1961.

Davies, William Cadwaladr and William Lewis Jones. *The University of Wales and Its Constituent Colleges*. London: F. E. Robinson & Co., 1905.

Demmon, Isaac N., ed. *History of the University of Michigan, by the Late Burke A. Hinsdale, with Biographical Sketches of the Regents and Members of the University Senate from 1837 to 1906*. Ann Arbor, MI: Published by the University, 1906.

De Montmorency, J. E. G. *The Progress of Education in England: A Sketch of the Development of English Educational Organization from Early Times to the Year 1904*. London: Knight & Co., 1904.

The Department of Education in the University of Manchester 1890–1911. Manchester: Manchester University Press, 1911.

"The Dermatology Society of Great Britain and Ireland." *The British Journal of Dermatology* (December 1905): 455–461.

Dexter, Edwin Grant. *A History of Education in the United States*. New York: Macmillan & Co., 1906.

Doherty, William T., Jr. and Festus P. Summers. *West Virginia University: Symbol of Unity In a Sectionalized State*. Morgantown: West Virginia University Press, 1982.

Donaldson, Gordon, ed. *Four Centuries: Edinburgh University Life, 1583–1983*. Edinburgh: Edinburgh University Press, 1983.

Dorland, William Alexander Newman. *The Sum of Feminine Achievement: A Critical and Analytical Study of Woman's Contribution to the Intellectual Progress of the World*. Boston: The Stratford Company, 1917.

Dorr, Rheta Childe. "Breaking Into the Human Race." *Hampton's Magazine* 27, 3 (September 1911): 317–329.

Durham Calendar with Almanack 1910–1911. Durham: Thomas Caldcleugh & Son and London: Whittaker & Co., 1910.

"Durham: University News and Notes." *The University Review* 1, 4 (August 1905): 424–425.

Dyhouse, Carol. *No Distinction of Sex? Women in British Universities 1870–1939*. London: UCL Press, 1995.

———. "Social Darwinistic Ideas and the Development of Women's Education in England, 1880–1920." *History of Education* 5, 1 (1976): 41–58.

Edwards, Bruce L. *C. S. Lewis: Life, Works, and Legacy*. Westport, CT: Praeger, 2007.

Elder, Glen H. "Appearance and Education in Marriage Mobility." *American Sociological Review* 34 (1969): 519–533.

Elder, John R. "Music in the University since 1898." *The Aberdeen University Review* 1, 3 (June 1914): 236–243.

Ellsworth, Adelaide. "The Woman Doctor in War." *The Pennsylvania Medical Journal: Official Organ of the Medical Society of the State of Pennsylvania* 22, 1 (October 1918): 23–25.

The Encyclopedia Americana. New York and Chicago: The Encyclopedia Americana Corporation, 1918.

Esarey, Logan. *A History of Indiana*. New York: Harcourt, Brace and Company, 1922.

Eschbach, Elizabeth Seymour. *The Higher Education of Women in England and America, 1865–1920*. New York: Garland, 1993.

Evans, George Eyre. *Aberystwyth and its Court Leet*. Aberystwyth: *Welsh Gazette*, 1902.

Evans, W. Gareth. *Education and Female Emancipation: The Welsh Experience, 1847–1914*. Cardiff: University of Wales Press, 1990.

"Event and Comment: Co-Educational Problems." *The Michigan Alumnus* 17, 1 (October 1910): 5–6.

"Event and Comment: The Michigan Ideal." *The Michigan Alumnus* 10, 90 (January 1904): 161–170.

Federal Writers' Project. *Indiana: A Guide to the Hoosier State.* New York: Oxford University Press, 1947.

Feldman, Jim. *The Buildings of the University of Wisconsin.* Madison, WI: University Archives, 1997.

Fiddes, Edward. *Chapters in the History of Owens College and of Manchester University, 1851–1914.* Manchester: Manchester University Press, 1937.

———. "The University Movement in Manchester (1851–1903)." In *Historical Essays in Honour of James Tait*, edited by J. G. Edwards, V. H. Galbraith, and E. F. Jacob, 97–109. Manchester: Printed for the subscribers, 1933.

"The First Co-Ed to Graduate from the University." *The Missouri Alumni Quarterly* (September 1905): 15.

Foley, Tadhg, ed. *From Queen's College to National University: Essays on the Academic History of QCG/UCG/NUI, Galway.* Dublin: Four Courts Press, 1999.

Folmsbee, Stanley J. "The Early History of the University of Tennessee: An Address in Commemoration of its 175th Anniversary." *The East Tennessee Historical Society's Publications* 42 (1970): 3–19.

———. *History of Tennessee, Volume 1.* New York: Lewis Historical Publishing Co., 1960.

Fowler, J. T. *Durham University: Earlier Foundations and Present Colleges.* London: F. E. Robinson, 1904.

Fraser, W. H. and R. J. Morris, eds. *People and Society in Scotland: Volume II, 1830–1914.* Edinburgh: John Donald Publishers, 1990.

Fred, E. B. "Women in Higher Education with Special Reference to the University of Wisconsin." *The Journal of Experimental Education* 31 (December, 1962): 158–172.

F. S. "The Irish University Question As Affecting Women." *The Westminster Review* 159 (January to June 1903): 610–624.

Fulton, William D. *Ohio General Statistics for the Fiscal Year Commencing July 1, 1917, and Ending June 30, 1918 Volume IV.* Springfield, OH: The Springfield Publishing Company, 1919.

General Alumni Association. *Alumni directory, University of Pittsburgh, Vol. 2, 1787–1916.* Pittsburgh: Aldine Printing Company; Smith Bros. Col., 1916.

"General College News." *The American Educational Review* 30, 4 (January 1909): 171–177.

Geyer-Kordesch, Johanna and Rona Ferguson. *Blue Stockings, Black Gowns, White Coats: A Brief History of Women Entering the Medical Profession in Scotland in Celebration of One Hundred Years of Women Graduates at the University of Glasgow.* Glasgow: University of Glasgow, Wellcome Unit for the History of Medicine, 1994.

Gilchrist, Marion. "Some Early Recollection of the Queen Margaret Medical School." *Surgo* (March 1948): 79–81.

Gissing, George. *The Odd Women*, edited by Arlene Young. Peterborough, ONT: Broadview Press, 1998.

The Glasgow University Calendar for the Year 1902–3. Glasgow: James Maclehose and Sons, 1902.

Gordon, Lynn D. *Gender and Higher Education in the Progressive Era.* New Haven, CT and London: Yale University Press, 1990.

———. "The Gibson-Girl Goes to College: Popular-Culture and Women's Higher Education in the Progressive Era, 1890–1920." *American Quarterly* 39, 2 (1987): 211–230.

Graham, Patricia Albjerg. "Expansion and Exclusion: A History of Women in American Higher Education." *Signs* 3 (Summer 1978): 759–773.

Green, Edwin L., ed. *A History of the University of South Carolina.* Columbia, SC: The State Company, 1916.

Gunn, Hugh. *The Distribution of University Centres in Britain: A Plea for the Highlands of Scotland.* Glasgow: Airlie Press, 1931.

Gwynn, Stephen. *The Famous Cities of Ireland.* Dublin and London: Maunsel & Co. and New York: The Macmillan Company, 1915.

Hague, Amy. "'What If the Power Does Lie Within Me?' Women Students at the University of Wisconsin, 1875–1900." *History of Higher Education Annual* (1984): 78–100.

Haight, Elizabeth Hazelton. "Pleasant Possibles in Lady Professors." *Journal of the Association of Collegiate Alumnae* (September 1917): 10–17.

Hall, Catherine, Keith McClelland, and Jane Rendall, eds. *Defining the Victorian Nation: Class, Race, Gender and the Reform Act of 1867.* Cambridge: Cambridge University Press, 2000.

Hall, John M. *England: An Account of Past and Contemporary Conditions and Progress.* Detroit: Bay View Reading Club, 1906.

Harding, Samuel Bannister. *Indiana University, 1820–1904: Historical Sketch, Development of the Course of Instruction, Bibliography.* Bloomington: Indiana University, 1904.

Harford, Judith. "An Experiment in the Development of Social Networks for Women: Women's Colleges in Ireland in the Nineteenth Century." *Paedagogica Historica* 43, 3 (June 2007): 365–381.

———. "The Movement for the Higher Education of Women in Ireland: Gender Equality or Denominational Rivalry?" *History of Education* 34, 5 (September 2005): 497–516.

———. *The Opening of University Education to Women in Ireland.* Dublin and Portland, OR: Irish Academic Press, 2008.

Harte, Negley B. *The University of London, 1836–1986: An Illustrated History.* London: Athlone, 1986.

Harvey, Charles M. "A Hundred Years of Ohio." *The World's Work: A History of Our Time* 5 (November 1902 to April 1903): 3229–3239.

Harwarth, Irene, Mindi Maline, and Elizabeth DeBra. *Women's College in the United States: History, Issues, and Challenges.* Darby, PA: Diane Publishing Company, 1997.

Hearnshaw, F. J. C. *The Centenary History of King's College, London.* London: G. G. Harrap & Company, 1929.

"Higher Education, University College, Lady Students' Claims." *Irish Independent*, March 22, 1905, 3.

Hill, Albert Ross. "Advantages and Disadvantages of Residential Halls for Women in Co-Educational Universities." In *Transactions and Proceedings of the National Association of State Universities in the United States of America No. 8, 1910*, 88–92. Hamilton, OH: Republican Publishing Company, 1910.

Historical Catalogue of the University of Mississippi 1849–1909. Nashville, TN: Marshall & Bruce Company, 1910.

Hogeland, Ronald W. "Coeducation of the Sexes at Oberlin College: A Study of Social Ideas in Mid-Nineteenth Century America." *Journal of Social History* 6 (1972): 160–176.

Hollis, Daniel Walker. *University of South Carolina Volume II. College to University.* Columbia: University of South Carolina Press, 1956.

Hollow, Betty. *Ohio University: The Spirit of a Singular Place, 1804–2004.* Athens: Ohio University Press, 2003.

Holmes, Rachel. *The Secret Life of Dr. James Barry: Victorian England's Most Eminent Surgeon.* Stroud, Gloucestershire: Tempus Publishing, 2007.

Hoover, Thomas N. *The History of Ohio University.* Athens: Ohio University Press, 1954.

Horner, Winifred Bryan. "Nineteenth-Century Higher Education: The Scottish-American Connection." In *Scottish Universities: Distinctiveness and Diversity*, edited by Jennifer J. Carter and Donald J. Witherington, 34–39. Edinburgh: Edinburgh University Press, 1992.

Horowitz, Helen Lefkowitz. *Alma Mater: Design and Experience in the Women's Colleges from Their Nineteenth-Century Beginnings to the 1930s.* Amherst: University of Massachusetts Press, 1993.

———. *Campus Life: Undergraduate Cultures from the End of the Eighteenth-Century.* Chicago: The University of Chicago Press, 1987.

———. "Does Gender Bend the History of Higher Education?" *American Literary History* 7, 2 (1995): 344–349.

Houston, R. A. "Scottish Education and Literacy, 1600–1800: an International Perspective." In *Improvement and Enlightenment: Proceedings of the Scottish Historical Studies Seminar, University of Strathclyde, 1987–1988*, edited by T. M. Devine, 43–61. Edinburgh: J. Donald, 1989.

Hughes, James Laughlin and Louis Richard Klemm. *Progress of Education in the Century.* Toronto and Philadelphia: The Linscott Publishing Company, 1907.

Hunt, Felicity, ed. *Lessons for Life: The Schooling of Girls and Women, 1850–1950.* Oxford: Blackwell, 1987.

Hutchins, H. B. "The University and Co-Education." *The Michigan Alumnus* 17, 160 (January 1911): 179–186.

Hutton, Lawrence. *Literary Landmarks of the Scottish Universities.* New York and London: G. P. Putnam's Sons, 1904.

Ingram, T. A. *The New Hazell Annual and Almanack.* London: Oxford University Press, 1917.

"Items of Interest." *The School World* 6, 64 (April 1904): 152–154.

Jarausch, Konrad H. *The Transformation of Higher Learning, 1860–1930: Expansion, Diversification, Social Opening, and Professionalization in England, Germany, Russia, and the United States.* Chicago: University of Chicago Press, 1983.

Johnson, James. "Women Inventors and Discoverers." *Cassier's Magazine: An Engineering Monthly* 36, 6 (October 1909): 548–552.

Johnson, Miss. "Higher Education of Women in the South." In *Proceedings of the Eleventh Conference for Education in the South*, 130–139. Nashville, TN: Published by the Executive Committee of the Conference, 1908.

Johnston, William, ed. *Roll of the Graduates of the University of Aberdeen 1860–1900, Aberdeen University Studies: No. 18.* Aberdeen: Aberdeen University Press, 1906.

Jones, David R. *The Origins of Civic Universities: Manchester, Leeds and Liverpool.* London: Routledge, 1988.

Jordan, Alison. *Margaret Byers: Pioneer of Women's Education and Founder of Victoria College, Belfast.* Belfast: The Institute of Irish Studies, The Queen's University of Belfast, 1991.

Jordan, T. W. "Report of the Dean of the College." In *University of Tennessee Register for 1901–1902 and Announcement for 1902–1903*, 196. Knoxville: The University of Tennessee Press, 1902.

Kaye, Elaine. *A History of Queen's College, London 1848–1972.* London: Chatto and Windus, 1972.

Kerber, Linda K. "The Republican Mother: Women and the Enlightenment—An American Perspective." *American Quarterly* 28 (1976): 187–205.

———. *Toward an Intellectual History of Women.* Chapel Hill: The University of North Carolina Press, 1997.

Kerr, John. *Scottish Education, School and University: From Early Times to 1908.* Cambridge: Cambridge University Press, 1910.

Kessenich, Henrietta Wood. "'Twas Long, Long Ago." *The Wisconsin Alumnus* (1938): 306–309.

Koch, Theodore W. *Handbook of the Libraries of the University of Michigan.* Ann Arbor, MI: George Wahr, 1910.

"Ladies' Notes." *QCG: a record of college life in the city of the tribes* 1, 1 (November 1902): 10–12.

"Ladies' Notes." *QCG: a record of college life in the city of the tribes* 1, 2 (February 1903): 48–50.

"Ladies' Notes." *QCG: a record of college life in the city of the tribes* 1, 3 (March 1903): 21–22.

"Ladies' Notes." *QCG: a record of college life in the city of the tribes* 2, 2 (February 1904): 54–56.

"Ladies' Notes." *QCG: a record of college life in the city of the tribes* 2, 3 (May 1904): 21–23.

Lasser, Carol, ed. *Educating Men and Women Together: Coeducation in a Changing World.* Urbana: University of Illinois Press, 1987.

La Vie: The Annual Publication of the Junior Class of the Pennsylvania State College Volume XXXIII. State College, PA: Published by the Class of Nineteen Twenty-Two, 1921.

Lavin, Deborah. "The Innocent Client Looking for Design Quality." In *Design Quality in Higher Education Buildings*, edited by the Royal Fine Art Commission, 30–33. London: Telford, 1996.

Leask, Wm. Keith. "A Notable Class Record." *The Aberdeen University Review* 1, 1 (October 1913): 55–66.

Leighton, Henry R., ed. *Memorials of Old Durham.* London: George Allen & Sons, 1910.

Levi, T. "Welsh Education." *The Cambrian Volume Twenty-Two* (Utica, NY: Thomas J. Griffiths, 1902), 11–13.

Levine, Philippa. *Prostitution, Race, and Politics: Policing Venereal Disease in the British Empire.* London: Routledge, 2003.

Lewis, C. S. *Collected Letters: Family Letters, 1905–1931*, edited by Walter Hooper. New York: HarperSanFrancisco, 2004.

"Literary Notes." In *Friends' Intelligencer and Journal.* Philadelphia: Friends' Intelligencer Association, 1902.

Little, L. M. "Women's Education: Forty Years Ago and Now." *Irish Independent*, June 1, 1906, 4.

Lloyd, Thomas, Julian Orbach, and Robert Scourfield. *The Buildings of Carmarthenshire and Ceredigion*. New Haven, CT: Yale University Press, 2006.

Luce, Morton. *A Handbook to the works of Alfred Lord Tennyson*. London: George Bell and Sons, 1906.

Lundberg, Emma O. "Women in the University of Wisconsin." *The Wisconsin Alumni Magazine* (April 1908): 263–269.

Lyon, A. B. and G. W. A. Lyon, eds. *Lyon Memorial: Massachusetts Families, Including Descendants of the Immigrants William Lyon, of Roxbury, Peter Lyon, of Dorchester, George Lyon, of Dorchester, with Introduction Treating of the English Ancestry of the American families*. Detroit: Wm. Graham Printing Co., 1905.

Mackay, Donald Iain. *Geographical Mobility and the Brain Drain. A Case Study of Aberdeen University Graduates, 1860–1960*. London: Allen & Unwin, 1969.

Mackinnon, Alison. "Male Heads on Female Shoulders? New Questions for the History of Women's Higher Education." *History of Education Review* (Australia) 19, 2 (1990): 36–47.

"Manchester." *The University Review* 2 (October 1905–March 1906): 420–422.

"Manchester." *The University Review* 4 (October 1906–March 1907): 146–149.

Manton, Jo. *Elizabeth Garrett Anderson*. London: Methuen, 1987.

Martens, Frederick H., Mildred W. Cochran, and W. Dermot Darby, eds. *The Art of Music: Volume Eleven, A Dictionary-Index of Musicians Book I A–L*. New York: The National Society of Music, 1917.

Martin, Ida Shaw. *The Sorority Handbook*. Menasha, WI: George Banta Publishing Company, 1918.

Martin, Jane Roland. *Reclaiming a Conversation: The Ideal of the Educated Woman*. New Haven, CT: Yale University Press, 1985.

Martzolff, Clement L. "Ohio University—The Historic College of the Old Northwest." *Ohio Archaeological and Historical Quarterly* 19, 2 (April 1910): 411–445.

Mathews, Jerry A. "Stephen B. Elkins." In *Twenty Years in The Press Gallery: A Concise History of Important Legislation from the 48th to the 58th Congress*, edited by O. O. Stealey, 270–274. New York: Publishers Printing Company, 1906.

McCandless, Amy Thompson. "Maintaining the Spirit and Tone of Robust Manliness: The Battle against Coeducation at Southern Colleges and Universities, 1890–1940." *NWSA Journal* 2, 2 (Spring 1990): 199–216.

McCorvey, Thomas Chalmers. "V. Henry Tutwiler, and the Influence of the University of Virginia on Education in Alabama." *Transactions of the Alabama Historical Society* 5 (1904): 83–106.

McCullough, Joseph A. "Alumni Address. South Carolina College and the State." In *Proceedings of the Centennial Celebration of South Carolina College, 1805–1905*, 187–212. Columbia, SC: The State Co., 1905.

McGuigan, Dorothy Gies. *A Dangerous Experiment: 100 Years of Women at the University of Michigan*. Ann Arbor, MI: Center for Continuing Education of Women, 1970.

McWilliams-Tullberg, Rita. *Women at Cambridge*. Cambridge: Cambridge University Press, 1998.

Melville, Frances H. *University Education for Women in Scotland: Its Effects on Social and Intellectual Life. A paper read at the conference of the National Union of Women Workers of Great Britain and Ireland, Edinburgh, October 1902*. St. Andrews, 1902.

Memorial Services Held in Honor of Major General William Crawford Gorgas By the Southern Society of Washington, D.C. Washington, DC: Government Printing Office, 1921.

A Memorial of the Seventy-Fifth Anniversary of the Founding of the University of Michigan: Held in Commencement Week June 23 to June 27, 1912. Ann Arbor, MI: Published by the University, 1915.

Miller-Bernal, Leslie and Susan L. Poulson, eds. *Going Coed: Women's Experiences in Formerly Men's Colleges and Universities, 1950–2000*. Nashville, TN: Vanderbilt University Press, 2004.

Mollenhoff, David V. *Madison: A History of the Formative Years*. Dubuque, IA: Kendall/Hunt Pub. Co., 1982.

Montaigu, The Comtesse De. "What Women Are Doing in America." *Womanhood: The Magazine of Woman's Progress and Interests—Political, Legal, Social, and Intellectual—and of Health and Beauty Culture* 11 (December 1903 to May 1904): 38–39.

Montgomery, James Riley, Stanley J. Folmsbee, and Lee Seifert Greene. *To Foster Knowledge: A History of The University of Tennessee 1794–1970*. Knoxville: The University of Tennessee Press, 1984.

Moore, Lindy. *Bajanellas and Semilinas: Aberdeen University and the Education of Women.* Aberdeen: Aberdeen University Press, 1991.

Morgan, Iwan. *The College by the Sea (A Record and a Review): "Nid Byd Byd Heb Wybodaeth."* Aberystwyth: Published by the Students' Representative Council in Collaboration with the College Council, 1928.

Morgan, John Vyrnwy. *A Study in Nationality.* London: Chapman & Hall, 1911.

Morgan, Kenneth O. *Rebirth of a Nation: A History of Modern Wales.* Oxford: Oxford University Press, 1989.

Morris, George Van Derveer. *A Man for a' That.* Cincinnati: Jennings & Pye, 1902.

Morrison, Sarah P. *Among Ourselves: To A Mother's Memory II: Catherine and Her Surroundings.* Plainfield, IN: Publishing Association of Friends, 1902.

Myers, Burton Dorr. *History of Indiana University Volume II: The Bryan Administration.* Bloomington: Published by Indiana University, 1952.

Myers, Christine D. "'The brilliant opening of a stubborn battle': The Queen Margaret College Bazaar and Women's Admission to Higher Education in Scotland, c. 1892." In *Leeds Working Papers in Victorian Studies, Volume 3: Platform-Pulpit-Rhetoric,* edited by Martin Hewitt, 150–165. Leeds: Leeds Centre for Victorian Studies, 2000.

———. "Gendering the 'Wisconsin Idea': The Women's Self-Government Association, c. 1898–1948." In *Gender, Politics and the Experience of Education: An International Perspective,* edited by Jane Martin and Jayne Goodman, 148–172. London: Woburn Press, 2002.

———. "The Glasgow Association for the Higher Education of Women, 1878–1883." *The Historian* 63, 2 (Winter 2001): 357–371.

———. "A Plea for the Highlands of Scotland: University Reform in the Early 20th Century." In *Contemporary Issues in Education,* edited by David Seth Preston, 141–158. Amsterdam and New York: Rodopi, 2004.

Nerad, Maresi. *The Academic Kitchen: A Social History of the Gender Stratification at the University of California, Berkeley.* Albany: State University of New York Press, 1988.

O'Dwyer, Frederick. *The Architecture of Deane & Woodward.* Cork: Cork University Press, 1997.

Ogren, Christine A. *The American State Normal School: "An Instrument of Great Good."* New York: Palgrave Macmillan, 2005.

———. "Where Coeds Were Coeducated: Normal Schools in Wisconsin, 1870–1920." *History of Education Quarterly* 25, 1 (Spring 1995): 1–26.

"Ohio University." *Journal of Pedagogy* 19, 1 (September 1906): 89.

Oldham, Bethenia McLemore. *Tennessee . . . and Tennesseans.* Clarksville, TN: W. P. Titus, 1903.

Olin, Helen R. *The Women of a State University: An Illustration of the Working of Coeducation in the Middle West.* New York and London: G. P. Putman, 1909.

Olson, James and Vera. *The University of Missouri: An Illustrated History.* Columbia, MO: University of Missouri Press, 1988.

O'Sullivan, Patrick. *The Irish in the New Communities.* Leicester: Leicester University Press, 1992.

Owen, Thomas McAdory. *History of Alabama and Dictionary of Alabama Biography Volume II.* Chicago: The S. J. Clarke Publishing Company, 1921.

"Oxford and Science." *Nature* 69, 1783 (December 31, 1903): 207–214.

Parkes, Susan M. and Judith Harford. "Women and Higher Education in Ireland." In *Female Education in Ireland 1700–1900: Minerva or Madonna,* edited by Deirdre Raftery and Susan M. Parkes, 105–144. Dublin and Portland, OR: Irish Academic Press, 2007.

Pearlman, Michael. "To Make the University Safe for Morality: Higher Education, Football and Military Training from the 1890s through the 1920s." *The Canadian Review of American Studies* 12, 1 (Spring 1981): 37–56.

Peckham, Howard H. *The Making of The University of Michigan 1817–1992,* edited and updated by Margaret L. Steneck and Nicholas H. Steneck. Ann Arbor: The University of Michigan Press, 1967, 1994.

Peers, Edgar Allison. *Redbrick University Revisited.* Liverpool: Liverpool University Press, 1996.

The Pennsylvania State College Alumni Directory, 1861–1935. State College, PA: Penn State Alumni Association, 1935.

Perkin, Joan. *Victorian Women*. New York: New York University Press, 1995.

Phi Beta Kappa: Catalogue of the Alpha of Missouri, 1901–1909. Columbia, MO: E. W. Stephens Publishing Company, 1909.

"Pieric Acid and Camphor Cure Ringworm." *The Eclectic-Medical Journal* 74, 5 (May 1914): 258.

Pointon, M. "Factors Influencing the Participation of Women and Girls in Physical Education, Physical Recreation and Sport in Great Britain during the Period, 1850–1920." *History of Education Society Bulletin* 24 (1979): 46–56.

Pope, Rhama D. and Maurice G. Verbeke. "Ladies' Educational Organizations in England, 1865–1885." *Paedagogica Historica* 3 (1976): 336–361.

President's Annual Report to the Board of Curators 1901–1902. Columbia, MO: E. W. Stephens, 1902.

Prochaska, Frank K. *Women and Philanthropy in Nineteenth-Century England*. Oxford: Clarendon Press, 1980.

Pyre, J. F. A. *Wisconsin*. New York: Oxford University Press, 1920.

Radke-Moss, Andrea G. *Bright Epoch: Women and Coeducation in the American West*. Lincoln: University of Nebraska Press, 2008.

Raffel, James A. *Historical Dictionary of School Desegregation: The American Experience*. Westport, CT: Greenwood Press, 1998.

Rainsford, George N. *Congress and Higher Education in the Nineteenth Century*. Knoxville: The University of Tennessee Press, 1972.

Rees, Gareth and David Istance. "Higher Education in Wales: The (Re-)emergence of a National System?" *Higher Education Quarterly* 51, 1 (January 1997): 49–67.

Report of the Chancellor to the Board of Trustees of Western University of Pennsylvania in Annual Session June 3rd, 1907. [Pittsburgh]: Western University of Pennsylvania, 1906.

Reynolds, Terry S. "The Education of Engineers in America before the Morrill Act." *History of Education Quarterly* 32, 4 (Winter 1992): 459–482.

Richards, Robert J. *Darwin and the Emergence of Evolutionary Theories of Mind and Behavior*. Chicago: University of Chicago Press, 1989.

Riley, Franklin L. *School History of Mississippi for Use in Public and Private Schools*. Richmond, VA: B. F. Johnson Publishing Company, 1915.

Robson, Ann P. and John M. Robson, eds. *Sexual Equality: Writings by John Stuart Mill, Harriet Taylor Mill and Helen Taylor*. Toronto: University of Toronto Press, 1994.

Rosenberg, Rosalind. *Beyond Separate Spheres: Intellectual Roots of Modern Feminism*. New Haven, CT and London: Yale University Press, 1982.

———. "The Limits of Access: The History of Coeducation in America." In *Women and Higher Education in American History*, edited by John M. Faragher and Florence Howe, 107–129. New York: Norton, 1988.

Rowland, Dunbar. *The Official and Statistical Register of the State of Mississippi 1912*. Nashville, TN: Brandon Printing Company, 1912.

Royal Commission on University Education in Wales. *Minutes of Evidence taken before the Royal Commissioners appointed to inquire Into the organisation and work of the University and its three constituent Colleges, and into the relations of the University to those Colleges and to other institutions in Wales providing education of a post-secondary nature, and to consider in what respects the present organisation of University Education in Wales can be improved and what changes, if any, are desirable in the constitution, functions and powers of the University and its three colleges*. Eleventh Day, Friday, 1st December, 1916.

The Royal University of Ireland. *The Calendar for the Year 1908*. Dublin: Alex. Thom & Co., 1908.

Rudolph, Frederick. *The American College and University: A History*. New York: Knopf, 1962.

———. *Curriculum: A History of the American Undergraduate Course of Study Since 1636*. San Francisco: Jossey-Bass Publishers, 1977.

Russett, Cynthia Eagle. *Sexual Science: The Victorian Construction of Womanhood*. Cambridge, MA: Harvard University Press, 1989.

The St. Andrews University Calendar for the Year 1904–1905. Edinburgh: William Blackwood and Sons, 1904.

Salmi, Jamil. *The Challenge of Establishing World Class Universities*. Washington, DC: World Bank, 2009.

Sanderson, Michael. *Education, Economic Change and Society in England 1780–1870*. Cambridge: Macmillan Press, 1995.

———. *The Universities in the Nineteenth Century*. London: Routledge and Kegan Paul, 1975.

Sansing, David G. *The University of Mississippi: A Sesquicentennial History*. Jackson: University Press of Mississippi, 1999.

Saunders, Ellen Virginia. "War-Time Journal of a 'Little Rebel'." *The Confederate Veteran Magazine* 28, 1 (January 1920): 11–12.

Sayer, George. *Jack: A Life of C. S. Lewis*. Wheaton, IL: Crossway Books, 1994.

Schakel, Peter J. *The Way into Narnia: A Reader's Guide*. Grand Rapids, MI: William B. Eerdmans Publishing Company, 2005.

Scott, Derek B. *The Singing Bourgeois: Songs of the Victorian Drawing Room and Parlour*. Milton Keynes and Philadelphia: Open University Press, 1989.

Scott, Joan Wallach. *Feminism and History*. Oxford and New York: Oxford University Press, 1996.

Sellers, James B. *History of the University of Alabama*. Tuscaloosa: University of Alabama Press, 1953.

S. E. P. "British Medical Association. Joint Discussion on the Treatment of Uterine Fibroids." *Medical Science: Abstracts & Reviews* 3, 3 (December 1920): 289.

Shafe, Michael. *University Education in Dundee 1881–1981: A Pictorial History*. Dundee: University of Dundee, 1982.

Shaw, T. Claye. "The Collegiate Training of Women." In *The Edinburgh Medical Journal Vol. XV*, edited by G. A. Gibson and Alexis Thomson, 444–448. Edinburgh and London: Young J. Pentland, 1904.

Shaw, Wilfred B. *A Short History of the University of Michigan*. Ann Arbor, MI: George Wahr, 1937.

Shimmin, A. N. *The University of Leeds: The First Half-Century*. London: Cambridge University Press, 1955.

Silver, Harold and John S. Teague. *The History of British Universities 1800–1969, excluding Oxford and Cambridge: A Bibliography*. London: Society for Research into Higher Education, 1970.

Silverstone, Rosalie and Audrey Ward. *Careers of Professional Women*. London: Croom Helm, 1980.

Simmons, Adele. "Education and Ideology in Nineteenth Century America: The Response of Educational Institutions to the Changing Role of women." In *Liberating Women's History: Theoretical & Critical Essays*, edited by Bernice A. Carroll, 115–126. Urbana: University of Illinois Press, 1976.

Skeffington, Florence. "Report of the Dean of the Woman's Department." In *University of Tennessee Register for 1901–1902 and Announcement for 1902–1903*, 196–197. Knoxville: The University of Tennessee Press, 1902.

Smith, A. Lapthorn. "Higher Education of Women and Race Suicide." *Popular Science Monthly* 66 (March 1905): 466–473.

Smith, Grace. "Indiana University." *Kappa Alpha Theta* 17, 2 (January 1903): 101–109.

Smith, Sarah J. "Retaking the Register: Women's Higher Education in Glasgow and Beyond, c. 1796–1845." *Gender & History* 12, 2 (2000): 310–335.

Smith-Rosenberg, Carroll. *Disorderly Conduct: Visions of Gender in Victorian America*. New York and Oxford: Oxford University Press, 1985, 1986.

Smith-Rosenberg, Carroll and Charles Rosenberg. "The Female Animal: Medical and Biological Views of Woman and Her Role in Nineteenth-Century America." *Journal of American History* 60 (1973): 332–356.

Snowden, Yates, ed. *History of South Carolina Volume II*. Chicago and New York: The Lewis Publishing Company, 1920.

Solomon, Barbara Miller. *In the Company of Educated Women*. New Haven, CT and London: Yale University Press, 1985.

Sonenklar, Carol. *We Are a Strong, Articulate Voice: A History of Women at Penn State*. University Park: The Pennsylvania State University Press, 2006.

Southgate, Donald. *University Education in Dundee: A Centenary History*. Edinburgh: Edinburgh University Press, 1982.

Staars, David. *The English Woman: Studies in Her Psychic Evolution*. Translated and edited by J. M. E. Brownlow. London: Smith, Elder, & Co., 1909.

Starrett, Agnes Lynn. *Through One Hundred and Fifty Years: The University of Pittsburgh*. Pittsburgh: University of Pittsburgh Press, 1937.

"Strong Women Remember." *Torchbearer: The Alumni Information Source of the University of Tennessee* 47, 2 (Summer 2008), http://www.utk.edu/torchbearer/4702/strong/index.shtml.

Summerfield, Carol J. and Mary Elizabeth Devine, eds. *International Dictionary of University Histories*. London: Taylor & Francis, 1988.

Sutherland, Gillian. "The Movement for the Higher Education of Women: Its Social and Intellectual Context in England, c. 1840–80." In *Politics and Social Change in Modern Britain*, edited by P. J. Waller, 91–116. Brighton, Sussex: Harvester Press, 1987.

Swanger, John E., comp. *Official Manual of the State of Missouri for the Years 1907–1908*. Jefferson City, MO: The Hugh Stephens Printing Company, 1907.

"'Sweet Girl Graduates.' Work of the Irish Association." *Irish Independent*, November 1, 1905, 5.

Talbot, Marion and Lois Kimball Mathews Rosenberry. *The History of the American Association of University Women, 1881–1931*. Boston: Houghton Mifflin, 1931.

Teicher, Barry and John W. Jenkins. *A History of Housing at the University of Wisconsin*. Madison, WI: UW History Project, 1987.

Thomas, M. Carey. "The Future of Woman's Higher Education." In *Mount Holyoke College: The Seventy-fifth Anniversary*, 100–104. South Hadley, MA: Mount Holyoke College, 1913.

Thompson, Edith. *Hockey As a Game for Women*. London: Edward Arnold, 1905.

Thompson, F. M. L. *The University of London and the World of Learning, 1836–1986*. London: Hambledon Press, 1990.

Thwing, Charles F. *A History of Higher Education in America*. New York: D. Appleton and Company, 1906.

Todd, Margaret. *The Life of Sophia Jex-Blake*. London: Macmillan and Co., 1918.

Trail, James W. H. "Natural Science in the Aberdeen Universities." In *Studies in the Development of the University, Aberdeen University Studies: No. 19*, edited by P. J. Anderson, 147–200. Aberdeen: Aberdeen University Press, 1906.

Tuke, M. J. *A History of Bedford College for Women, 1849–1937*. London: Oxford University Press, 1939.

Turk, Diana B. *Bound by a Mighty Vow: Sisterhood and Women's Fraternities, 1870–1920*. New York and London: New York University Press, 2004.

Turner, A. Logan, ed. *History of the University of Edinburgh 1883–1933*. Edinburgh: Oliver and Boyd, 1933.

Tylecote, Mabel. *The Education of Women at Manchester University 1883 to 1933*. Manchester: Manchester University Press, 1941.

"The Universities." *The University Review* 5, 24 (April 1907): 38–70.

The Universities (Scotland) Act, 1889 together with Ordinances of the Commissioners under the said Act, with relative Regulations & Declarations and University Court Ordinances made and approved subsequent to the expiry of the Powers of the Commissioners. With an Appendix containing the Universities (Scotland) Act, 1858. Glasgow, 1915.

"The University." In *UCC Honours Degree Programmes 09*. Cork: University College Cork, Ireland.

University of Alabama Bulletin; Centennial Celebration 1831–1931 90 (June 1931).

The University of Glasgow Through Five Centuries. [Glasgow]: Published by the University in commemoration of the Fifth Centenary, 1951.

University of Manchester: Register of Graduates and Holders of Diplomas and Certificates 1851–1958. Manchester: Manchester University Press, 1958.

"U.W. One of First to Admit Women." *The Wisconsin State Journal*, December 31, 1919.

Van de Warker, Ely. *Woman's Unfitness for Higher Coeducation*. New York: Grafton Press, 1903.

Vernon, Keith. *Universities and the State in England, 1850–1939*. Abingdon, Oxon and New York: RoutledgeFalmer, 2004.

Vicinus, Martha. *Independent Women: Work and Community for Single Women, 1850–1920*. Chicago: University of Chicago Press, 1985.

———, ed. *Suffer and Be Still: Women in the Victorian Age*. Bloomington: Indiana University Press, 1972.

———, ed. *A Widening Sphere*. Bloomington: Indiana University Press, 1977.

Vickery, Margaret Birney. *Buildings for Bluestockings: The Architecture and Social History of Women's Colleges in Late Victorian England.* Newark: University of Delaware Press, 1999.

The Victoria University of Manchester Medical School. Manchester: Manchester University Press, 1908.

The Victoria University of Manchester: Register of Graduates up to July 1st, 1908. Manchester: Manchester University Press, 1908.

Vinovskis, Maris and Richard Bernard. "Beyond Catharine Beecher: Female Education in the Antebellum Period." *Signs* 3 (1978): 856–869.

Wake, Jehanne. *Princess Louise: Queen Victoria's Unconventional Daughter.* London: Collins, 1988.

Wallace, Alfred Russel. *Darwinism: An Exposition of the Theory of Natural Selection with Some of Its Applications.* London: Macmillan and Co., 1912.

Waller, P. J., ed. *Politics and Social Change in Modern Britain.* Brighton, Sussex: Prentice Hall/ Harvester Wheatsheaf, 1987.

Ward, Herbert. *The Educational System of England and Wales and Its Recent History.* Cambridge: Cambridge University Press, 1935.

Watson, E. J. *Handbook of South Carolina: Resources, Institutions and Industries of the State.* Columbia, SC: The State Company, 1908.

Watson, Foster, ed. *The Encyclopaedia and Dictionary of Education in Four Volumes, Volume I.* London, Bath, Melbourne, Toronto, and New York: Sir Isaac Pitman & Sons, 1921.

Waugh, Lillian J. and Judith G. Stitzel. "'Anything But Cordial': Coeducation and West Virginia University's Early Women." *West Virginia History* 49 (1990): 69–80.

Welsh, Beatrice. *After the Dawn. A Record of the Pioneer Work in Edinburgh for the Higher Education of Women.* Edinburgh: Oliver and Boyd, 1939.

"What the Board of Regents Have to Say of Dr. Twombly, and What the President Has to Say of Them." *The Madison Daily Democrat*, January 21, 1909, 1.

Wharton, Annabel. "Gender, Architecture, and Institutional Self-Preservation: The Case of Duke University." *South Atlantic Quarterly* 90, 1 (Winter 1991): 175–217.

Whitehill, A. R. *History of Education in West Virginia.* Washington, DC: Government Printing Office, 1902.

Who's Who, 1904: An Annual Biographical Dictionary, Fifty-Sixth Year of Issue. London: Adam and Charles Black and New York: The Macmillan Company, 1904.

Williams, Gwyn A. *The Welsh in Their History.* London: Croom Helm, 1982.

Wilson, A. N. *C. S. Lewis: A Biography.* New York: Norton, 2002.

Wilson, Calvin Dill. *Working One's Way through College and University: A Guide to Paths and Opportunities to Earn an Education at American Colleges and Universities.* Chicago: A. C. McClurg & Co., 1912.

"Woman at Wisconsin: A Chronology." *The Wisconsin Alumni Magazine* (March, 1916): frontispiece.

"Women Graduates' Association." *Irish Independent*, February 21, 1905, 7.

"Women Graduates. Royal University Status." *Irish Independent*, February 20, 1906, 7.

Women in Scotland Bibliography Group. *Women in Scotland: An Annotated Bibliography.* Edinburgh: Open University of Scotland, 1988.

"Women and Their Work We Ought to Know About." *The Woman's Medical Journal* 17, 4 (April 1907): 226.

Woods, Robert. *The Population of Britain in the Nineteenth Century.* Cambridge: Cambridge University Press, 1995.

Woody, Thomas K. *A History of Women's Education in the United States*, 2 Volumes. New York: Octagon Books, 1929.

Wronker, Charlotte "Co-Education in the 'Varsity." *The Missouri Alumni Quarterly* (December 1905): 27–33.

WVU Women: The First Century. Morgantown: WVU Women's Centenary Project, West Virginia University, 1989.

Yellin, Jean Fagan. *Women & Sisters: The Antislavery Feminists in American Culture.* New Haven, CT and London: Yale University Press, 1989.

Yeo, Eileen Janes. *Radical femininity: Women's Self-Representation in the Public Sphere.* Manchester and New York: Manchester University Press, 1998.

INDEX

GPSR Compliance
The European Union's (EU) General Product Safety Regulation (GPSR) is a set of rules that requires consumer products to be safe and our obligations to ensure this.

If you have any concerns about our products, you can contact us on

ProductSafety@springernature.com

In case Publisher is established outside the EU, the EU authorized representative is:

Springer Nature Customer Service Center GmbH
Europaplatz 3
69115 Heidelberg, Germany

www.ingramcontent.com/pod-product-compliance
Lightning Source LLC
LaVergne TN
LVHW011809200926
848484LV00005B/1117
* 9 7 8 0 2 3 0 6 2 2 3 7 1 *